147
Advances in Biochemical Engineering/Biotechnology

Aims and Scope

This book series reviews current trends in modern biotechnology and biochemical engineering. Its aim is to cover all aspects of these interdisciplinary disciplines, where knowledge, methods and expertise are required from chemistry, biochemistry, microbiology, molecular biology, chemical engineering and computer science.

Volumes are organized topically and provide a comprehensive discussion of developments in the field over the past 3–5 years. The series also discusses new discoveries and applications. Special volumes are dedicated to selected topics which focus on new biotechnological products and new processes for their synthesis and purification.

In general, volumes are edited by well-known guest editors. The series editor and publisher will, however, always be pleased to receive suggestions and supplementary information. Manuscripts are accepted in English.

In references, Advances in Biochemical Engineering/Biotechnology is abbreviated as *Adv Biochem Eng Biotechnol* and cited as a journal.

More information about this series at http://www.springer.com/series/10

Joydeep Mukherjee
Editor

Biotechnological Applications of Biodiversity

With contributions by

Karen Amaya · Jacobo Arango · Lakshmi Balachandran
Beata Dedicova · John R. J. French · R. Andrew Hayes
Guangrong Hu · Shiqi Ji · D. İpek Kurtböke
Fuli Li · Girish Mahajan · Prem Narain Mathur
Danilo E. Moreta · Ronald J. Quinn · Samir Kumar Samanta
Barindra Sana · Michael Gomez Selvaraj
Tuhinadri Sen · Maarten van Zonneveld · Shi'an Wang
Yanchong Yu · Gongke Zhou

Springer

Editor
Joydeep Mukherjee
School of Environmental Studies
Jadavpur University
Kolkata, West Bengal
India

ISSN 0724-6145 ISSN 1616-8542 (electronic)
ISBN 978-3-662-51230-2 ISBN 978-3-662-45097-0 (eBook)
DOI 10.1007/978-3-662-45097-0

Springer Heidelberg New York Dordrecht London

Softcover reprint of the hardcover 1st edition 2015

Printed on acid-free paper

Springer is part of Springer Science+Business Media (www.springer.com)

Preface

Biodiversity is the outcome of successive periods of evolution for over three billion years. From simple unicellular microbes to the complex human body, all are equally important components of biodiversity, interacting to form functional ecosystems. Biological resources have sustained human society over thousands of years and the diversity of these resources has been exploited for three basic necessities: food, clothes, and shelter by pre-historic people as well as modern mankind. Recognizing the enormous value of biodiversity for present and future generations, the United Nations Conference on Environment and Development (the Rio "Earth Summit") proclaimed the Convention on Biological Diversity (CBD) in 1992. Through this global agreement, 193 nations aspire to the "conservation of biological diversity, the sustainable use of its components and the fair and equitable sharing of benefits arising from the use of genetic resources" [1]. Furthermore, the United Nations have declared the present decade (2011–2020) as the "United Nations Decade on Biodiversity." With the objective of stopping biodiversity loss and in the long run regaining the lost biodiversity, governments agreed to the "Strategic Plan for Biodiversity 2011–2020 and the Aichi Targets." Among the five targets, "enhancing the benefits to all from biodiversity and ecosystem services" is one important strategic goal. Against this backdrop, I consider the publication of this book very well-timed.

The Article 2 of the CBD defines biotechnology as "any technological application that uses biological systems, living organisms, or derivatives thereof, to make or modify products or processes for specific use" [2]. The major facets of our lifestyle that have been touched by biotechnology are agriculture, medicine, bulk products, environment, and energy. Biodiversity is intricately linked with the provision of services by biotechnology. The diversity of food and fiber crops is crucial to feed, clothe, and house the growing population, particularly in the developing world. According to the World Health Organization, better knowledge of the earth's biodiversity is vital for future medical and pharmacological discoveries that will keep us away from death and disease. The bioprocess industry is looking for new enzymes and metabolites that resist harsh industrial manufacturing conditions like extremes of temperature, pH, and pressure. Biodiversity holds the key. The rate and extent of bioremediation can be significantly increased by the application of novel organisms, hence the importance of biodiversity in environmental protection. As predicted by the "International Energy Outlook 2013," the world's energy consumption would increase by a massive 56 % between 2010 and 2040 and the demand would be highest in China and India. Biodiversity has the potential to provide novel bioresources to meet this demand.

Accordingly, the chapters of this book have been selected to cover the spectrum of biotechnological applications of biodiversity. In "Current Issues in Cereal Crop Biodiversity" Danilo E. Moreta and colleagues write about the biodiversity of cereal crops such as rice, wheat, maize, millets and an emerging staple food, quinoa. In "Biodiversity in Production of Antibiotics and Other Bioactive Compounds" Girish Mahajan and Lakshmi Balachandran have highlighted the importance of the diversity of microbes in providing leads for the development of new drugs. In "Medicinal Plants, Human Health and Biodiversity: A Broad Review," Tuhinadri Sen and Samir Samanta emphasize the role of plant biodiversity in affording botanical drugs and herbal medicines on which the majority of the world's population (particularly in the developing countries) are dependant. İpek Kurtböke and co-authors, in "Eco-Taxonomic Insights into Actinomycete Symbionts of Termites for Discovery of Novel Bioactive Compounds" review the microbial diversity of a very small ecosystem, the termite gut and its potential to deliver a wide range of useful bioproducts. In "Bioresources for Control of Environmental Pollution" Barindra Sana describes the diversity of plants, microbes, and lower eukaryotes and their application in bioremediation of environmental pollutants. In "Organisms for Biofuel Production: Natural Bioresources and Methodologies for Improving Their Biosynthetic Potentials," Guangrong Hu and colleagues write about the diverse plants, algae, yeasts, and bacteria as producers of biodiesel, gasoline, jet fuel, alkanes, and hydrogen. Taxonomical listing of species currently used or being explored vis-à-vis the bases of their selection for biotechnological applications have been presented by the authors. Modern approaches to discover new biodiversity have also been discussed. Conservation strategies form an important part of the chapters. Commercial biotechnological processes exploiting biodiversity have also been focused.

Legal and policy issues in biodiversity are gaining importance alongside the scientific and technological innovations for its exploitation. Unfortunately, a north-south conflict exists on the utilization of biological diversity. The global south (comprising mostly of developing nations) is rich in biodiversity but has limited access to advanced technology, while the global north (consisting of developed countries) is bioresource poor but possesses the economic power and scientific technology required for commercialization of bioresources. Repeatedly, the south has accused the industrialized north of biopiracy [3]. To prevent the commercialization of biodiversity without paying rational compensation to the rightful owners, the "Nagoya Protocol on Access to Genetic Resources and the Fair and Equitable Sharing of Benefits Arising from their Utilization" was adopted by the governing body of the CBD in 2010. This international agreement strives for "sharing the benefits arising from the utilization of genetic resources in a fair and equitable way, including by appropriate access to genetic resources and by appropriate transfer of relevant technologies, taking into account all rights over those resources and to technologies, and by appropriate funding, thereby contributing to the conservation of biological diversity and the sustainable use of its components" [4]. It is hoped that successful implementation of the Nagoya Protocol will ease the north-south conflict and promote congruous biotechnological applications of biodiversity not only for us but also for the generations to come.

I thank the Managing Editor, Prof. Dr. Thomas Scheper and the Publishing Editor, Elizabeth Hawkins for giving me the opportunity to edit this book on a very important global issue. I thank the authors for spending their valuable time preparing their excellent contributions. My sincere thanks also go to all the reviewers for their meticulous corrections that vastly improved the manuscripts. I hope the readers will find every chapter interesting and informative.

Kolkata, India, April 2014 Joydeep Mukherjee

References

1. Convention on Biological Diversity (2014) http://www.cbd.int/history. Accessed 28 Apr 2014
2. Convention on Biological Diversity (2014) http://www.cbd.int/convention/articles/?a=cbd-02. Accessed 28 Apr 2014
3. Chen J (2005) Biodiversity and biotechnology: a misunderstood relation. Available via http://www.cbd.int/doc/articles/2005/a-00468.pdf. Accessed 28 Apr 2014
4. Convention on Biological Diversity (2014) http://www.cbd.int/abs/. Accessed 28 Apr 2014

Contents

Current Issues in Cereal Crop Biodiversity 1
Danilo E. Moreta, Prem Narain Mathur, Maarten van Zonneveld, Karen Amaya, Jacobo Arango, Michael Gomez Selvaraj and Beata Dedicova

Biodiversity in Production of Antibiotics and Other Bioactive Compounds 37
Girish Mahajan and Lakshmi Balachandran

Medicinal Plants, Human Health and Biodiversity: A Broad Review 59
Tuhinadri Sen and Samir Kumar Samanta

Eco-Taxonomic Insights into Actinomycete Symbionts of Termites for Discovery of Novel Bioactive Compounds 111
D. İpek Kurtböke, John R. J. French, R. Andrew Hayes and Ronald J. Quinn

Bioresources for Control of Environmental Pollution 137
Barindra Sana

Organisms for Biofuel Production: Natural Bioresources and Methodologies for Improving Their Biosynthetic Potentials 185
Guangrong Hu, Shiqi Ji, Yanchong Yu, Shi'an Wang, Gongke Zhou and Fuli Li

Index 225

Adv Biochem Eng Biotechnol (2015) 147: 1–35
DOI: 10.1007/10_2013_263

Published Online: 20 December 2013

Current Issues in Cereal Crop Biodiversity

Danilo E. Moreta, Prem Narain Mathur, Maarten van Zonneveld, Karen Amaya, Jacobo Arango, Michael Gomez Selvaraj and Beata Dedicova

Abstract The exploration, conservation, and use of agricultural biodiversity are essential components of efficient transdisciplinary research for a sustainable agriculture and food sector. Most recent advances on plant biotechnology and crop genomics must be complemented with a holistic management of plant genetic resources. Plant breeding programs aimed at improving agricultural productivity and food security can benefit from the systematic exploitation and conservation of genetic diversity to meet the demands of a growing population facing climate change. The genetic diversity of staple small grains, including rice, maize, wheat, millets, and more recently quinoa, have been surveyed to encourage utilization and prioritization of areas for germplasm conservation. Geographic information system technologies and spatial analysis are now being used as powerful tools to elucidate genetic and ecological patterns in the distribution of cultivated and wild species to establish coherent programs for the management of plant genetic resources for food and agriculture.

Keywords Biotechnology · Climate change · Crop improvement · Food security · Genetic resources

List of Abbreviations

AVRDC	The World Vegetable Center
BNI	Biological Nitrification Inhibition
CIAT	International Center for Tropical Agriculture
CGIAR	Consultative Group on International Agricultural Research
CGRP	Canadian Genetic Resources Programme

D. E. Moreta (✉) · J. Arango · M. G. Selvaraj · B. Dedicova
International Center for Tropical Agriculture (CIAT), Cali, Colombia
e-mail: d.moreta@cgiar.org

P. N. Mathur
Bioversity International (Office for South Asia), New Delhi, India

M. van Zonneveld · K. Amaya
Bioversity International (Americas Office), Cali, Colombia

DNA	Deoxyribonucleic acid
FAO	Food and Agriculture Organization of the United Nations
GBIF	Global Biodiversity Information Facility
GCDT	Global Crop Diversity Trust
GEF	Global Environment Facility
GHG	Greenhouse gas
GM	Genetically modified
GMO	Genetically modified organism
IBPGR	International Board for Plant Genetic Resources
ICRISAT	International Crops Research Institute for the Semi-Arid Tropics
IFAD	International Fund for Agricultural Development
ILRI	International Livestock Research Institute
IRD	Institut de Recherche pour le Développement
ITPGRFA	International Treaty on Plant Genetic Resources for Food and Agriculture
IUCN	International Union for Conservation of Nature
JIRCAS	Japan International Research Center for Agricultural Sciences
MLS	Multilateral system
NBPGR	National Bureau of Plant Genetic Resources (India)
NUS	Neglected and underutilized species
ORSTOM	Office de la Recherche Scientifique et Technique d'Outre-Mer
SINGER	System-wide Information Network for Genetic Resources
SMTA	Standard Material Transfer Agreement
T-DNA	Transfer-deoxyribonucleic acid
TALENs	Transcription activator-like effector nucleases
UNEP	United Nations Environment Programme
UNU	United Nations University
USDA-ARS	United States Department of Agriculture, Agricultural Research Service
WEMA	Water Efficient Maize for Africa
WHO	World Health Organization

Contents

1 Introduction 3
2 Agricultural Biodiversity: Concept, Importance and Scope 3
2.1 Conservation of Agrobiodiversity in Small Grains: Status and Applications 4
2.2 Genetic Diversity of Pearl Millet and Its Wild Relatives: Distribution, Conservation, and Use 10
2.3 Quinoa Diversity and Its Potential in the New Millennium 17
2.4 A Practical Use of Agrobiodiversity in Cereal Crops to Mitigate Climate Change through Regulation of Soil Nitrification 21

2.5 Impact of Nitrogen Pollutants on Biodiversity 23
2.6 Transgenic Crops and Food Security 24
3 Summary and Conclusions 25
References 26

1 Introduction

Agricultural biodiversity is the key to successful biotechnological approaches for crop improvement. Plant breeding programs aimed at increasing crop productivity and improving food security rely on traits that must be efficiently managed and exploited for the sustainable delivery of cultivars without compromising genetic diversity. Exploration of new biodiversity in the face of climate change and the threat of genetic erosion must complement current trends in crop diversity. Enhanced conservation strategies of important agricultural biodiversity need to be deployed to take advantage of cutting-edge technologies on spacial analysis to monitor the patterns of plant diversity and distribution and prioritize areas for conservation. Most recent advances in the genetic modification of crop plants (i.e. transgenics) are also an important issue for agricultural production. Next-generation crops need to be created using transdiciplinary strategies, with the goal of making farming more productive and more profitable. Major staple crops, such as rice, maize, wheat, millets, and most recently quinoa, play an important role for millions of people worldwide because these crops sustain the lives of the poorest rural farmers. Therefore, diversity and conservation issues involving the utilization of these staple crops will significantly contribute to a household's food security and nutrition of smallholders, mainly in the tropics.

2 Agricultural Biodiversity: Concept, Importance and Scope

Agricultural biodiversity, also known as agrobiodiversity, can be defined as all of the components of biological diversity that are relevant to food and agriculture, including agricultural ecosystems [1]. From a pragmatic perspective, agrobiodiversity is the result of the interaction between the environment, genetic resources, and management systems and the practices used by people from diverse cultural backgrounds.

Agrobiodiversity is an integral part of overall biodiversity; it comprises the variety and variability of animals, plants, and microorganisms at the genetic, species, and ecosystem levels that are used for food and agriculture, including crops, livestock, forestry, and fisheries. Culture and local knowledge are regarded as essential parts of agrobiodiversity because it is the human activity of agriculture that affects and shapes this biodiversity. In other words, agrobiodiversity is the result of natural selection and human intervention over millennia, and it plays a key role in sustainable development, including processes for, and in support of, food production and food security [2, 3]. Some part of this biodiversity is directly managed to supply the goods and services that people need; however, most of it is not directly intended for production purposes and remains important as a source of materials for its contributions to ecosystem services, such as pollination, control of greenhouse gas emissions, and soil dynamics [4].

The diversity in crops consists of the crops, landraces, and cultivars grown by farmers [5]. The world is currently facing serious environmental problems due to loss of biological diversity at alarming rates. Scientists have estimated that by 2025, 60,000 plant species could be lost. The Food and Agriculture Organization of the United Nations (FAO) has also stated that, since 1900, approximately three quarters of the genetic diversity of domestic agricultural crops has already been lost [6].

It is well known that modern, intensive agriculture reduces agricultural biodiversity [4, 5]; this loss of genetic diversity in agriculture is known as genetic erosion [7]. It is commonly assumed that plant breeding efforts are an important cause of genetic erosion of crops. However, the effects of urbanization and modern agricultural practices are important factors as well. Climate change and environmental degradation can also contribute to changes in cropping patterns and the disappearance of traditional varieties [5].

The spread and adoption of modern crop varieties has implications for genetic erosion and a decline in crop genetic diversity. Conservation of crop genetic resources is therefore a prerequisite for future generations. Organized and well-focused exploitation and conservation strategies of biodiversity will allow users to breed crop varieties for improved food security and face new challenges in the era of climate change.

2.1 Conservation of Agrobiodiversity in Small Grains: Status and Applications

It is generally accepted that the modernization of agriculture and land use changes negatively affect economic plant diversity, both on farmers' fields and in home gardens [5]. This might eventually lead to the genetic erosion of cultivars and crop wild relatives with potentially useful traits for current and future human use. Therefore, many collecting missions have been organized in the past decades to establish extensive international and national genebank collections for important food crops, including cereals [8]. More than 3 million cereal accessions (i.e. samples of living plant material collected from particular locations) are stored *ex situ* worldwide [9]. They account for almost half of all genetic materials conserved globally in genebank collections [9]. This confirms the importance of cereals for global food security and agricultural production. The three crops with the most accessions conserved *ex situ* at a global level are the cereals rice (*Oryza* spp.), wheat (*Triticum* spp.), and barley (*Hordeum vulgare* subsp. *vulgare*) [9]. Of other cereal crops, such as maize (*Zea mays* L.), sorghum (*Sorghum* spp.), oat (*Avena sativa* L.), and millets (e.g. *Pennisetum glaucum*), less but still huge amounts of materials are being conserved *ex situ* [9]. To facilitate access to genetic material for evaluation, breeding, and direct use, some of these collections are put in the Multilateral System (MLS) under the conditions defined in the International Treaty on Plant Genetic Resources for Food and Agriculture (www.planttreaty.org). Several of the collections that are in the open domain, such as those held in trust by the Consultative Group on International Agricultural Research (CGIAR), can be consulted through the GENESYS Web portal (http://www.genesys-pgr.org), developed by Bioversity International.

The genetic integrity of accessions is maintained as much as possible in *ex situ* gene bank collections to conserve the specific characteristics of each material for evaluation, breeding, and direct use. For example, most accessions of the cereal collections held in trust by CGIAR have been characterized (88 %) [9]. This percentage is higher than all other types of crops conserved by CGIAR centers and The World Vegetable Center [9]. However, considering all reported national collections around the world, the amount of characterized cereal genetic resources is lower and similar to levels of genetic resource characterization of other crop types [9]. As a drawback, *ex situ* collections do not maintain the continued process of interactions between plants, humans, and environmental factors that take place on farms and between farmers [10]. *In situ* conservation of crops such as cereals is thought to be important to maintain the adaptive genetic variation of crop populations through interactions with their environment, including human management and selection. Several newly sampled barley materials in Morocco, for example, had more disease resistance than accessions collected several decades ago at the same location [11]. This supports the need of cereal genetic resources in situ conservation to allow evolution of adaptive genetic variation to important diseases for overall crop production. On the other hand, some historic *ex situ* accessions included rare genes to resistance that were extinct in the current species populations due to genetic erosion [11]. This highlights the importance of *ex situ* conservation as well as the necessity to develop complementary strategies of *ex situ* and in situ conservation.

Therefore, there is a need to assess the diversity status and dynamics of in situ plant genetic resources and develop complementary *ex situ* and in situ conservation strategies [9, 12–14]. At the same time, these types of analyses are useful to identify remaining gaps of diversity that are missing in existing genebank collections and that should therefore be targeted for germplasm collecting [9, 13]. Of course, periodic monitoring activities are required to measure the effectiveness of in situ conservation over time and to check the viability of seed material in *ex situ* collections.

The main purpose of in situ conservation is to maintain genetic variation in cultivated crop populations for phenotypic selection by farmers and/or natural processes [15]. This allows maintenance of the processes of microevolution and continuous adaptation of crop populations to their environments, including farmer management. The genetic structure of populations can change when phenotypic traits are heritable and selection is sufficiently strong. Following Darwin's concepts of selection, this allows cumulative directional genetic response over generations—that is, microevolution of these populations to natural and human selection [15, 16]. Microevolution in plant populations is further driven by factors such as random mutation, recombination, and genetic drift [17, 18]. As an additional factor, many smallholders in all parts of the world periodically introduce new materials from neighbors and other localities into their systems to sustain productivity [19]. These factors and activities together make on-farm plant genetic resources management a dynamic system of crop genetic diversity use. Farmers may select for changing preferences as well choose to maintain desired phenotypic traits [15]. The variety of traits that is maintained and evolving under farmers' care is often unknown to conventional breeders, entrepreneurs, and consumers. This makes on-farm conservation areas potential sources of untapped diversity for the development of new crop varieties for local and wider use. Even genetic diversity itself in cultivated populations may be a trait of farmers' selection for ecosystem services, such as pest and disease control [20].

The status and trends of intraspecific crop diversity are traditionally being assessed and monitored through varietal diversity, either through farmer interviews or morphological or botanical classifications. For example, classification on the basis of traits that are important for farmers suggests that sorghum and pearl millet varietal diversity in Niger remained at similar levels during the last three decades of the twentieth century [21]. A reason for the maintenance of varietal diversity could be that the areas under study are marginal terrains where the cultivation of traditional crop and varieties outperforms cash crops and/or crop genetic diversity is being maintained as a risk management strategy [21]. Taxonomic keys have been used to monitor traditional maize varieties (www.biodiversidad.gob.mx/genes/proyectoMaices.html). Taxonomic studies are the basis for understanding variation in plant genetic resources. However, assessment of varietal diversity according to morphological or botanical characterization may still lead to a substantial degree of misidentification [22], and taxonomic keys may not exist for botanical varieties of specific crops or from particular geographic areas. This limits intraspecific diversity studies.

Molecular tools that identify polymorphisms have created novel opportunities for assessing crop genetic diversity, particularly when these markers are linked to adaptive traits and applied in combination with new geospatial methods of geographic and environmental analysis [23–25]. In addition, geographic information systems (GIS) can contribute significantly to improving the understanding and monitoring of spatial and temporal patterns of crop diversity [26]. Application of spatial analyses on georeferenced diversity data allows the formulation and implementation of better-targeted, and hence more effective, conservation strategies of inter and intra-specific plant diversity. For example, geospatial analyses combined with molecular marker characterization data have been used to support conservation strategies for African rice (*O. glaberrima* Steud.) genetic resources [27]. This study found the highest African rice genetic diversity in intermediate humid conditions and a decrease of genetic diversity under semiarid conditions and humid conditions [27]. This may have implications for the genetic resource conservation of this crop under changing climate conditions [27].

As an example of how to apply geospatial analysis to support plant genetic resources conservation, maize microsatellite diversity was mapped in the Americas. A dataset freely available by the Genetic Architecture of Maize and Teosinte project at www.panzea.org was used. It consists of molecularly characterized maize accessions from different genebanks. This dataset has been used by Vigouroux et al. [22] to understand the genetic structure (*beta* diversity) in America's maize distribution. Four big genetic groups can be distinguished, consisting of material from (1) temperate zones in the United States and South America; (2) Mexican highlands; (3) tropical lowland; and (4) the Andes [22]. This data was used to map maize *alpha* diversity. A total of 1,145 georeferenced accessions that had microsatellite data for 92 markers were selected. Following van Zonneveld et al. [28], a 20 min grid layer (corresponding to approximately 33 km in the study area) was constructed for all genetic parameters, applying a circular neighborhood with a diameter of two degrees (corresponding to approximately 222 km) to improve visualization and group geographically isolated germplasm accessions. Calculations were done in the R program version 2.15.1 with the packages Raster [29] and Adegenet [30]. To standardize comparability of these parameters between cells, sample bias was corrected through resampling without replacement after Leberg [31] to a sample size of six trees. Per parameter, the average value was calculated from six subsamples following the bootstrap method developed by Thomas et al. [32].

The highest allelic richness is found in the central highlands of Mexico (Fig. 1). This indicates a high variation of maize genetic resources, and this area therefore should be prioritized for conservation actions. Nevertheless, the center of maize domestication is thought to be located in the southern lowlands, which thus is also an important area for conservation [33]. Also, a high number of alleles is found in these areas (Fig. 1). In a few scattered areas in northern Mexico, high levels of diversity are also observed.

Allelic richness is lower in South America, where maize was introduced later. However, high levels of diversity can be found in Ecuador in the Andean region, Venezuela, northern Colombia, and Bolivia in the tropical lowlands (Fig. 1). These materials belong to different genetic groups than the material from the Mexican highlands [22]. These clusters may reflect different groups of evolution and adaptation to different environments. To maximize conservation of plant genetic resources, areas from these clusters that harbor high genetic variation should be prioritized for conservation.

The high genetic maize diversity in the Mexican highlands can be explained by the high levels of introgression between crop wild relatives and maize in that area [33, 34]. Indeed, gene flow between cultivated plants and their relatives in overlapping areas of distribution can cause elevated levels of intraspecific cultivated plant diversity. Such insights allow a better understanding of the role of evolutionary processes in the development of current species distributions and, where relevant, their domestication [35]. Similar phenomena have also been observed in other areas. Higher levels of molecular diversity of domesticated emmer wheat (*Triticum turgidum* L. subsp. *dicoccon* (Schrank) Thell.) and bread wheat (*T. aestivum* L.) have been found in the eastern Mediterranean and Turkey, respectively, and are located south and west of their putative centers of domestication due to crossing between domesticates and their wild ancestors [36].

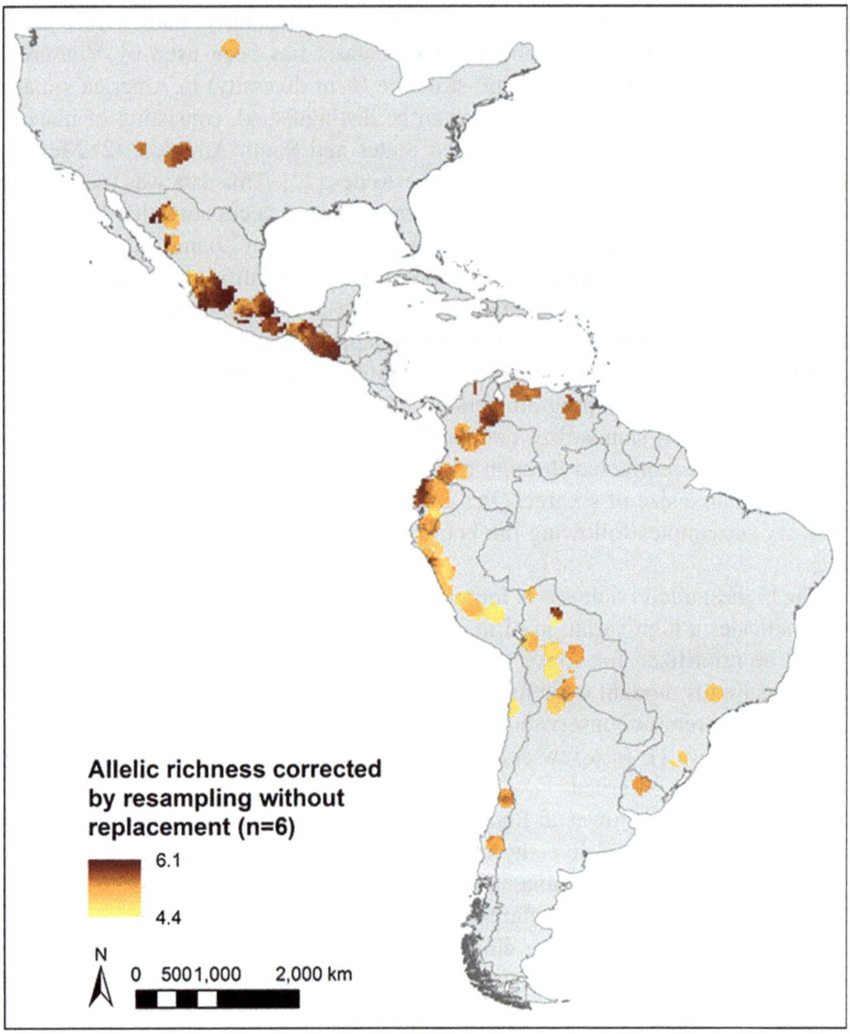

Fig. 1 Maize microsatellite diversity in the Americas constructed from geospatial analysis as a tool for genetic resources conservation. The average number of alleles per locus is shown

In addition to studies across crop distribution ranges to target areas for in situ conservation and germplasm collecting, local spatial studies in prioritized traditional rural communities are important to provide input to the development of appropriate on-farm PGR management strategies [25]. These studies help to (1) increase the understanding of how farmers manage and conserve crop diversity within a community and; (2) identify the optimal geographic scale for interventions and crop diversity monitoring, as well as the social context in which in situ conservation should be implemented [37]. Estimates of distribution and levels of crop diversity in rural communities also help to determine the need to introduce new varieties into local seed systems and improve seed distribution systems accordingly [38].

Several case studies illustrate how genetic analysis helps to support in situ conservation interventions. A microsatellite marker study of diversity and structure of local rice varieties (*Oryza sativa* L., *O. glaberrima* Steud.) in the Republic of Guinea revealed genetic differences between two different agro-ecosystems, but no differentiation was shown between villages or farms within each of the contrasting agro-ecosystems [37]. This suggests that most genetic diversity can be conserved within just a few farms of a village [37]. Simple sequence repeat (SSR) markers also detected high seed exchange within villages of a traditional rice variety in Thailand [39]. In addition, high genetic differentiation between villages was found [39]. This indicates long periods of local adaptation and selection. Also, low genetic differentiation between seed lots of different farmers of the specific Mexican maize variety, Jala, suggests high seed exchange between farmers [40]. This implies that within communities only a few farmer fields would be necessary to target for in situ conservation and that collection for *ex situ* conservation of many individuals in a few farmer fields is preferred to collection of a limited number of individuals in many fields [40].

Climate change will certainly impact landraces conserved in situ, such as the native maize races in Mexico [18]. Nevertheless, it remains difficult to predict whether a local landrace goes extinct or can adapt [18]. This depends partly on the magnitude of the climate alterations, the novelty of new climates, and the amount of genetic variation present in landrace populations [18]. Seed exchange between farmers remains an important means to adapt local seed material to changing climates. Ecogeographic analysis of traditional maize systems in Mexico shows that mid-elevation communities can adapt fairly easy their production systems to climate change through seed exchange with farmers within a 10 km radius, where a wide range of different micro-climates can be found [41]. In contrast, highland and lowland systems that have less local micro-climate diversity require seed material from geographically more distant locations. The latter would require active support from governmental and non-governmental organizations [41].

During the domestication process of crops, overall genetic diversity is generally reduced, whereas phenotypic diversity at specific parts of the genome related to traits of interest are increased. However, many unknown traits of interest may have been lost during that domestication process. Crop wild relatives and progenitors may still include many potentially interesting traits, such as adaptive traits to heat stress and other climate change–related traits [42]. Therefore, recently more emphasis has been placed on the conservation of crop wild relatives. Microsatellite studies on rice in Vietnam confirm that wild rice populations contain much more diversity than cultivated populations [43]. A good example is mapping the genetic diversity of the wild subspecies of barley (*Hordeum vulgare* subsp. *spontaneum*) in Eastern Mediterranean and Middle East (Israel, Palestine, Syria, and Turkey), and Central Asia [44]. In these locations, barley here has higher genetic diversity in the eastern Mediterranean than elsewhere in its distribution. This area should therefore be a focus of conservation activities for barley genetic resources.

2.2 Genetic Diversity of Pearl Millet and Its Wild Relatives: Distribution, Conservation, and Use

2.2.1 Background

Pearl millet [*Pennisetum glaucum* (L.) R. Br.] is the fifth most important cereal crop in the world after rice, wheat, maize, and sorghum. It is a widely grown rainfed cereal crop in the arid and semi-arid regions of Africa and Southern Asia, and it can be grown in areas where rainfall is not sufficient (200 to 600 mm/year) for the cultivation of maize and sorghum. In other countries, it is grown under intensive cultivation as a forage crop. Pearl millet accounts for almost half of global millet production, with 60 % of the cultivation areas in Africa, followed by 35 % in Asian countries. European countries represent 4 % of millet cultivation and North America only 1 %, mainly for forage. Today, millet is a staple for more than 500 million people. Areas planted with pearl millet are estimated at 15 million ha annually in Africa and 14 million ha in Asia. Global production exceeds 10 million tons a year [45]. In sub-Saharan Africa, pearl millet is the third major crop, with the major producing countries being Nigeria, Niger, Burkina Faso, Chad, Mali, Mauritania, and Senegal in the West and Sudan and Uganda in the East. In Southern Africa, maize has partially or completely displaced millet cultivation because of commercial farming. India is the largest producer of pearl millet, both in terms of area (9.3 m ha) and production (9.3 mt), with an average productivity of 1044 kg/ha during the last 5 years (2007–2011).

The trends in area, production, and productivity of pearl millet suggest that area has increased marginally (2 %) during the last 2 years (2010–2011) and productivity has gone up by 19 % [46]. The major pearl millet growing states in India are Rajasthan, Maharashtra, Gujarat, Uttar Pradesh, Haryana, Karnataka, Madhya Pradesh, Tamil Nadu and Andhra Pradesh. However, productivity is the highest in Haryana, followed by Gujarat, Uttar Pradesh, Madhya Pradesh, Tamil Nadu, Andhra Pradesh, Rajasthan, Maharashtra, and Karnataka. It is mainly cultivated during the *Kharif* (rainy) season across the country. However, it is also grown to a lesser extent during the *Rabi* (post-rainy) season in Andhra Pradesh, producing high yields and excellent grain quality. Outside Africa and India, millets are also grown in Australia, China, Canada, Mexico, Russia, and the United States, primarily grown as a forage crop for livestock production [45]. Pearl millet is endowed with enormous genetic variability for various morphological traits, yield components, adaptation, and quality traits. Pearl millet is also nutritionally superior compared to maize and rice. The protein content of pearl millet is higher than maize and it has a relatively high vitamin A content.

2.2.2 Taxonomy, Origin, and Distribution

Pearl millet, *Pennisetum glaucum,* is an annual, allogamous, cross-pollinated, diploid cereal belonging to the *Poaceae* family, subfamily *Panicoideae*, tribe

Paniceae, subtribe *Panicinae*, section *Penicillaria,* and genus *Pennisetum.* The genus *Pennisetum* contains about 140 species. The important wild relatives of cultivated pearl millet include the progenitor, *Pennisetum glaucum* subsp. *monodii* Maire, *P. purpureum* K. Schumach, *P. pedicellatum* Trin., *P. orientale* Rich, *P. mezianum* Leeke, and *P. squamulatum* Fresen. Previous names are *P. typhoideum* L.C. Rich and*P. americanum* (L.) Leeke. The four cultivated forms of pearl millet are *typhoides* (found mainly in India and Africa), *nigritarum* (dominant in eastern Sahel), *globosum* (dominant in the western Sahel, probably originating in Sahelian Africa in a diffuse belt stretching from western Sudan to Senegal), and *leonis* (dominant on the West African coast) [47–49].

Linnaeus [50] originally placed pearl millet cultigens into two separate species (*P. glaucum* and *P. americanum*) of the genus *Panicum*. Later, he moved several of these elements to the genus *Holcus* [51]. Rechard [52] grouped pearl millet along with a number of species previously listed under both *Panicum* L. and *Cenchrus* L. in a new genus, *Pennisetum.* Willdenow [53], however, established the genus *Penicillaria* to include pearl millet, but Steudel [54] reduced it to its present status as a section in *Pennisetum.* He merged many variants of pearl millet into a single polymorphic species, recognized as *P. typhoideum* L. Rich. The limits of the section were expanded by Leeke [55] to include all those wild species of *Pennisetum* having penicillate anther tips and involucral bristles. The generic name *Pennisetum* has been derived from two Latin words, *Penna* and *Seta*, meaning feather and bristles (i.e., feathery bristles). The most extensive treatment of the genus *Pennisetum* was contributed by Stapf and Hubbard [56], who divided the genus into five sections: *Gymnothrix*, *Brevuvalvula*, *Penicillaria*, *Heterostachya*, and *Eu-pennisetum.* Cultivated pearl millet and its wild and weedy relatives were included in section *Penicillaria*, which included 14 cultivated, 6 wild, and 13 intermediate species. Brunken [47] further reduced the number of species in section *Penicillaria* to two, on morphological and cytological grounds; *P. purpureum* was maintained as a separate species because of its tetraploid chromosome number and perennial lifecycle.

All the diploid cultivated, weedy, and wild taxa that frequently hybridize without genetic barriers are classified under a single species, *P. americanum.* Based on the morphology and adaptive strategies to domestication, *P. americanum* was further divided into three subspecies: *americanum,* including the cultivated forms; *monodii,* including the wild forms; and *stenostachyum,* with the weedy forms. Clayton and Renvoize [57] demonstrated that the taxonomically correct name for cultivated pearl millet is *P. glaucum.* They recognized the weedy forms (colloquially called *shibra*) as *P. sieberanum* and their wild progenitor as *P. violaceum. P. violaceum* differs from pearl millet in having involucres that are sessile, deciduous at maturity, and always contain a single spikelet.

The geographical origin and the center of domestication of pearl millet are situated in western Africa. The plant was subsequently introduced into India, where the earliest archaeological records date back to 2000 BC [48, 58–60]. Records exist for cultivation of pearl millet in the United States in the 1850s, and the crop was introduced into Brazil in the 1960s. The oldest findings of wild and domesticated pearl millet were recorded at about 3500 BC in Dhar Tichitt, a Saharan site in Mauritania [61]. Birimi in northern Ghana has laid claim to one of the earliest findings of domesticated pearl millet, dated at about 1459 BC [62, 63].

These archaeobotanical findings in the Sahara and Sahel confirm the hypothesis of original distribution and widespread utilization of wild and cultivated pearl millet across sub-Saharan Africa [61, 62]. However, there is dispute among scholars as to whether pearl millet has a single center of origin or more than one place of origin (the so-called non-centers), which would have resulted from domestication processes occurring independently in several regions. According to the latter hypothesis, the whole Sahel, from Mauritania to western Sudan, was originally covered with these non-centers [60, 64–67]. Whether domestication took place as multiple parallel processes in the non-centers in several places along the Sahelian distribution belt of the wild progenitors or at one specific place [60, 68], the ultimate center of origin of the wild progenitors, *P. monodii* and *P. mollissimum*, is most likely to be situated in the Sahara desert [62, 63, 69].

Based on the distribution of pearl millet throughout the continent, the uniform cradle of domestication is likely to be the regions of Mauritania, Senegal, and western Mali [48, 68, 70]. Today's cultivated forms developed out of this domestication cradle [71]. Next, these first early-maturing forms of domesticated pearl millet were carried eastwards, facilitated by their efficient adaptation to arid conditions [62]. About 3,000 years BC, the first translocation carried the crop to eastern Africa [68, 71] and then to India, where 3000-year-old carbonized pearl millet was detected at a site on the eastern coast [63, 69].

Another diffusion took place in the region near Lake Chad, more precisely on the Nigerian side [72], where a secondary center of diversity developed at about 2010 BC [68, 73]. There, photoperiod-sensitive cultivars were selected, which adapted to the more humid conditions in the southern Sudanian zone [62, 71]. These late-maturing cultivars were transported further towards the Sudanian zone of southwestern Africa, from northern Nigeria to southern Senegal, as evidenced by the above-mentioned findings in northern Ghana [68, 73]. The third and last major translocation took pearl millet towards the plateaus of southern Africa, across Uganda and towards Namibia, at about 1000 BC [68, 71].

2.2.3 Overview of Pearl Millet Collections

The International Crops Research Institute for the Semi-Arid Tropics (ICRISAT) has the single largest consolidated collection of pearl millet in the world, which comprises a total of 22,211 accessions. Of these, 750 are of wild species (24 species), 19,377 are landraces, 132 are improved cultivars, 1,943 are breeding/research materials, and 25 are others [74]. India contributed a significant number of pearl millet accessions to the global collection maintained at ICRISAT (6,647 accessions). The remaining accessions were collected from about 51 countries. The major diversity centers of pearl millet are considered to be relatively well represented in the collection at ICRISAT. In addition to ICRISAT, the Institut de

Recherche pour le Développement in France also maintains 3,968 accessions of pearl millet from 16 countries. Collection of these accessions was supported by Bioversity International and Office de la Recherche Scientifique et Technique d'Outre-Mer. The Canadian Genetic Resources Programme in Saskatoon, Canada maintains 3,821 accessions covering a few species, with emphasis on *Pennisetum glaucum* (3390 accessions). Accessions of other species include: *P. violaceum* (221), *P. macrourum* (1), *P. purpureum* (12), *P. orientale* (1), *P. pedicellatum* (11), *P. polystachion* (8), *P. ramosum* (3), *P. unisetum* (1), and other species (14). The number of seeds maintained in these collections is moderate for long-term conservation; it is intended as safety duplication but not for distribution. In addition to these global collections, the Agricultural Research Station of the U.S. Department of Agriculture (USDA) at Griffin, Georgia maintains 1,314 accessions from 32 countries, of which only 1 is a wild relative, 290 are of breeding/research material, and 552 are for other purposes.

Among the national collections, the largest was recorded from the Indian genebank at the National Bureau of Plant Genetic Resources (NBPGR), which maintains 8,913 accessions under long-term conservation. Most of the accessions are indigenous (8,827), with only 168 accessions from other countries. The Indian collection also includes 221 advanced improved varieties and 272 accessions of breeding/research material. No other countries in South Asia, except Pakistan (193 accessions), have reported pearl millet collections. Among African countries, the collections were reported from gene banks based in Algeria, Benin, Botswana, Burkina Faso, Burundi, Cameroon, Central African Republic, Chad, Eritrea, Ethiopia, Ghana, Kenya, Malawi, Mali, Mauritania, Mozambique, Namibia, Niger, Nigeria, Senegal, Sierra Leone, Sudan, Democratic Republic of the Congo, Uganda, Zambia, and Zimbabwe. In total, 56,580 pearl millet accessions were recorded from various sources. Landraces represent the largest proportion of pearl millet germplasm conserved in gene banks worldwide (49,973 accessions). Of these, only 3 % are wild relatives collections (1,630), 0.80 % are advanced improved varieties (452), 6 % are breeding/research materials (3,600), and 2 % are of unknown description (947).

Some progress has been made in the recent past in mapping the pearl millet diversity collected worldwide. Global databases show that georeference data have been assigned to 16,855 accessions. These collections are being maintained at ICRISAT (13542 accessions), USDA Agricultural Research Service (ARS) (472 accessions) and the International Livestock Research Institute (13 accessions). With support from Bioversity International, 2,828 accessions were collected and became part of the global collections being maintained by ICRISAT and USDA-ARS. Not much information is available for any of the national collections. Based on the georeference information from the global database, the distribution pattern of these accessions is shown in Fig. 2, whereas the mapping of *Pennisetum* species (excluding *Pennisetum glaucum*) is shown in Fig. 3.

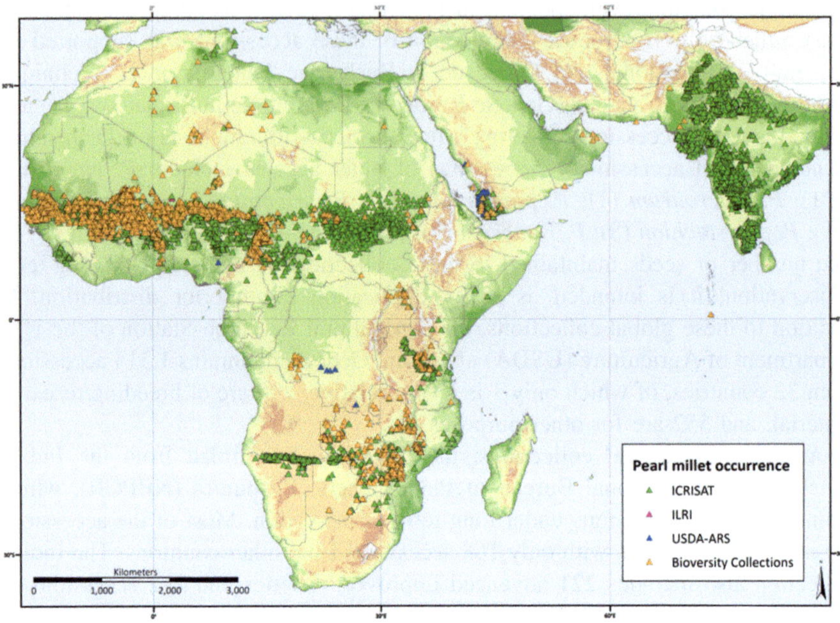

Fig. 2 Mapping of pearl millet accessions based on information available from the global database

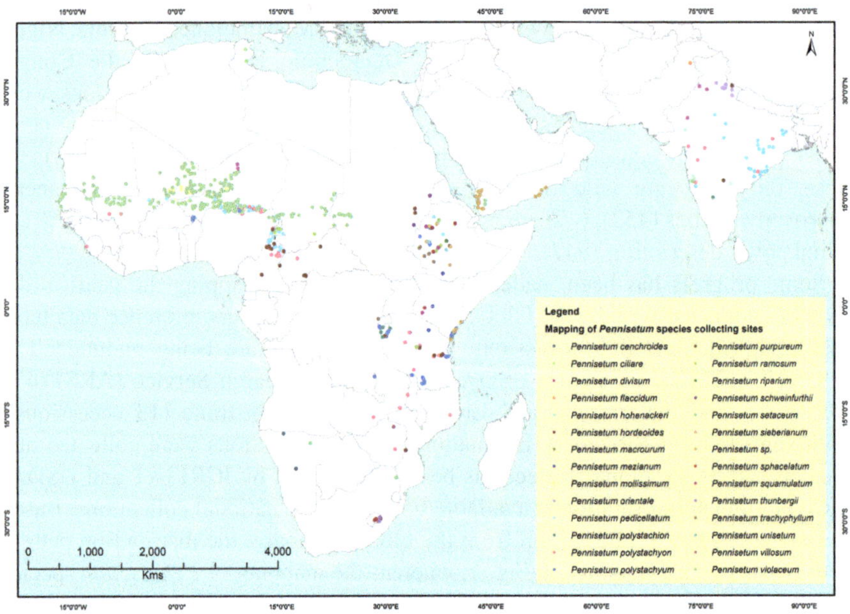

Fig. 3 Mapping of *Pennisetum* species (excluding *Pennisetum glaucum*) accession collecting sites based on information obtained from the global accession level information portal GENESYS, developed by Bioversity International (http://www.genesys-pgr.org/)

2.2.4 Characterization and Evaluation

From a global survey undertaken by Bioversity International, it appears that although a modest number of accessions have been assembled and maintained in many countries, systematic characterization and evaluation activities are not sufficient. One reason provided by most gene banks is lack of adequate human and financial resources. Evaluation activities, especially in Africa, have been fewer than hoped. In the ICRISAT gene bank, all cultivated accessions have been characterized and evaluated for 23 morphoagronomic characters following the descriptors for pearl millet [75]. Selected pearl millet germplasm accessions of Indian and African origin were evaluated by NBPGR for important agronomic characters at different locations in India in collaboration with ICRISAT and catalogues were published [76, 77]. Considerable phenotypic diversity was observed for almost all quantitative traits. Distribution of qualitative traits indicates occurrence of nine panicle shapes (cylindrical, conical, spindle, club, candle, dumb bell, lanceolate, oblanceolate, and globose), five seed shapes (obovate, oblanceolate, elliptical,hexagonal, globular) and ten seed colors (ivory, cream, yellow, grey, dark grey, grey-brown, brown, purple, purplish-black, and mixture of white and grey) in the entire collection. Accessions with candle-shaped panicles, short bristled panicles, globular seed shape, grey seed color, and seeds with partly corneous endosperm texture are predominant in the collection maintained at ICRISAT [74].

Based on the characterization and evaluation information contained in the database, sources of resistance to biotic and abiotic stresses, adaptation, and nutritional qualitative traits have been identified. It was also reported that some of the landraces have wide adaptation and are therefore very useful in light of the changing climate scenario and for use in crop improvement programs. Much of this diversity is still available in areas of early cultivation in Africa and regions of early introduction in Asia.

2.2.5 Utilization

Identification of useful germplasm for crop improvement is the first step in encouraging utilization. From the information obtained through the Bioversity International survey questionnaire, it was difficult to obtain a good comparison of utilization activities being carried out in various genebanks, especially in Africa. However, in India, modest efforts have been undertaken in the last three decades to exploit pearl millet germplasm with useful genes for crop improvement, especially in developing composites. The *iniadi* germplasm from the Togo-Ghana-Burkina Faso-Benin region of Western Africa is most commonly used in pearl millet breeding programs worldwide [78]. At ICRISAT, a small seed sample of each accession is available on request to all research workers under the Standard Material Transfer Agreement of the International Treaty on Plant Genetic Resources for Food and Agriculture [74]. To further enhance the utilization of pearl millet germplasm, ICRISAT evaluated sets of selected germplasm accessions at different locations in India and several other countries in Africa; trait-specific genepools (early maturing, high tillering, large panicle, and large grain) were developed to provide partially conserved genotypes to the breeders. There has been a general lack of interest in using wild species because of the large genetic variability in pearl millet landraces. However, some wild species are very useful in pearl millet improvement programs, notably *P. glaucum* subsp. *monodii* for new source of cytoplasmic-nuclear male sterility (CMS); *P. purpureum* for forage, stiff stalk, and restorer genes of the A1 CMS system; *P. orientale* for drought tolerance and forage; *P. schweinfurthii* for large seed; *P. pedicellatum* and *P. polystachion* for downy mildew resistance, and *P. squamulatum* for apomictic gene [48].

In general, the Indian pearl millet landraces have contributed to earliness, high tillering, high harvest index, and local adaptation, whereas African materials have been a good source of high head volume, large seed size, and disease resistance. In order to enhance the use of these genetic resources, Bioversity International, in consultation with various partners, has developed a comprehensive list of descriptors for pearl millet [79]. This strategic set of descriptors, together with passport data, are an integral part of the information available through the global accession-level information portal GENESYS. It will facilitate access to and utilization of pearl millet accessions held in gene banks worldwide.

2.2.6 Distribution of *Pennisetum* Species and Gaps in World Collection

A study of the distribution of *Pennisetum* species and of gaps in world collections was undertaken by Bioversity International with support from the Global Crop Diversity Trust (GCDT) and the World Bank, using datasets of herbarium collections, as well as the germplasm collections available from Global Biodiversity Information Facility and the System-wide Information Network for Genetic Resources and the climate database available at WorldClim. Based on the available records, 53 wild species and 2 infraspecific taxa have been identified, accounting for a total of 55 taxa for the genus *Pennisetum*. These different taxa are classified as follows, according to their closeness to the cropped species *P. glaucum,* using the Maxted and Kell [80] model. The analysis dataset contained 4,326 observations (http://gisweb.ciat.cgiar.org/GapAnalysis/) with 3,364 (78 %) being herbarium specimens and 962 (22 %) being gene bank accessions. The average number of total samples per taxon was 79, indicating that available data is not particularly limited, although it is concentrated in certain taxa (i.e. *P. ciliare*, *P. polystachion*, *P. purpureum*, *P. violaceum*, *P. clandestinum*, *P. villosum*). Other taxa such as *P. domingense*, *P. lanatum,* and *P. sieberianum* present a very limited sampling and/or data availability and thus need further characterization and sampling in order to obtain a reliable ecogeographic evaluation.

The gap analysis of the *Pennisetum* genepool showed that there are 47 out of 55 taxa under analysis that are either underrepresented or not represented in gene banks; these were therefore flagged as high-priority species. Twenty-six of these taxa presented only 10 data points, which indicates that these species in particular need to be further collected. Only species *P. violaceum* was found to be adequately represented in gene banks, while *P. ciliare*, P. *flaccidum*, *P. orientale,* and *P. pedicellatum* were found to be relatively underrepresented and thus flagged as medium-priority species. Based on the analysis, all these species have been identified as high priority for conservation.

2.3 Quinoa Diversity and Its Potential in the New Millennium

Quinoa (*Chenopodium quinoa* Willd.) is an ancient native Andean grain that was extensively cultivated in the Andean region by pre-Columbian cultures, such as the Tiahuanacota and Inca, from around 5,000 years ago. This grain was used in the diet of the settlers from the inter-Andean valleys and the high plateaus along with other native species such as potato (*Solanum tuberosum* L.), oca (*Oxalis tuberosa* Molina), amaranth (*Amaranthus caudatus* L.), chili peppers (*Capsicum* sp.), among others [81, 82]. Quinoa is an herbaceous annual Chenopod that has played not only a vital role in family food security and farmer livelihoods, but also in economic, social, ecological, nutritional, and cultural contexts [83, 84]. Quinoa plants are part of various Andean ecosystems; the grains and leaves of this crop are used for food while its subproducts are used for forage, fuel wood, as well as in rituals and handicrafts [85].

This crop, however, began to be marginalized in the sixteenth century with the introduction of cereals such as barley and wheat [81, 86], leading to a significant decrease of quinoa-cultivated areas in the Andean countries [86]. Nevertheless, quinoa cultivation continued in marginal areas. For many decades, quinoa has been considered as a food for the poor and the peasants, but now it is considered to be today's "golden grain". In fact, quinoa was catalogued by FAO in 1996 as "one of the most promissory crops for humanity not only due to its high nutritional value and versatility, but also because it offers alternatives to solve the increasingly serious problems of human nutrition" [86].

Quinoa′s key asset lies in its potential as a high quality source food. Quinoa has high concentrations of some essential amino acids (lysine, methionine, threonine, and tryptophan) that usually limit the quality of human diet. Quinoa is also rich in oligo elements, vitamins, and is gluten free, just to mention some of its nutritional properties [87]. The high content of these amino acids plus the high-quality protein found in its grain can cover the nutritional requirements for schoolchildren and even adults [88], which are serious problems in some populations around the world.

The Andean region is one of eight centers of origin and diversity of cultivated plants in the world described by Vavilov in 1953; that is, it is one of the regions that has the highest diversity of cultivated crops and their wild relatives. In this region and specifically around Lake Titicaca, a high plateau region in Bolivia and Peru, the highest genetic diversity of wild and cultivated quinoa genotypes are found in farmer fields [81, 89, 90].

Agriculture in the Andean highlands is characterized by a high degree of risk due to a range of harsh climatic factors such as frost, hail, wind, drought, high radiation, and poor and saline soils [91]. It is highly probable that during domestication, Andean farmers selected certain genotypes based on their use and their tolerance to adverse biotic and abiotic factors, thereby obtaining current plants and ecotypes that possess high diversity of traits. Some of these ecotypes have been strictly selected as a source of food (finding up to hundred different recipes), while others have been selected based upon their tolerance to salinity, poor soils, cold climate, frost and hail, drought, and flooding. It is important to remark that some genotypes have also been selected due to their high yield potential and precocity [89, 92, 93].

During the domestication of quinoa by Andean farmers, a wide range of morphological modifications have occurred to the plant, such as the condensation of the inflorescence in the highest part of the plant, an increase in plant and seed size, reduction of the testa, loss of seed dormancy and dispersion mechanisms, and high variation in levels of pigmentation, among others. Today, we can find quinoa plants with high production of larger seeds and clear colors; these traits reflect the long time man has been using, selecting, and cultivating this species [93]. Although this species has been completely domesticated, the seeds still contain saponin, which must be extracted before consumption [86]. Quinoa ancestors and relatives still exhibit the wild traits of the crop [93].

The genetic variability of quinoa is huge [91]. For example, one of the most popular varieties grown today in Bolivia (*Quinua real)* has at least 73 ecotypes. This variety is extremely well adapted to saline environments and has big grains. In addition, 47 local landraces and approximately 10 improved varieties are also found (Table 1).

In other quinoa-producing countries within the Andean region, a considerable diversity of local and improved varieties can also be found [86, 92]. The existence of this vast amount of ecotypes, improved and local varieties, and/or landraces are the result of the intense breeding processes, domestication, and use and conservation practices adopted by ancient and contemporary Andean farmer communities based on their traditional knowledge [86].

Currently, the demand for quinoa in the national and international markets has increased; quinoa production areas have also shown this tendency due to high market prices. However, this demand has preference for larger seed size, uniform color and white color, and saponin-free seeds [93, 94]. This has progressively caused the domination of a group of cultivated varieties with these traits, such as various ecotypes of *Quinua real* [89, 94]. Although markets have preferred these varieties, Bonifacio et al. [94] reported that, in the last couple of years, those with red and brown seeds have also been accepted now in some markets, although not to the same extent. Nevertheless, this market preference can be risky as it can further lead to genetic erosion and the loss of some varieties. Unfortunately, this process is in progress, and it has been stated that three varieties of quinoa and a

Table 1 Common names of varieties, landraces, or ecotypes found in Bolivia

Varieties or landraces	Common name
Quinua Real	Achachino[a], achachino rojo, café chullpa, canchis/qanchis blanca, canchis roja, carequimeña, ch'illpi amapola, ch'ullpi blanco, ch'ullpi rojo, ch'ullpi rosado, chachahua, challamoko, challamuro, chhuku puñete, chillpi, chipaya, hilo, huallata/wallata, imilla, intinayra[a], jiskitu, kairoja, kellu/q'illu[a], lipeña, mañiqueña, mañiqueña nor lípez, mañiqueña palaya, manzano, mok'o rosado/moqu rosado, moqu, moqu chacala, mururata, negra, negra blanquita, negra blanquita planta roja, pandela/rosada[a], pandela amarilla, perlasa, phisanqalla 3 hermanos, phisanqalla amarantiforme, phisanqalla hembra, phisanqalla macho, pisankalla (ayrampu)/pasankalla[a], pucauya, puñete, punta blanca, q'illu puñete, q'uitu/koitu, q'uitu rojo, qanchis amarillo, qanchis anaranjado, qanchis rosado, qhaslala blanca, quinua roja, real blanca/real[a], romerilla, rosa blanca, santa maría, señora, sorata, tacagua, timsa/timza 1, timza nor lípez, toledo[a], toledo anaranjado, toledo rojo, tres hermanos/siete hermanos, tupita, ucaya, utusaya, utusaya local chacala, wila jipiña, wilalaca
Other local landraces	Acu juira, ajara (wild), amarilla, amarilla maranganí, arroz jupha, blanca, blanca de july, choq'e pito, chuchi jiura, churo iri, cochasqui, coytu, cuntur naira, elva, granadilla, imbaya, ingapirca, iry, janko cayun juira, janko jhupa, juchuy mojo, kaslala/kaslali matizada, katamari, kcancolla, kelly juira, llulluchi, mezcla, misa jupa, mixtura, negra, negra de oruro, noventona, palco, ploma, pureja, quilliwillu, roja, roja coporaque, sallami, siki, tunkahuan, waca misu, waranta, wila cayun janq'o, wilacoimi, witulla, yubi
Improved varieties	Chuca paca[a], jacha grano, jilata, kamiri, ratuqui, robura, sajama[a], samaranti, sayaña[a], surumi[a]

[a] Most commonly used by communities
Sources [89, 93, 94, 195–197]

wild relative have already been included in the International Union for Conservation of Nature red list categories as endangered (EN), near threatened (NT) and least concern (LC) [95].

Apaza et al. [96] reported that the highest diversity of quinoa is found in *aynokas, mandas*, and *laymes*, a group of traditional organization systems or communal fields maintained by farmer communities in field borders and sacred places (all called *Gentil wasi* or *Phiru*) using local cultivation methods and traditional knowledge. If these places continue being eroded and the traditional practices and knowledge become lost by the adoption of modern culture practices, it is highly probable that the diversity present in these sites will disappear.

To conserve the enormous genetic diversity of quinoa and some of its wild relatives, various germplasm banks were created in the 1960s throughout the South American region. These banks are being managed by public or private entities interested in the *ex situ* conservation of plant genetic resources, thereby finding more facilities in countries with higher diversity (Table 2). Collections of quinoa germplasm began in this period with the help and sponsorship of various private and public organizations, donors, and projects that still subsist today.

Table 2 Approximate number of quinoa accessions conserved in *ex situ* facilities across South America

Country	Number of entities that conserve quinoa germplasm	Number of accessions conserved (countries of origin)	Access to the material
Argentina	2	63 (Argentina, Bolivia, Chile, Ecuador, Peru, U.S.A.)	Yes
Bolivia	8	More than 7,077 (Bolivia, Peru, Ecuador, Colombia, Argentina, Chile, Mexico, U.S.A., Denmark, Holland and England)	Yes
Chile	3	152 (Chile)	Yes
Colombia	3	More than 328 (Colombia, other)	Yes
Ecuador	1	608 (Ecuador, Bolivia, Peru, Colombia, Argentina)	Yes
Peru	12	More than 4,431 (Peru, Bolivia, Colombia, Ecuador)	Yes

Sources [86, 198, 199]

Although these banks keep an important number of quinoa accessions, the variability stored in these collections do not represent all the Andean diversity [86], especially when it comes to wild populations and wild relatives [97]. There are at least ten wild species or wild relatives and a couple of subspecies of the *Chenopodium* genus, known as wild quinoa or *ajara*, that are used for certain recipes by Andean farmers. These wild species are not being properly conserved.

Since 2001, some quinoa varieties and wild relatives, such as *cañahua* (*Chenopodium pallidicaule* Aellen), have been evaluated for yield, harvest index, postharvest, market, industrialization potential, new products and uses, among other fields, within the framework of the International Fund for Agricultural Development Neglected and Underutilized Species–Bioversity International Project 2001–2014. Additionally, other wild relatives are being evaluated for their potential to improve the tolerance to biotic and abiotic stresses as part of a United Nations Environment Programme/Global Environment Facility project for nutritive properties [90].

To date, results show that there is a huge potential in the conservation and use of quinoa diversity and its wild relatives. Studies also show that there is considerable variation among cultivars for a wide range of traits. This fact allows users to take advantage of quinoa's versatility for use in more than 100 different preparations and products. Other than its cosmetic and industrial and pharmaceutical uses, quinoa can also be used as forage, medicine, pesticide, ornament, and fuel wood [91, 93, 98]. In addition, quinoa has a promising market potential for high-value grain, subproducts, and byproducts that still remain unexploited [99].

Both quinoa and its wild relatives possess an enormous reservoir of genetic variation that can be exploited through plant breeding. They are also an essential resource to meet the challenge to improve food security, enhance agricultural production, and sustain productivity in the context of a growing world population and the threats of climate change [90, 100]. Likewise, quinoa diversity can be greatly exploited through biotechnological approaches to the benefit of agricultural industries [98].

2.4 A Practical Use of Agrobiodiversity in Cereal Crops to Mitigate Climate Change through Regulation of Soil Nitrification

It has been shown repeatedly how the exploration of the natural diversity in cereals -and in crops in general- has been the solution for multiple challenges that naturally arise as a result of agricultural practices (tolerance to biotic and abiotic stresses, yield increments, adaptation to different conditions other than the ones in the center of origin, etc.). Along with the need of yield increments in the main staples cereals -as the global population is exponentially growing-, climate change is a prominent issue that is challenging agriculture and even mankind. Agriculture is an important source of anthropogenic emissions of the greenhouse gases (GHG) methane (CH_4) and nitrous oxide (N_2O) associated with nitrogen (N) fertilizer production [101]. It is now known that N_2O has a higher ozone-depletion potential than any other reactive chemical, including carbon dioxide [102], and also that agricultural practices are the major source of N_2O emissions to the atmosphere [103]. Under this scenario, a disjunctive stands, either to continue practicing agriculture for food and feed in the way we are currently doing or to look into alternatives to implement a "climate-smart" agriculture, where we as humans not only aim to feed the world but also care for the environment. Once again, researchers have found that natural diversity in certain crops confer characteristics to some lines that are "eco-friendly" and that, for example, reduce the emission of GHG as CH_4 and N_2O [104].

Much of the N_2O produced by agriculture is generated from the use of N fertilizers that, after application to soil, feed the nitrification reaction; in this process, nitrifying microorganisms take ammonium and convert it into nitrite and finally into nitrate, a compound susceptible to leaching, thereby contaminating bodies of water. Nitrate is eventually reduced by denitrification, releasing N_2O gas as a byproduct [105]. Researchers from the International Center for Tropical Agriculture and the Japan International Research Center for Agricultural Sciences observed that the tropical pasture *Brachiaria* spp. has the ability to inhibit the soil nitrification process by releasing chemical compounds from its root system to the soil; a compound with major inhibition capacity was identified and named brachialactone [106]. This phenomenon was termed biological nitrification inhibition (BNI) [107]. The BNI trait would decrease costs for the farmer because the N applied will stay longer in the soil and the plants would have a better chance to intake it before it is converted into forms that are prone to leaching and losses via gaseous forms, thereby improving the ecoefficiency of agricultural practices by diminishing the amount of N_2O emissions that are released to the atmosphere. Subbarao et al. [106] demonstrated under field conditions that plots of *Brachiaria* pastures significantly reduced N_2O emissions when compared to bare soil, soybean, and guinea grass plots.

After the BNI concept was brought to light, researchers identified the great potential that this phenomenon has in agricultural practices by reducing the N contaminants (i.e., nitrate and N_2O); therefore, this strategy can be applied to mitigate global warming largely occasioned by GHG. For that reason, Subbarao et al. [108] achieved and attempt to identify in which other plant species the BNI potential was presented, as well as to what extent there was genetic variability of the BNI potential in cereal crops. The outcome of this investigation showed that BNI is widely existent in pasture grasses, but the investigation also reported that the roots of two cereal crops *Sorghum bicolor* (L.) Moench var. hybrid sorgo and *Pennisetum glaucum* (L.) R. Br. var. CIVT (Pearl millet) release chemicals to the soil that reduce the population of nitrifier microorganism and therefore inhibit the nitrification rates. Other cereals, such as rice, maize, barley, and wheat, were also tested in this preliminary screening with negative results (no BNI activity detected), but in most cases using single lines; therefore, it is recommended to test a diversity panel for each species to really inspect the BNI trait on each cereal crop. Tanaka et al. [109] tested the BNI activity in the root exudates of 36 different rice genotypes and found substantial genotypic variation for BNI, with the upland cultivar IAC25 expressing high BNI activity, whereas lowland varieties such as Nipponbare or IR64 exhibited low BNI activity.

Although initial evaluations in wheat showed that the roots of this cereal do not exudate chemical compounds to the soil that successfully control the nitrification process, further analysis of *Leymus racemosus* (mammoth wild rye, a wild wheat relative), resulted in the discovery of a high BNI capacity that efficiently suppressed soil nitrification [110]. The exploration of genetic diversity of wild predecessors of wheat showed how the BNI trait has been lost in the course of decades of breeding and selection of "desired" (at a given time and condition) agronomic conditions that inevitable lead to accidental loss of other valuable traits. As a result, the chromosomal location of the genes conferring the BNI trait has been identified and wheat varieties expressing the BNI trait have been formed by the production of wheat–*Leymus racemosus* chromosome addition lines [111].

The cereal crop in which significant BNI research has been conducted is sorghum, and both hydrophilic and hydrophobic root exudates with BNI activity have been identified [112, 113]. An ample natural diversity has also been identified among sorghum lines for the release of sorgoleone (a major hydrophobic root exudate with the highest BNI capacity). Nimbal et al. [114] evaluated sorgoleone production among 25 sorghum lines, finding nearly a 30-fold variation. Current efforts will focus on the exploration of sorghum diversity to exploit the BNI trait and to identify sorghum genetic stocks with high potential to release chemicals to the soil that suppress nitrification, reduce N_2O emissions, and improve nitrogen use efficiency in sorghum-based production systems [113].

Cereal production uses the most of N-fertilizers; wheat itself accounts for a third of the global production [115] to cope with the food demand of a growing population. The indiscriminate use of N fertilizers results in N contaminants (e.g. nitrate and nitrite) that are causing major problems to bodies of water and are also contributing to global warming. BNI function represents a novel opportunity to naturally establish agricultural systems with improved N use efficiency and reduced N-contaminants. Therefore, the exploration of diversity of major cereal crops is imperative to promote the ecoefficiency of agricultural systems.

2.5 Impact of Nitrogen Pollutants on Biodiversity

Biodiversity is declining at an exceptional rate and on a worldwide scale. Indeed, loss of ecosystem functions and services associated with such declines has generated international debate [116–118]. Agricultural crops can be injured when exposed to high concentrations of various air, water, and soil pollutants. Air pollution affects plants in many ways, which have implications for overall biodiversity and ecology. There is evidence that air pollution can reduce some plants' ability to reproduce, thus causing long-term changes to population ecology [119]. Of the different kinds of pollutants damaging the environment, nitrate pollution is a major problem along with the pollution of the atmosphere by ammonia and oxides of nitrogen. Nitrogen is a beneficial plant fertilizer in small amounts, but large amounts cause negative impacts on ecosystems and serious threat to biodiversity of many groups of organisms, including diversity of plants [120–122]. Nitrogen deposition refers to the input of reactive nitrogen species from the atmosphere to the biosphere. At the global scale, current N emission scenarios project that most regions will have increased rates of atmospheric N deposition in 2030 [123], which is causing concern about significant impacts on global plant biodiversity [116, 117, 124].

Even though low-to-medium levels of N addition ($\leq$100 kg N/ha/year) generally did not alter plant diversity through time, high levels of N application significantly reduced species diversity [125]. The declines of diversity appeared to arise from N-related changes in soil properties, such as significant decreases in pH and extractable calcium (Ca) and increases in extractable aluminum (Al). Research reports revealed that N deposition may have shifted plant communities towards species composition typical of high-N availability. This shift has often been associated with a loss in diversity of plant species, particularly in areas with high deposition rates [126]. Mechanisms underlying the declines of diversity include competitive exclusion of more N-efficient dominant species by relatively fast-growing nitrophilic species, as a result of high-N availability induced by N deposition. Other such mechanisms include increased susceptibility to secondary stress and disturbance factors and species invasions [127, 128]. The research review of N-addition experiments across the tropics and subtropics have shown that N deposition may potentially affect plant diversity in some ecosystems more than originally thought, and because atmospheric N loads are gradually increasing in some tropical areas such as Asia, research on this topic is now urgently required [127].

Although improved agronomic approaches are one way to enhance nitrogen use efficiency (NUE) and reduce the N inputs, there is a growing interest in understanding the genetics of NUE in crop plants. Integration of these approaches may reduce both N deposition and the rapid loss of biodiversity.

2.6 Transgenic Crops and Food Security

The development of technologies for plant transformation started in the early 1980s following the production of chimeric genes [129–132], transformation vectors [133, 134], and DNA delivery systems [135–139], combined with plant tissue culture regeneration systems, which were pioneered by Murashige and Skoog [140]. This continued over the decades with the development of tissue culture for the three most important cereals: rice [141, 142], maize [143, 144], and wheat [145, 146]. The transformation of cereals since the 1990s was achieved with *Agrobacterium*-based protocols for rice [147], maize [148], and wheat [149], usually with a low-copy insertion of transgenes [150–152] and the *gus* reporter gene [153] with antibiotic selectable marker genes [154, 155]. These techniques were the most widely used in earlier transgenic plant research and crop development. Currently, more than 50 selectable marker genes are available for transgenic research and commercialization, and they can be divided into several categories depending on the mode of action and substrates used [156–158].

An important milestone in plant transgenic technology is the ability to generate plants free of the selectable marker gene using co-transformation with multiple T-DNAs [159, 160]. Among the major advances in plant transformation technology, we can consider the understanding of the plant cell–*Agrobacterium* relationship [161] and the molecular mechanisms of T-DNA transfer [162, 163] together with integration into the plant genome [164, 165].

Vain [166, 167] reviewed publication trends in and scientific knowledge on plant transgenic science and technologies available worldwide. In addition, the security of transgenic plants and products has been scrutinized [168–172], including the economic, environmental, and social dimensions [173].

The newest discoveries describing gene silencing [174], gene targeting [175], and RNA interference [176, 177], with alteration of the transcriptional activity of genes by using zinc-fingers nucleases [178, 179], together with the possibility of genome editing by TALENs (transcription activator-like effector nucleases) [180, 181], including the technology of using site-directed nucleases, are providing an opportunity to develop a new generation of transgenic crops with both improved traits and minimized potential for unintended effects that can impact safety [182].

It is very well known that agricultural production in the twenty-first century is going to face a number of new challenges [183–186]. A staple crop such as rice, for which more than half of the global population already depends on its production, will be under more and more pressure to increase its yield steadily; this has happened during the last four decades in some countries, where rice yield doubled or tripled [187]. New and multi-integrated strategies, including functional genomics, phenomics, and transgenics, coupled with conventional rice breeding, can help to develop new rice cultivars (referred to by Zhang [188] as Green Super Rice) and can meet future demand for world rice production. Take into consideration the fact that, from 1996 to 2012, millions of farmers in 30 countries worldwide adopted biotech/genetically modified organism (GMO) crops at an unprecedented level, going from 1.7 million ha in 1996 to 170 million ha in 2012 [189], a 100-fold increase. In addition, Golden Rice [190] is expected to be released in the Philippines in 2014, together with drought-tolerant sugarcane in Indonesia, biotech maize [191], and rice [192] in China. As a result, up to a billion poor people in rice households in Asia could benefit. Beyond 2015, it is difficult to predict what will come to the market, but expectations are also high in Africa, where the private/public partnership Water Efficient Maize for Africa [193] will release the first biotech drought-tolerant maize in the sub-Saharan region, where the need for drought-tolerant crops is greatest.

Commercialized crops could help to reduce the impacts of agriculture on biodiversity by alleviating pressure to convert additional land into agricultural use [194].

3 Summary and Conclusions

The management of plant genetic resources for food and agriculture is becoming a central issue in the context of sustainable agriculture, climate change, and food security. As the world faces unprecedented challenges, the exploration, use, and conservation of agricultural biodiversity will play a major role to address the most critical socioeconomical and environmental issues concerning agricultural research and food security strategies. The agricultural biodiversity of staple small grains, such as rice, maize, wheat, millets, and quinoa, is currently being studied from several perspectives to offer information to researchers and breeders worldwide.

Only a few accessions of pearl millets have been maintained so far in several countries. In-depth characterization and evaluation of these accessions remain insufficient due to lack of human and financial resources. Nevertheless, significant phenotypic diversity has been observed for most quantitative traits in a set of selected pearl millet germplasm accessions. This diversity is therefore very useful for use in crop improvement programs. Although the genetic variability in pearl millet landraces is large, some wild species have also been considered as a source of genes controlling agronomically important traits.

In the light of land change uses, modernization of agricultural practices, and climate change, which eventually affect plant diversity patterns, transnational efforts have focus on the establishment of extensive genebank collections for cereal crops. Both in situ and *ex situ* conservation of plant genetic resources are complementary strategies for the preservation of agricultural biodiversity. Geospatial analysis tools can be applied to map microsatellite diversity of cereal accessions characterized molecularly in order to identify priority geographic areas for conservation.

Although quinoa and its wild relatives are a reservoir of rich genetic variation that can be used by breeders for further improvement of the species, quinoa diversity is being lost. Market demand has encouraged the cultivation of commercially valuable varieties with certain characteristics, neglecting the immense variation found within the species. Some local quinoa genebanks have been created, containing many known quinoa varieties with a wide range of diverse agronomic and nutritional properties. However, current quinoa genebanks still do not represent the Andean diversity of the cultivated and wild relatives.

Agrobiodiversity can provide additional sources of new traits. Some cereal crops have the ability to inhibit soil nitrification and reduce the emissions of greenhouse gases (BNI activity) by releasing root exudates. The screening of some cereal accessions revealed genotypic variation for BNI. The exploration of agrobiodiversity for the BNI trait offers a novel strategy to improve the nitrogen use efficiency and increase crop productivity while reducing the environmental impacts of agriculture.

Chemical pollutants that are released into the environment are posing a threat to plant biodiversity. Nitrogen pollution is a major problem worldwide due to the indiscriminate use of synthetic nitrogen fertilizers in agricultural systems. This practice affects the natural balance of soil ecosystems, leading to significant losses of plant biodiversity as a result of changes in soil properties and plant community structure. Research addressing the genetics of nitrogen use efficiency in crop plants plus the adoption of sustainable agronomic practices offer an integrated approach to minimize the negative impact of reactive nitrogen on plant biodiversity and the environment.

Integration of transgenics, genomics, proteomics, and conventional plant breeding strategies can greatly accelerate the development of new cereal cultivars. Other than increasing crop productivity, commercialized genetically modified crops could help to reduce the impacts of agriculture on biodiversity by alleviating pressure to convert additional land into agricultural use.

References

1. CBD (2000) Agricultural biological diversity: review of phase I of the programme of work and adoption of a multi-year programme. http://www.cbd.int/decision/cop/?id=7147. Accessed 25 Feb 2013
2. Bioversity International (2009) Learning agrobiodiversity the importance of agricultural biodiversity and the role of universities. http://www.bioversityinternational.org/fileadmin/bioversityDocs/Training/Agrobiodiversity_Education/Learning_agrobiodiversity.pdf. Accessed 20 Feb 2013
3. FAO (1999) What is agrobiodiversity? http://www.fao.org/docrep/007/y5609e/y5609e01.htm#TopOfPage. Accessed 25 Feb 2013

4. Frison EA, Cherfas J, Hodgkin T (2011) Agricultural biodiversity is essential for a sustainable improvement in food and nutrition security. Sustainability 3:238–253. doi:10.3390/su3010238
5. van de Wouw M, Kik C, van Hintum T et al (2009) Genetic erosion in crops: concept, results and challenges. Plant Genet Res: Charact Utilization 8:1–15. doi:10.1017/S1479262109990062
6. Singh A, Singh HN, Singh J (2008) Rice biodiversity and its social implication. Int J Rural Stud 15(2)
7. Pistorius R (1997) Scientists, plants and politics—A history of the plant genetic resources movement. International Plant Genetic Resources Institute, Rome
8. Thormann I, Gaisberger H, Mattei F et al (2012) Digitization and online availability of original collecting mission data to improve data quality and enhance the conservation and use of plant genetic resources. Genet Resour Crop Evol 59:635–644
9. FAO (2010) The second report on the state of the world's plant genetic resources for food and agriculture. http://www.fao.org/agriculture/seed/sow2/en/. Accessed 17 Apr 2013
10. Altieri MA, Merrick LC (1987) *In situ* conservation of crop genetic resources through maintenance of traditional farming. Econ Bot 41:86–96
11. Jensen HR, Dreiseitl A, Sadiki M et al (2012) The Red Queen and the seed bank: pathogen resistance of *ex situ* and in situ conserved barley. Evol Appl 5:353–367
12. Dulloo ME, Hunter D, Borelli T (2010) *Ex situ* and in situ conservation of agricultural biodiversity: major advances and research needs. Notulae Botanicae Horti Agrobotanici Cluj-Napoca 38:123–135
13. FAO (2011a) Draft updated global plan of action for the conservation and sustainable utilization of plant genetic resources for food and agriculture. In: Fifth session of the Intergovernmental Technical Working Group on Plant Genetic Resources for Food and Agriculture, Rome, 27–29 Apr 2011
14. Pereira HM, Ferrier S, Walters M et al (2013) Essential biodiversity variables. Science 339:227–278
15. Cleveland DA, Soleri D (2007) Extending Darwin's analogy: bridging differences in concepts of selection between farmers, biologists, and plant breeders. Econ Bot 61:121–136
16. Thomas M, Dawson JC, Goldringer I et al (2011) Seed exchanges, a key to analyze crop diversity dynamics in farmer-led on-farm conservation. Genet Resour Crop Evol 58:321–338
17. Briggs D, Walters M (1999) Evolution: some general considerations. Plant variation and evolution. 3rd edn. Cambridge University Press, Cambridge, pp 367–398
18. Mercer KL, Perales HR (2010) Evolutionary response of landraces to climate change in centers of crop diversity. Evol Appl 3:480–493
19. Zeven AC (1999) The traditional inexplicable replacement of seed and seed ware of landraces and cultivars: a review. Euphytica 110:181–191
20. Hajjar R, Jarvis DI, Gemmill-Herren B (2008) The utility of crop genetic diversity in maintaining ecosystem services. Agric Ecosyst Environ 123:261–270
21. Bezançon G, Pham JL, Deu M et al (2009) Changes in the diversity and geographic distribution of cultivated millet (*Pennisetum glaucum* (L.) R. Br.) and sorghum (*Sorghum bicolor* (L.) Moench) varieties in Niger between 1976 and 2003. Genet Resour Crop Evol 56:223–236
22. Vigouroux Y, Glaubitz JC, Matsuoka Y et al (2008) Population structure and genetic diversity of new world maize races assessed by DNA microsatellites. Am J Bot 95:1240–1253
23. Chan LM, Brown JL, Yoder AD (2011) Integrating statistical genetic and geospatial methods brings new power to phylogeography. Mol Phylogenet Evol 59:523–537
24. Kozak KH, Graham CH, Wiens JJ (2008) Integrating GIS-based environmental data into evolutionary biology. Trends Ecol Evol 23:141–148
25. van Zonneveld M, Dawson I, Thomas E, et al (2013) Application of molecular markers in spatial analysis to optimize in situ conservation of plant genetic resources. In: Tuberosa. R,

Adler A, Frison E (eds) Advances in genomics of plant genetic resources. Springer, New York
26. Guarino L, Jarvis A, Hijmans RJ et al (2002) Geographic information systems (GIS) and the conservation and use of plant genetic resources. In: Engels JMM, Ramanatha Rao V, Brown AHD, Jackson MT (eds) Managing plant genetic diversity. International Plant Genetic Resources Institute (IPGRI), Rome, pp 387–404
27. Kiambi DK, Newbury HJ, Maxted N et al (2008) Molecular genetic variation in the African wild rice *Oryza longistaminata* A. Chev. et Roehr. and its association with environmental variables. Afr J Biotechnol 7:1446–1460
28. van Zonneveld M, Scheldeman X, Escribano P et al (2012) Mapping genetic diversity of cherimoya (*Annona cherimola* Mill.): application of spatial analysis for conservation and use of plant genetic resources. PloS ONE 7:e29845
29. Hijmans RJ, van Etten J (2012) Geographic analysis and modelling with raster data. R package "Raster". http://cran.r-project.org/web/packages/raster/raster.pdf. Accessed 17 Apr 2013
30. Jombart T (2008) Adegenet: a R package for the multivariate analysis of genetic markers. Bioinformatics 24:1403–1405
31. Leberg PL (2002) Estimating allelic richness: effects of sample size and bottlenecks. Mol Ecol 11:2445–2449
32. Thomas E, van Zonneveld M, Loo J et al (2012) Present spatial diversity patterns of *Theobroma cacao* L. in the Neotropics reflect genetic differentiation in Pleistocene refugia followed by human-influenced dispersal. PLoS ONE 7:e47676
33. van Heerwaarden J, Doebley J, Briggs WH et al (2010) Genetic signals of origin, spread, and introgression in a large sample of maize landraces. Proc Natl Acad Sci. doi:10.1073/pnas.1013011108
34. Hufford MB, Lubinksy P, Pyhäjärvi T et al (2013) The genomic signature of crop-wild introgression in maize. PLoS Genet 9:e1003477
35. Russell J, Dawson IK, Flavell AJ et al (2011) Analysis of >1000 single nucleotide polymorphisms in geographically matched samples of landrace and wild barley indicates secondary contact and chromosome-level differences in diversity around domestication genes. New Phytol 191:564–578
36. Dvorak J, Luo M-Ch, Akhunov ED (2011) N.I. Vavilov's theory of centres of diversity in the light of current understanding of wheat diversity, domestication and evolution. Czech J Genet Plant Breed 47:S20–S27
37. Barry MB, Pham JL, Courtois B et al (2007) Rice genetic diversity at farm and village levels and genetic structure of local varieties reveal need for in situ conservation. Genet Resour Crop Evol 54:1675–1690
38. Jarvis D, Hodgkin T, Sthapit BR et al (2011) An heuristic framework for identifying multiple ways of supporting the conservation and use of traditional crop varieties within the agricultural production system. Crit Rev Plant Sci 30:125–176
39. Pusadee T, Jamjoda S, Chiang Y-C et al (2009) Genetic structure and isolation by distance in a landrace of Thai rice. Proc Nat Acad Sci 106:13880–13885
40. Rice EB, Smith ME, Mitchell SE et al (2006) Conservation and change: a comparison of in situ and *ex situ* conservation of Jala maize germplasm. Crop Sci 46:428–436
41. Bellon MR, Hodson D, Hellin J (2011) Assessing the vulnerability of traditional maize seed systems in Mexico to climate change. Proc Natl Acad Sci 108:13432–13437
42. Tanksley SD, McCouch SR (1997) Seed banks and molecular maps: unlocking genetic potential from the wild. Science 227:1063–1066
43. Ishii T, Hiraoka T, Kanzaki T et al (2011) Evaluation of genetic variation among wild populations and local varieties of rice. Rice 4:170–177
44. Russell J, van Zonneveld M, Dawson IK, et al (2013) Genetic diversity and ecological niche modeling of wild barley: refugia, large-scale post-LGM range expansion and limited mid-future climate threats? PloS ONE (accepted)

45. National Research Council (NRC) (1996) Pearl millet. Lost Crops of Africa, vol 1, Grains. National Academy Press, Washington, pp 77–126
46. Yadav OP (2011) Project Coordinator's Review (2010-11) on Pearl Millet Research. http://www.aicpmip.res.in/pcr2011.pdf. Accessed 02 May 2013
47. Brunken JN, de Wet JMJ, Harlan JR (1977) The morphology and domestication of pearl millet. Econ Botan 31:163–174
48. Rai KN, AppaRao S, Reddy KN (1997) Pearl Millet In: Fuccillo D, Sears L, Stapleton P (eds) Biodiversity in Trust. Cambridge University Press, Cambridge, pp 243–258
49. The Syngenta Foundation for Sustainable Agriculture (2006) Harnessing Modern Science in Africa to Sustain Sorghum and Pearl Millet Production for Resource poor Farmers. www.syngentafoundation.com/millet.html. Accessed 02 May 2013
50. Linnaeus C (1753) Species plantarum. Laurentius Salvius, Stockholm
51. Linnaeus C (1759) Systema naturae, 10th edn, vol 2. Laurentius Salvius, Stockholm
52. Rechard L (1805) *Pennisetum.* In: Person CH (ed) Synopsis plantarum,vol 1. Cotta, Tubingen
53. Willdenow KL (1809) Enumeratio plantarum horti regni botanici Berolinensis. Berolini, Berlin
54. Steudel EG (1855) Synopsis *plantarum glumaceum.* Stuttgart. B65.10.01. M. inscription vol. 1: Ferd Muller, M.D., Ph.D. 1987
55. Leeke P (1907) Untersuchunguber Abstammung and Heimat der Negerhirse (*Pennisetum americanum* (L.)L schum.). Z Naturwissenchaften 79:1–108
56. Stapf O, Hubbard CE (1934) Pennisetum. In: Prain D (ed) Flora of tropical Africa, Part 9. Crown Agents, London
57. Clayton WD, Renvoize SA (1982) Gramineae. In: Polhill RM (ed) Flora of tropical East Africa, Part 3. Balkema, Rotterdam
58. Gari JA (2002) Review of the African millet diversity. In: Paper for the International workshop on fonio, food security and livelihood among the rural poor in West Africa, IPGRI/IFAD, Bamako, Mali. Programme for Neglected and Underutilized Species. IPGRI, Rome, Italy, 19–22 Nov 2001
59. Hanna WW (1987) Utilization of wild relatives of pearl millet. In: Proceedings of the international pearl millet wokshop, ICRISAT Center, India. ICRISAT, Patancheru, Andhra Pradesh, India, 7–11 Apr 1986
60. Oumar I, Mariac C, Pham JL et al (2008) Phylogeny and origin of pearl millet (*Pennisetum* glaucum [L.] R. Br) as revealed by microsatellite loci. Theor Appl Genet 117:489–497
61. Amblard S, Pernes J (1989) The identification of the cultivated pearl millet (*Pennisetum*) amongst plant impressions on pottery from Oued Chebbi (Dhar Oualata, Mauritania). Afr Archaeol Rev 7:117–126
62. D'Andrea AC, Casey J (2002) Pearl millet and kintampo subsistence. Afr Archaeol Rev 19:147–173
63. D'Andrea AC, Klee M, Casey J (2001) Archaeological evidence for pearl millet (*Pennisetum glaucum*) in sub-saharan West Africa. Antiquity 75:341–348
64. Clark JD (1962) The spread of food production in sub-Saharan Africa. J Afr Hist III 2:211–228
65. Harlan JR (1971) Agricultural origins: centres and non-centres. Science 174:468–474
66. Marchais L (1994) Wild pearl millet population (*Pennisetum glaucum,* Poaceae) integrity in agricultural Sahelian areas. An example from Keita (Niger). Plant Syst Evol 189:233–245
67. Portères R (1962) Berceaux agricolos primaries sur le continent africain. J Afr Hist III 2:195–210
68. Tostain S, Marchais L (1993) Wild pearl millet population (*Pennisetum glaucum*, Poaceae) integrity in agricultural Sahelian areas. Plant Syst Evol 189:233–245
69. Khairwal IS, Rai KN, Diwakar B et al (2007) Pearl millet: crop management and seed production manual. ICRISAT, Patancheru, Andhra Pradesh, India, pp 108
70. Tostain S (1992) Enzyme diversity in pearl millet (*Pennisetum glaucum* L.). Theor Appl Genet 83:736–742

71. Tostain S (1998) Le mil, une longue histoire: hypothèses sur sa domestication et ses migrations. In: Chastenet M (ed) Plantes et Paysages d'Afrique: Une Histoire à Explorer. Karthala and Centre de Recherches Africaines, Paris, pp 461–490
72. Klee M, Zach B, Stika HP (2004) Four thousand years of plant exploitation in the Lake Chad basin (Nigeria), part III: plant impressions in postherds from the final stone age Gajiganna culture. Veg Hist Archaeobotony 13:131–142
73. Tostain S, Riandey MF, Marchais L (1987) Enzyme diversity in pearl millet (*Pennisetum glaucum*) I. West Africa. Theor Appl Genet 74:188–193
74. Upadhyaya HD, Reddy KN, Gowda CLL (2007) Pearl millet germplasm at ICRISAT genebank—status and impact. International Crops Research Institute for the Semi-Arid Tropics (ICRISAT), Patancheru 502 324, Andhra Pradesh, India
75. IBPGR and ICRISAT (1993) Descriptors for Pearl Millet. International Plant genetic resources Institute, Rome
76. Mathur PN, Rana RS, Aggarwal RC (1993) Evaluation of pearl millet germplasm part I. National Bureau of Plant Genetic Resources, New Delhi
77. Mathur PN, Rana RS, Aggarwal RC (1993) Evaluation of pearl millet germplasm part II. National Bureau of Plant Genetic Resources, New Delhi
78. Andrews DJ, Anand Kumar K (1996) Use of the West African pearl millet landrace Iniadi in cultivar development. Plant Genet Resour Newsl 105:15–22
79. Bioversity International (2010) Key access and utilization descriptors for pearl millet genetic resources. Bioversity International, Rome, Italy. http://www.bioversityinternational.org/index.php?id=19&user_bioversitypublications_pi1%5BshowUid%5D=3376. Accessed 02 May 2013
80. Maxted N, Kell SP (2009) Establishment of a global network for the *In Situ* conservation of crop wild relatives: status and needs. FAO Commission on Genetic Resources for Food and Agriculture, Rome, p 266
81. Mujica A (1992) Granos y leguminosas andinas. In: Hernandez J, Bermejo J, Leon J (eds) Cultivos marginados: otra perspectiva de 1492. Organización de la Naciones Unidas para la Agricultura y la Alimentación (FAO), Roma
82. Rojas W, Pinto M, Soto JL (2010a) Distribución geográfica y variabilidad genética de los granos andinos. In: Rojas W, Soto JL, Pinto M, Jäger M, Padulosi S (eds) Granos Andinos. Avances, logros y experiencias desarrolladas en quinua, cañahua y amaranto en Bolivia. Bioversity International, Roma
83. Aroni G, Pinto M, Rojas W (2012) Small-scale quinoa processing technology in the southernAltiplano of Bolivia. In: Giuliani A, Hintermann F, Rojas W, Padulosi S (eds) Biodiversity of Andean grains: balancing market potential and sustainable livelihoods. Bioversity International, Rome
84. Carrasco E, Soto JL (2010) Importancia de los granos andinos. In: Rojas W, Soto JL, Pinto M, Jäger M, Padulosi S(eds) Granos Andinos. Avances, logros y experiencias desarrolladas en quinua, cañahua y amaranto en Bolivia. Bioversity International, Roma
85. Pinto M, Alarcón V, Soto JL, Rojas W (2010a) Usos tradicionales, no tradicionales e innovaciones agroindustriales de los granos andinos. In: Rojas W, Soto JL, Pinto M, Jäger M, Padulosi S (eds) Granos Andinos. Avances, logros y experiencias desarrolladas en quinua, cañahua y amaranto en Bolivia. Bioversity International, Roma
86. FAO (2011b) La Quinua: cultivo milenario para contribuir a la seguridad alimentaria mundial. FAO, Oficina Regional para América Latina y el Caribe, La Paz
87. Rojas W, Pinto M, Soto JL, Alcocer E (2010b) Valor nutricional, agroindustrial y funcional de los granos andinos. In: Rojas W, Soto JL, Pinto M, Jäger M, Padulosi S (eds) Granos Andinos. Avances, logros y experiencias desarrolladas en quinua, cañahua y amaranto en Bolivia. Bioversity International, Roma
88. FAO, WHO, UNU (1985) Necesidades de energía y proteínas. Informes Técnicos. OMS, Ginebra

89. Astudillo D (2012) Livelihoods of quinoa producers in southern Bolivia. In: Giuliani A, Hintermann F, Rojas W, Padulosi S (eds) Biodiversity of Andean grains: balancing market potential and sustainable livelihoods. Bioversity International, Italy
90. Hunter D, Heywood V (eds) (2011) Crop wild relatives: a manual of in situ conservation, 1st edn. Earthscan, London
91. Jacobsen SE (2003) The Worldwide Potential for Quinoa (*Chenopodium quinoa* Willd.). Food Rev Int 19:167–177
92. Apaza V, Estrada R, Quispe MG (2010b) Identificación y selección participativa de material genético promisorio. In: Bravo R, Valdivia R, Andrade K, Padulosi S, Jäger M (eds) Granos Andinos. Avances, logros y experiencias desarrolladas en quinua, cañihua y kiwicha en Perú. Bioversity International, Roma
93. Mujica A, Izquierdo J, Marathee JP (2001) Origen y Descripción de la Quinua. In: Mujica A, Jacobsen SE, Izquierdo J, Marathee JP. Quinua (*Chenopodium quinoa* Willd.). Ancestral Cultivo Andino, Alimento del Presente y Futuro. FAO. Santiago
94. Bonifacio A, Aroni G, Villca M (2012) Catálogo Etnobotánico de la Quinua Real. PROINPA, Cochabamba
95. Ministerio de Medio Ambiente y Agua (MMAyA), Viceministerio de Medio Ambiente, Biodiversidad y Cambios Climáticos (VMABCC), Bioversity International (2009) Libro Rojo de Parientes Silvestres de Cultivos de Bolivia. VMABCC-Bioversity International, La Paz
96. Apaza V, Catacora P, Quispe MG (2010a) Distribución geográfica y variabilidad genética de los granos andinos. In: Bravo R, Valdivia R, Andrade K, Padulosi S, Jäger M (eds). Granos Andinos. Avances, logros y experiencias desarrolladas en quinua, cañihua y kiwicha en Perú. Bioversity International, Roma
97. Rojas W, Pinto M (2010) Colecta de germoplasma. In: Rojas W, Soto JL, Pinto M, Jäger M, Padulosi S (eds) Granos Andinos. Avances, logros y experiencias desarrolladas en quinua, cañahua y amaranto en Bolivia. Bioversity International, Roma
98. Bravo R, Valdivia R, Andrade K, Padulosi S, Jäger M (eds) (2010) Granos Andinos. Avances, logros y experiencias desarrolladas en quinua, cañihua y kiwicha en Perú. Bioversity International, Roma
99. Jaeger M (2012) Novel products, markets and partnerships in value chains for Andean grains in Peru and Bolivia. In: Giuliani A, Hintermann F, Rojas W, Padulosi S (eds) Biodiversity of Andean grains: balancing market potential and sustainable livelihoods. Bioversity International, Rome
100. Bravo R, Catacora P (2010) Situación actual de los bancos nacionales de germoplasma. In: Bravo R, Valdivia R, Andrade K, Padulosi S, Jäger M (eds) Granos Andinos. Avances, logros y experiencias desarrolladas en quinua, cañihua y kiwicha en Perú. Bioversity International, Roma
101. Reay SD, Davidson EA, Smith KA et al (2012) Global agriculture and nitrous oxide emissions. Nat Clim Chang 2:410–416. doi:10.1038/nclimate1458
102. Ravishankara AR, Daniel JS, Portmann RW (2009) Nitrous oxide (N_2O): the dominant ozone-depleting substance emitted in the 21st century. Science 326:123–125
103. Wuebbles DJ (2009) Nitrous oxide: no laughing matter. Science 326:56–57
104. Philippot L, Hallin S (2011) Towards food, feed and energy crops mitigating climate change. Trends Plant Sci 16:476–480
105. Canfield DE, Glazer AN, Falkowski PG (2010) The evolution and future of Earth's nitrogen cycle. Science 330:192–196
106. Subbarao GV, Nakahara K, Hurtado MP et al (2009) Evidence for biological nitrification inhibition in *Brachiaria* pastures. Proc Natl Acad Sci 106:17302–17307
107. Subbarao GV, Sahrawat KL, Nakahara K et al (2012) A paradigm shift towards low-nitrifying production systems: the role of biological nitrification inhibition (BNI). Ann Bot. doi:10.1093/aob/mcs230
108. Subbarao GV, Rondon M, Ito O et al (2007) Biological nitrification inhibition (BNI)—Is it a widespread phenomenon? Plant Soil 294:5–18

109. Tanaka JP, Nardi P, Wissuwa M (2010) Nitrification inhibition activity, a novel trait in root exudates of rice. AoB Plants 2010:1–11
110. Subbarao GV, Ban T, Masahiro K et al (2007) Can biological nitrification inhibition (BNI) genes from perennial Leymus racemosus (Triticeae) combat nitrification in wheat farming? Plant Soil 299:55–64
111. Zahn LM (2007) A boost from wild wheat. Science 318:171. doi:10.1126/science.318.5848.171c
112. Hossain AKMZ, Subbarao GV, Pearse SJ et al (2008) Detection, isolation and characterization of a root-exuded compound, methyl 3-(4-hydroxyphenyl) propionate, responsible for biological nitrification inhibition by sorghum (Sorghum bicolor). New Phytol 180:442–451
113. Subbarao GV, Nakahara K, Ishikawa T et al (2013) Biological nitrification inhibition (BNI) activity in sorghum and its characterization. Plant Soil 366:243–259
114. Nimbal CI, Pedersen JF, Yerkes CN et al (1996) Phytotoxicity and distribution of sorgoleone in grain sorghum. J Agric Food Chem 44:1343–1347
115. Dobermann A, Cassman KG (2005) Cereal area and nitrogen use efficiency are drivers of future nitrogen fertilizer consumption. Sci China Life Sci 48:745–758
116. Phoenix GK, Hicks WK, Cinderby S et al (2006) Atmospheric nitrogen deposition in world biodiversity hotspots: the need for a greater global perspective in assessing N deposition impacts. Glob Change Biol 12:470–476
117. Sala OE, Chapin FS III, Armesto JJ et al (2000) Global biodiversity scenarios for the year 2100. Science 287:1770–1774
118. Zhou Z, Sun OJ, Huang J et al (2006) Land-use affects the relationship between species diversity and productivity at the local scale in a semi-arid steppe ecosystem. Funct Ecol 20:753–762
119. Cox RM (1992) Air pollution effects on plant reproductive processes and possible consequences to their population ecology. In: Barker JR, Tingey DT (eds) Air pollution effects on biodiversity. Springer, New York, pp 131–158
120. Clark CM, Tilman D (2008) Loss of plant species after chronic low-level nitrogen deposition to prairie grasslands. Nature 451:712–715
121. Nordin A, Strengbom J, Ericson L (2006) Responses to ammonium and nitrate additions by boreal plants and their natural enemies. Environ Pollut 41:167–174
122. Strengbom J, Nordin A, Näsholm T, Ericson L (2002) Parasitic fungus mediates change in nitrogen-exposed boreal forest vegetation. J Ecol 90:61–67
123. Dentener F, Drevet J, Lamarque JF et al (2006) Nitrogen and sulfur deposition on regional and global scales: a multi-model evaluation. Glob Biogeochem Cycles 20:GB4003. doi: 10.1029/2005GB002672
124. Vitousek PM, Aber JD, Howarth RW et al (1997) Human alteration of the global nitrogen cycle: sources and consequences. Ecol Appl 7:737–750
125. Lu X, Mo J, Gilliam FS, Zhou G et al (2010) Effects of experimental nitrogen additions on plant diversity in an old-growth tropical forest. Glob Chang Biol 16:2688–2700
126. Bobbink R, Hornung M, Roelofs JGM (1998) The effects of air-borne nitrogen pollutants on species diversity in natural and semi-natural European vegetation. J Ecol 86:717–738
127. Bobbink R, Hicks K, Galloway J et al (2010) Global assessment of nitrogen deposition effects on terrestrial plant diversity: a synthesis. Ecol Appl 20:30–59
128. Gilliam FS, Hockenberry AW, Adams MB (2006) Effects of atmospheric nitrogen deposition on the herbaceouslayer of a central Appalachian hardwood forest. J Torrey Bot Soc 133:240–254
129. Bevan MW, Flavell RB, Chilton MD (1983) A chimeric antibiotic-resistance gene as a selectable marker for plant cell transformation. Nature 304:184–187
130. Fraley RT, Rogers SG, Horsch RB et al (1983) Expression of bacterial genes in plant cells. Proc Natl Acad Sci 80:4803–4807
131. Herrera-Estrella L, Depicker A, Van Montagu M et al (1983) Expression of chimaeric genes transferred into plant cells using a Ti-plasmid-derived vector. Nature 303:209–213

132. Horsch RB, Fry JE, Hoffmann NL et al (1985) A simple and general method for transferring genes into plants. Science 227:1229–1231
133. Bevan M (1984) Binary Agrobacterium vectors for plant transformation. Nucleic Acids Res 12:8711–8721
134. Hoekema A, Hirsch PR, Hooykaas PJJ et al (1983) A binary plant vector strategy based on separation of Vir-region and T-region of the Agrobacterium tumefaciens Ti-plasmid. Nature 303:179–180
135. Chilton MD, Tepfer DA, Petit A et al (1982) Agrobacterium rhizogenes inserts T-DNA into the genomes of the host plant root cells. Nature 295:432–434
136. Draper J, Davey MR, Freeman JP et al (1982) Ti plasmid homologous sequences present in tissues from Agrobacterium plasmid transformed petunia protoplasts. Plant Cell Physiol 23:451–458
137. Krens FA, Molendijk L, Wullems GJ et al (1982) *In vitro* transformation of plant protoplasts with Ti-plasmid DNA. Nature 296:72–74
138. Paszkowski J, Shillito RD, Saul M et al (1984) Direct gene transfer to plants. EMBO J 3:2717–2722
139. Sanford JC, Klein TM, Wolf ED et al (1987) Delivery of substances into cells and tissues using a particle bombardment process. J Part Sci Technol 5:27–37
140. Murashige T, Skoog F (1962) A revised medium for rapid growth and bio assays with tobacco tissue cultures. Physiol Plant 15:473–497
141. Bajaj YPS (1991) Biotechnology in rice improvement. In: Bajaj YPS (ed) Biotechnology in Agriculture and Forestry, vol 14. Rice. Springer, Berlin, pp 1–15
142. Shimammoto K, Tareda R, Izawa H (1989) Fertile transgenic rice plants regenerated from transformed protoplast. Nature 338:274–276
143. Bajaj YPS (1994) Biotechnology in maize improvement. In: Bajaj YPS (ed) Biotechnology in Agriculture and Forestry vol. 25. Maize. Springer, Berlin, pp 1–17
144. Gordon-Kamm WJ, Spencer TM, Mangano ML et al (1990) Transformation of maize cells and regeneration of fertile transgenic plants. Plant Cell 2:603–618
145. Bajaj YPS (ed) (1990) Biotechnology in agriculture and forestry, vol. 13. Wheat. Springer, Berlin
146. Vasil V, Castillo AM, From ME et al (1992) Herbicide resistant fertile transgenic wheat plants obtained by microprojectil bombardment of regenerable embryogenic callus. Biotechnol 10:667–674
147. Hiei Y, Ohta S, Komari T et al (1994) Efficient transformation of rice (*Oryza sativa* L.) mediated by Agrobacterium and sequence analysis of the boundaries of the T-DNA. Plant J 6:271–282
148. Ishida V, Saito H, Ohta O et al (1996) High efficiency transformation of maize (*Zea mays* L.) mediated by Agrobacterium tumefaciens. Nat Biotechnol 6:745–750
149. Cheng M, Fry JE, Pang S et al (1997) Genetic Transformation of Wheat Mediated by *Agrobacterium tumefaciens*. Plant Physiol 115:971–980
150. Afolabi AS, Worland B, Snape JW et al (2004) A large-scale study of rice plants transformed with different T-DNAs provides new insights into locus composition and T-DNA linkage configurations. Theor Appl Genet 109:815–826
151. Ingham DJ, Beer S, Money S et al (2001) Quantitative Real-Time PCR assay for determining transgene copy number in transformed plants. Biotechniques 31:132–140
152. Vain P, Afolabi AS, Worland B et al (2003) Transgene behavior in populations of rice plants transformed using a new dual binary vector system: pGreen/pSoup. Theor Appl Genet 107:210–217
153. Jefferson RA, Burgess SM, Hirsh D (1986) Beta-glucuronidase from *Escherichia coli* as a gene-fusion marker. Proc Natl Acad Sci 83:8447–8451
154. Carrer H, Hockenberry TN, Svab Z et al (1993) Kanamycin resistance as a selectable marker for plastid transformation in tobacco. Mol Gen Genet 241:49–56
155. Waldron C, Murphy EB, Roberts JL et al (1985) Resistance to hygromycin B. Plant Mol Biol 5:103–108

156. Miki B, McHugh S (2003) Selectable marker genes in transgenic plants: application, alternatives and biosafety. J Biotechnol 107:193–232
157. Padilla IMG, Burgos L (2010) Aminoglycoside antibiotics: structure, functions and effects on in vitro plant culture and genetic transformation protocols. Plant Cell Rep 29:1203–1213
158. Sundar IK, Sakthitel N (2008) Advances in selectable marker genes for plant transformation. J Plant Physiol 165:1698–1716
159. Komari T, Hiei Y, Saito Y et al (1996) Vectors carrying two separate T-DNAs for co-transformation of higher plants mediated by *Agrobacterium tumefaciens* and segregation of transformants free from selection markers. Plant J 10:165–174
160. Miller M, Tagliani L, Wang N et al (2002) High efficiency transgene segregation in co-transformed maize plants using an *Agrobacterium tumefaciens* 2 T-DNA binary system. Transgenic Res 11:381–396
161. Gelvin SB (2003) Improving plant genetic engineering by manipulating the host. Trends Biotechnol 21:95–98
162. Tzfira T, Citovsky V (2006) Agrobacterium-mediated genetic transformation of plants: biology and biotechnology. Curr Opin Biotechnol 17:147–154
163. Zambryski P (1992) Chronicles from the Agrobacterium–plant cell DNA transfer story. Ann Rev Plant Physiol Mol Biol 43:4645–4690
164. Gelvin SB (2010) Plant proteins involved in Agrobacterium mediated genetic transformation. Ann Rev Phytopatol 48:45–68
165. Gelvin SB (2010b) Finding the way to the nucleus.Curr Opin Microbiol 13:53–58
166. Vain P (2005) Plant transgenic science knowledge. Nat Biotechnol 23:1348–1349
167. Vain P (2007) Thirty years of plant transformation technology development. Plant Biotechnol J 5:221–229
168. Areal FJ, Riesgo L, Rodríguez-Cerezo E (2011) Attitudes of European farmers towards GM crop adoption. Plant Biotechnol J 9:945–957
169. EU Commission (2010) A decade of EU funded GMO research (2001–2010). ftp://cordis.europa.eu/pub/fp7/kbbe/docs/a-decade-of-eu-funded-gmo-research_en.pdf. Accessed 02 May 2013
170. European Policy Evaluation Consortium (EPEC) (2011) The Evaluation of the EU legislative framework in the field of cultivation of GMOs under Directive 2001/18/EC. http://ec.europa.eu/food/food/biotechnology/evaluation/docs/gmo_cultivation_report_en.pdf. Accessed 02 May 2013
171. Lusser M, Rodríguez-Cerezo E (2012) Comparative regulatory approaches for new plant breeding techniques. Workshop proceedings European Commission, JRC Technical Report EUR 25237 EN (2012)
172. Phillips PWB (2002) Biotechnology in the global agri-food systems. Trends Biotechnol 20:376–381
173. Park JR, McFarlane I, Phipps RH et al (2011) The role of transgenic crops in sustainable development. Plant Biotechol J 9:2–21
174. Dunoyer P, Himber C, Voinnet O (2006) Induction, suppression and requirement of RNA silencing pathways in virulent *Agrobacterium tumefaciens* infections. Nat Genet 38:258–263
175. Shaked H, Melamed-Bessudo C, Levy AA (2005) High frequency gene targeting in Arabidopsis plants expressing the yeast RAD54 gene. Proc Natl Acad Sci 12:265–269
176. Miki D, Itoh R, Shimamoto K (2005) RNA silencing of single and multiple members in a gene family of rice. Plant Physiol 138:1903–1913
177. Reynolds A, Leake D, Boese Q et al (2004) Rational sRNAi design for RNA interterference. Nat Biotechnol 22:326–330
178. Carroll D (2011) Genome engineering with zinc-finger nucleases. Genetics 188:773–782
179. Shukla VK, Doyon Y, Miller JC et al (2009) Precise genome modification in the crop species *Zea mays* using zinc-finger nucleases. Nature 459:437–441
180. Boch J, Scholze H, Schornack S (2009) Breaking the code of DNA binding specificity of TAL-type III effectors. Science 326:1509–1512

181. Bogdanove AJ, Voytas JF (2011) TAL effectors: customizable proteins for DNA targeting. Science 333:1843–1846
182. Podevin N, Davies HV, Hartung F (2013) Site-directed nucleases: a paradigm shift in predictable, knowledge-based plant breeding. Trends Biotechnol. doi:10.1016/j.tibtech.2013.03.004
183. Chen H, Lin Y, Zhang Q (2010) Rice. In: Kempker F, Jung Ch (eds) Genetic modification of plants. Biotechnology in Agriculture and Forestry, vol 64. Springer, Berlin, p 423–451
184. Hirano HY, Hirai A, Sano Y, Sasaki T (eds) (2008) Rice Biology in the Genomics era. Biotechnol in Agriculture and Forestry, vol 62. Springer, Berlin
185. Serraj R, Bennett J, Hardy B (eds) (2008) Drought frontiers in rice: crop improvement for increased rainfed production. World Scientific Publishing, Singapore
186. Sheehy JE, Mitchell PL, Hardy B (eds) (2007) Charting new pathways to C4 rice. IRRI, World Scientific, Los Banos
187. FAOSTAT (2013) http://faostat.fao.org. Accessed 05 Mar 2013
188. Zhang Q (2007) Strategies for developing Green Super Rice. Proc Natl Acad Sci 16:16402–16409
189. James C (2012) Global status of commercialized Biotech/GMO crops: 2012. ISAAA Briefs No. 44. ISAAA, Ithaca, NY
190. Datta SK, Datta K, Parkhi V et al (2007) Golden rice: introgression, breeding, and field evaluation. Euphytica 154:271–278
191. D'Halluin K, Vanderstraeten Ch, Stals E et al (2008) Homologous recombination: a basis for targeted genome optimization in crop species such as maize. Plant Biotechnol J 6:93–102
192. Duan Y, Zhai Ch, Li H (2012) An efficient and high-throughput protocol for Agrobacterium mediated transformation based on phosphomannose isomerase positive selection in Japonica rice (*Oryza sativa* L.). Plant Cell Rep 31:1611–1624
193. Water Efficient Maize for Africa (2011) Water efficient maize for Africa: pushing GMO crops onto Africa. The African Center for Biosafety. http://www.monsanto.com/improvingagriculture/Pages/water-efficient-maize-for-africa.aspx. Accessed 10 May 2013
194. Carpenter JE (2011) Impact of GM crops on biodiversity. GM Crops 2:7–23
195. Aroni J, Aroni G, Quispe R, Bonifacio A (2003) Catálogo de Quinoa Real. Fundación PROINPA, La Paz
196. Pinto M, Marin W, Rojas W (2010b) Estrategias para la conservación y promoción de los granos andinos: ferias y concursos. In: Rojas W, Soto JL, Pinto M, Jäger M, Padulosi S (eds) Granos Andinos. Avances, logros y experiencias desarrolladas en quinua, cañahua y amaranto en Bolivia. Bioversity International, Roma
197. Soto JL (2010) Tecnología del cultivo de granos andinos. In: Rojas W, Soto JL, Pinto M, Jäger M, Padulosi S (eds) Granos Andinos. Avances, logros y experiencias desarrolladas en quinua, cañahua y amaranto en Bolivia. Bioversity International, Roma
198. Knudsen H (ed) (2000) Directorio de Colecciones de Germoplasma en América Latina y el Caribe, 1st edn. International Plant Genetic Resources Institute (IPGRI), Roma
199. Mazón N, Rivera M, Peralta E, Estrella J, Tapia C (2002) Catálogo del banco de germoplasma de quinua (*Chenopodium quinoa* Willd.) del INIAP—Ecuador. Programa Nacional de Leguminosas y Granos Andinos, Departamento Nacional de Recursos Fitogenéticos y Biotecnología, Estación Experimental Santa Catalina. Quito

Adv Biochem Eng Biotechnol (2015) 147: 37–58
DOI: 10.1007/10_2014_268

Published Online: 20 May 2014

Biodiversity in Production of Antibiotics and Other Bioactive Compounds

Girish Mahajan and Lakshmi Balachandran

Abstract Microbes continue to play a highly considerable role in the drug discovery and development process. Nevertheless, the number of new chemical entities (NCEs) of microbial origin that has been approved by the Food and Drug Administration (FDA) has been reduced in the past decade. This scarcity can be partly attributed to the redundancy in the discovered molecules from microbial isolates, which are isolated from common terrestrial ecological units. However, this situation can be partly overcome by exploring rarely exploited ecological niches as the source of microbes, which reduces the chances of isolating compounds similar to existing ones. The use of modern and advanced isolation techniques, modification of the existing fermentation methods, genetic modifications to induce expression of silent genes, analytical tools for the detection and identification of new chemical entities, use of polymers in fermentation to enhance yield of fermented compounds, and so on, have all aided in enhancing the frequency of acquiring novel compounds. These compounds are representative of numerous classes of diverse compounds. Thus, compounds of microbial origin and their analogues undergoing clinical trials continue to demonstrate the importance of compounds from microbial sources in modern drug discovery.

Keywords Actinomycetes · Fungi · Myxobacteria · Biodiversity · Antibiotics · Antitumor · Anticancer

G. Mahajan (✉)
Piramal Enterprises Limited, 1-Nirlon Complex, Goregaon,
Mumbai 400 063, India
e-mail: girish.mahajan@piramal.com

L. Balachandran
Karmic Lifesciences Inc., 802, Bldg No 3, Raheja Mindspace,
Airoli, Navi Mumbai, Mumbai 400 708, India

Contents

1 General Introduction 38
1.1 Biodiversity 39
2 Actinomycete-Derived Compounds 40
2.1 Streptomyces: A Sustained Gold Mine of Bioactive Compounds 41
2.2 Rare Actinomycetes: Future Gold Mine of Bioactive Compounds 45
3 Myxobacteria-Derived Compounds 46
4 Eubacteria-Derived Compounds 50
5 Fungal-Derived Compounds 50
5.1 Slime-Molds–Derived Compounds 51
6 Pipeline of Microbial Bioactive Compounds 52
References 53

1 General Introduction

Natural product compounds (NPCs), especially those mined from microbes (bacteria and lower eukaryotes) are established resources for a variety of remedial agents. Such drugs of microbial origin have been classified as (i) original microbial products, (ii) products derived or chemically synthesized from microbial products, or (iii) synthetic products based on microbial product structures [1]. Early scientific observations on the antagonism among soil microflora led scientists to speculate on the existence of some compound that held the key to their survival [1, 2]. The soil, despite teeming with billions of microbes, enables only some of them to endure the struggle for existence. In recent times, many compounds used in the treatment and management of cancer, infections due to drug-resistant microbes (bacteria, fungi, and viruses), and immunosuppressive disorders, have been derived from microbial sources.

Investigations and observations gave way to the concept of the term "antibiosis" (against life). The renowned Nobel Laureate, Selman Abraham Waksman, coined the word "antibiotic". The discovery and launch of microbial antibiotics such as penicillin and streptomycin, were early evidence that microbes could be further explored for novel bioactive compounds for human use. The pharmaceutical industry owes an immense amount of its early success to the development of antibacterial drugs, and as an upshot the market is abundant with old drug scaffolds. Essentially the scaffold of a molecule is taken to be its framework, defined as all its ring systems and all the linkers that connect them [3]. For the past seven decades, the need for new antibiotics has relied largely upon semisynthetic tailoring of natural product scaffolds, discovered in the middle of the twentieth century. During the past decade, however, advances in technology such as high-throughput screening facility, the launch of high-resolution NMR facilities, upgraded separation systems, and, moreover, recent molecular techniques for the investigation of marine metagenomes have revealed a large number of new phylogenetic lines of groups of bacteria and archaea [4, 5], which has sparked a resurgence in the discovery of natural product antibiotics from microbial sources.

Microbial metabolites are among the most important chemotherapeutic agents in oncology. This aspect of microbes was identified as early as 1940 with the discovery of actinomycin from *Streptomyces* [2]. Since then, many compounds with anticancer properties have been isolated from microorganisms. More than 60 % of the current compounds with antineoplasic activity have been originally isolated as natural products or are their derivatives. Among the approved products deserving special attention are actinomycin D, anthracyclines (daunorubicin, doxorubicin, epirubicin, pirirubicin, and valrubicin), bleomycin, mitomycin C, anthracenones (mithramycin, streptozotocin, and pentostatin), enediynes (calcheamycin), taxol, and epothilones [2]. Several of these compounds were discovered by (i) understanding the genetics of secondary metabolism in Actinomycetes, myxobacteria, other eubacteria, fungi, and slime molds (ii) exploring the marine environment, and (iii) applying modern screening technologies. In quite a few cases, the discovery of a novel natural derived product has been reported to be used as a tool to better understand compound targets and new pathways in the disease process [6].

This review describes the current role of biodiversity in drug discovery and pharmaceuticals from microbial sources, and aims to take the reader through a journey of recent advances in the role of biodiversity in the synthesis of novel scaffolds, having an unreported framework of chemical rings. We have focused essentially on those bioactive compounds from microorganisms, which are reported and being used as antibiotic and other bioactive compounds without any further chemical modifications.

1.1 Biodiversity

For researchers involved in the discovery of novel bioactive microbial products, microbial diversity is a key factor for the novelty of the molecules. Although wide diversity is observed among the microbes with reference to their habitat, metabolism, and extremity tolerance, microbes with an established record of synthesis of novel pharmaceutically important lead compounds are very limited. Actinomycetes, fungi, and myxobacteria are the leaders among these microbes.

Prior to the discovery of antibiotics in the nineteenth and twentieth centuries, natural remedies and herbal treatments were used for the treatment of most infectious diseases (or medical conditions). The serendipitous discovery of penicillin (from *Penicillium rubens*), followed by streptomycin (from *Streptomyces griseus*) transformed the lives of millions of people. Since then, natural habitats have been continuously explored for new antibiotics and other bioactive compounds in order to combat the onslaught of new infections and other diseases.

Different communities of microbes coexist in extreme terrestrial regions and oceans, and they constitute an untapped source of bioactive compounds. Advances in basic research have enabled scientists to understand the course of disease and the way a drug works at the molecular level. Continuous improvements in isolation techniques for screening, separation, and isolation have aided the identification of over one million natural compounds, of which 50–60 % are of plant origin and over 5 % are of microbial origin [2]. Around 25 % of these compounds are reported to be biologically active, of which 10 % are derived from microbial sources [2]. There have been approximately 22,500 biologically active compounds [2] obtained thus far from microbes. Of these, 45 % are produced by Actinomycetes, 38 % by fungi, and 17 % by unicellular bacteria [2, 7, 8]. This highlights the immense contribution of these microbes in the production of antibiotics.

Natural habitats, especially the soil—and plant-associated environments, are teeming with microbes that produce bioactive metabolites that shield them against extreme environmental conditions. Such bioactive entities presumably confer an ecological advantage to the producer, by prolonging their survival in an environment challenged by predators and competitors. Secondary metabolites at subinhibitory concentrations also influence developmental changes in the producer. Critical processes such as nutrient supply, developmental changes, survival rate under stressful conditions, and complex interactions are presumably affected by these metabolites [9].

The euphoria over the discovery of a new drug is often short-lived due to the development of resistance in microbes or the tumor cells in addition to the drug's toxicity. This results in limiting the optimum use of a drug. Hence, there is a compelling need for discovering new drugs with newer mechanisms to tackle the menace of drug resistance. Identification of new molecules for disease management mandates the exploration of diverse ecosystems [2].

The marine ecosystem houses most of the animals from the 28 major animal phyla, thus comprising nearly half of the total biodiversity for the discovery of useful therapeutic agents [10]. Soils from the Antarctic regions, extreme cold deserts, playa regions, geothermal vents, hot spring outlets, high pH lakes, acidic water bodies, metal mining areas, sugarcane bagasse, marine sediment soils, and soil from the areas of radionuclear (heavy metal) waste depositions, among others are some of the unique regions for isolation of diverse microbes.

In terms of microbial diversity, not all the microbial phyla have been cultivated, and among the cultivated microbes, not all of them produce secondary metabolites. In this scenario, targeting bacterial phyla, renowned for antibiotic production seems to be a viable option. Hence, Actinomycetales and fungi are attractive targets, inasmuch as they produce most of the antibiotics currently in use [11].

2 Actinomycete-Derived Compounds

Actinomycetes are prokaryotes whose growth (prothallus) consists of branching threads, and rods, and occasionally give rise to a typical mycelium, which is unicellular, during the early stages of growth. The hyphae, which are generally nonseptate, have a tendency to turn septate under special conditions, such as while growing in solid culture media for a long time [12]. Actinomycetes with prostate

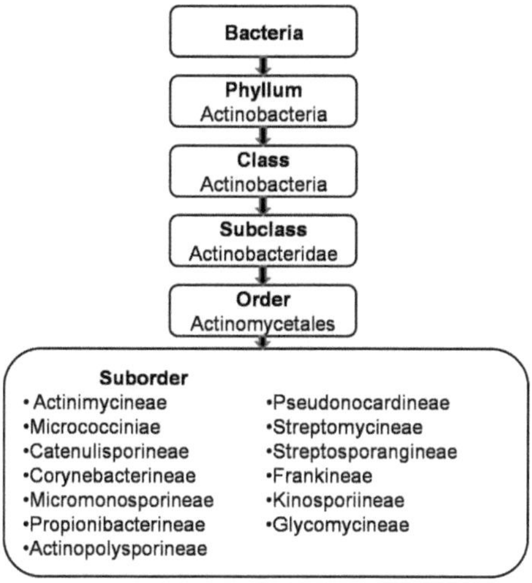

Fig. 1 Systemic classification of Actinomycetes

mycelium grow on the substrate, and those with aerial mycelium grow above the vegetative growth [13]. Currently, Actinomycetes are classified as actinobacteria and include Gram-positive bacteria with their DNA high in guanine-plus-cytosine content (69–73 mol %); and extensive branching substrates and aerial mycelia [14, 15]. The complete taxonomy of Actinomycetes (Fig. 1) and details of each genus can be observed elsewhere [1, 16, 17].

The historical discovery of streptomycin in 1945 was preceded by decades of research on the slender filamentous bacteria, the Actinomycetes. The pioneering work of Selman Waksman inspired many researchers to pursue research on Actinomycetes, which was a new microbe then, hovering for an identity between fungi and bacteria. The most notable feature of these bacteria was the production of secondary metabolites, most of which possessed antimicrobial properties. Many antibacterial compounds including tetracyclines, cephalosporins, aminoglycosides, and macrolides were derived, which addressed concerns regarding disease management in the early twentieth century [18, 19]. A representative list of the different classes of antibiotics produced by Actinomycetes is presented in Table 1.

2.1 *Streptomyces: A Sustained Gold Mine of Bioactive Compounds*

Streptomyces, a well-explored genus of Gram-positive bacteria, is included in the phylum Actinobacteria. These prokaryotes present a strikingly similar lifestyle to that of filamentous fungi and, as do fungi, most Streptomycetes live as saprophytes in the soil [53]. In fact, almost half of all known natural products (NPs) are produced by Actinomycetes (mainly *Streptomyces*) [19, 53]. Nearly two thirds of

Table 1 Antibacterial agents from Actinomycetes

Class	Mechanism of action	Actinomycetes	Antibiotic	Reference
Aminoglycoside	Inhibition of protein synthesis and increase in translation errors	*Streptomyces kanamyceticus*	Bekanamycin	[20]
	Inhibition of protein synthesis and increase in translation errors	*Streptomyces kanamyceticus*	Kanamycin	[21]
	Inhibits protein synthesis by binding to L6 protein of 50S ribosomal subunit	*Micromonospora purpurea*	Gentamicin	[22]
	Binds to 30S and in some cases the 50S subunit causing miscoding; inhibits initiation and elongation during protein synthesis	*Streptomyces fradiae*	Neomycin	[23]
	Inhibits bacterial protein synthesis	*Streptomyces griseus*	Streptocin	[24]
	Inhibits protein synthesis by binding to S12 protein of 30S ribosomal subunit, causing miscoding or inhibiting initiation	*Streptomyces griseus*	Streptomycin	[25]
Acetamide	Inhibits protein biosynthesis by impairing translation on the 50S ribosomal subunit	*Streptomyces venezuelae*	Chloramphencol	[26]
Aminocoumarin	Inhibits DNA synthesis by inhibiting the DNA polymerization	*Streptomyces niveus/ S. spheroids*	Novobiocin	[27]
Aminocyclitol	Disrupts bacterial protein synthesis	*Streptomyces spectabilis*	Spectinomycin	[28]
Cyclic hexapeptide	Inhibits seryl-t-RNA synthetase and impairs protein biosynthesis	*Streptomyces griseus*	Grisein0 (albomycin)	[29]
Cyclic oligopeptide	Impairment of the coupling of the 30-S initiation complex to the 50-S ribosomal subunit	*Streptomyces azureus* and *Streptomyces laurentii*	Thiostrepton	[30]
Galactoocto-pyranoside	Inhibits bacterial protein synthesis	*Streptomyces lincolnensis*	Lincomycin	[31]
Galactoocto-pyranoside	Inhibits bacterial protein synthesis	*Streptomyces sp*	Clindamycin	[32]
Glycolipo-depsipeptide	Inhibits transglycosylation in peptidoglycan synthesis	*Streptomyces azureus* and *Streptomyces laurentii*	Thiostrepton	[33]
Glycolipo-depsipeptide	Inhibits transglycosylation in peptidoglycan synthesis	*Actinoplanes sp* ATCC 33706	Ramoplanin (INN)	[34]

(continued)

Table 1 (continued)

Class	Mechanism of action	Actinomycetes	Antibiotic	Reference
Glycopeptide	Inhibits cell wall synthesis	*Amycolatopsis orientalis*	Vancomycin	[35]
Glycopeptide	Binds to D-ALA-D-ALA terminal end of peptidoglycan precursors and inhibits cell-wall synthesis	*Actinoplanin teichomyceticus*	Teicoplanin	[36]
Imidazo pyridine-4-one	Inhibits polypeptide synthesis via interaction with the ribosome	*Streptomyces lavendulae, Streptomyces noursei*	Streptothricin	[37]
Lipopeptide	Bactericidal activity by disrupting plasma membrane function without penetrating into the cytoplasm	*Streptomyces roseosporus*	Daptomycin	[38]
Macrolide	Inhibits bacterial protein synthesis	*Streptomyces halstedii*	Carbomycin	[39]
Macrolide	Inhibits elongation at transpeptidation step of protein biosynthesis	*Saccharopolyspora erythrea*	Erythromycin	[40]
Macrolide	Inhibits bacterial protein synthesis	*Streptomyces antibioticus*	Oleandomycin	[41]
Macrolide	Inhibits protein biosynthesis by rapid breakdown of polyribosome's by binding 50S unit	*Streptomyces ambofaciens*	Spiramycin	[42]
Natural polycyclicpolyketide	PABA pathway inhibitor	*Verrucosispora AB-18-032*	Abyssomicins	[43]
Naphthalene (ansamycins subclass)	Inhibits bacterial DNA-dependent RNA-polymerase	*Amycolatopsis rifamycinica*	Rifamycin	[44]
Peptide	Inhibits bacterial protein synthesis	*Streptomyces pyridomyceticus*	Pyridomycin	[45]
Thiopeptide	Inhibits bacterial protein synthesis	*Kocuria sp*	PM181104	[45, 46]
Polyketide-Streptogramin	Inhibits protein biosynthesis by binding to 50S ribosome unit	*Streptomyces virginiae*	Streptogramin A	[47]
Polyene lactam macrolides antibiotic	Inhibits bacterial protein synthesis	*Micromonospora sp*	Micromonosporin	[48]

(continued)

Table 1 (continued)

Class	Mechanism of action	Actinomycetes	Antibiotic	Reference
Tetracyclines	Inhibits protein synthesis (elongation) by preventing binding of aminoacyl-tRNA to the 30S subunit	*Streptomyces aureofaciens*	Chlortetracycline	[49]
Tetracycline	Inhibits protein synthesis (elongation) by preventing binding of aminoacyl-tRNA to the 30S subunit	*Streptomyces rimosus*	Oxy tetracycline	[50]
Thiolactone	Inhibition of fatty acid synthesis	*Nocardia sp*	Thiolactomycin	[51]
Unknown	–	*Streptomyces sp*	Bonactin	[52]

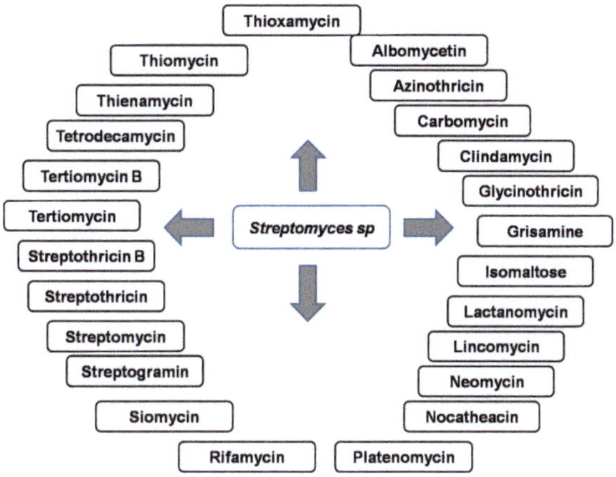

Fig. 2 Representative metabolites by *Streptomyces sp*

Table 2 Recently discovered bioactive compounds from *Streptomyces* sp

Antibiotic	Actinomycetes	Potential use	Reference
BE43472A	*Streptomyces* strain (N1-78-1)	Antibacterial	[55]
Citreamicin delta	*Streptomyces vinaceus*	Antitumor	[56]
Dynemicin	*Micromonospora chersina*	Anticancer	[57]
Lenticulone	*Streptomyces sp* JP 95	Antibacterial	[58]
Lucensimycin D	*Streptomyces lucencis* MA 7349	Antibacterial	[59]
Mediomycin B	*Streptomyces mediocidicus*	Antifungal	[60]
Rapamycin	*Streptomyces hygroscopicus*	Immunosuppressant	[61]
Sansanmycin A	*Streptomyces sp SS*	Antibacterial	[62]

all known antibiotics are produced by these Actinomycetes. The secondary metabolites expressed by *Streptomyces* also find application in the treatment of cancer and autoimmune diseases [19, 53, 54] (Fig. 2). Currently it is reported that there are more than 2,400 different secondary metabolites produced by *Streptomyces* sp. [53]. Scientists and researchers believe that there could be many more such metabolites with therapeutic potential to be discovered and explored [54].

Some of the recently discovered antibiotics from *Streptomyces* are listed in Table 2.

2.2 Rare Actinomycetes: Future Gold Mine of Bioactive Compounds

Streptomyces and other common Actinomycetes have since been exploited so often that the prospects of a new strain often seem remote. Very similar strains most often produce the same or similar compounds, thus hampering the rationale for discovery of new antibiotics. In the quest for new strains and products, marine Actinomycetes home to novel genera and have resulted in some new leads.

Table 3 List of representative bioactive compounds from rare actinomycetes

Antibiotic	Actinomycetes	Potential use	Reference
EHA–2	*Actinomadura spadix*	Antimicrobial	[64]
Teichoplanin	*Actinoplanes teichomyceticus*	Antibiotic	[65]
Vancomycin	*Amycolatopsis orientalis*	Antibiotic	[66]
Pyridomycin	*Dactylosporangium falvum*	Antibiotic	[67]
Aridicins A, B and C	*Kibdelosporangium aridum*	Antimicrobial	[68]

The rare Actinomycetes are usually regarded as strains of Actinomycetes whose frequency of isolation by conventional methods is lower than that of Streptomycete strains and usually comprises those genera other than *Streptomyces*. Notable producers of secondary bioactive metabolites from this class of Actinomycetes are from genera such as *Actinomadura, Actinoplanes, Amycolatopsis, Dactylosporangium, Kibdelosporangium, Kitasatospora, Microbiospora, Planomonospora, Planobispora, Salinispora, Streptosporangium, and Verrucosispora* (Table 3) [63].

Marinospora, affiliated with the *Streptomycetaceae* family has yielded some secondary metabolites named marinimycins, with potential antibacterial and cytotoxic activity [69]. Novel compounds of the napyradiomycin class have also been identified from the "MAR4" lineage [70].

Among the terrestrial sources, *Ktedenobacteria, Actinospica,* and *Catenulispora* also appear to possess the ability to produce secondary metabolites.

3 Myxobacteria-Derived Compounds

Myxobacteria, the gliding, Gram-negative bacteria, produce highly colored macroscopic fruiting bodies on decomposed wood and other substrates. Myxobacteria are unique, with a lifestyle differing from all other prokaryotes. They are capable of excreting hydrolytic enzymes and decomposing various and complex biopolymers but can also lyse and destroy other prokaryotes, and even eukaryotic cells [71]. It has been reported that myxobacteria form a phylogenetically coherent group and constitute the order Myxococcales in the class Deltaproteobacteria. They are subdivided into the three suborders Cystobacterineae, Sorangiineae, and Nannocystineae [71, 72]. They produce a large number of unusual secondary metabolites, with potential antibiotic activity [73]. Myxobacteria have been regarded as "microbe factories" for active secondary metabolites because they have great potential as producers of new drugs [72]. They move by an axonal cellular motion (i.e., gliding) and form fruiting bodies when resources are scarce. Individual cells of myxobacteria organize themselves as waves during cooperative feeding. As the cells collide, they aggregate in mounds that grow in size, forming fruiting bodies that can harbor up to 10^5 individuals. Cells within these structures become myxospores, which germinate to new swarms when nutrients are available. Diverse proteins and metabolites mediate these signaling processes [73, 74]. Their secondary metabolites are unusual hybrids

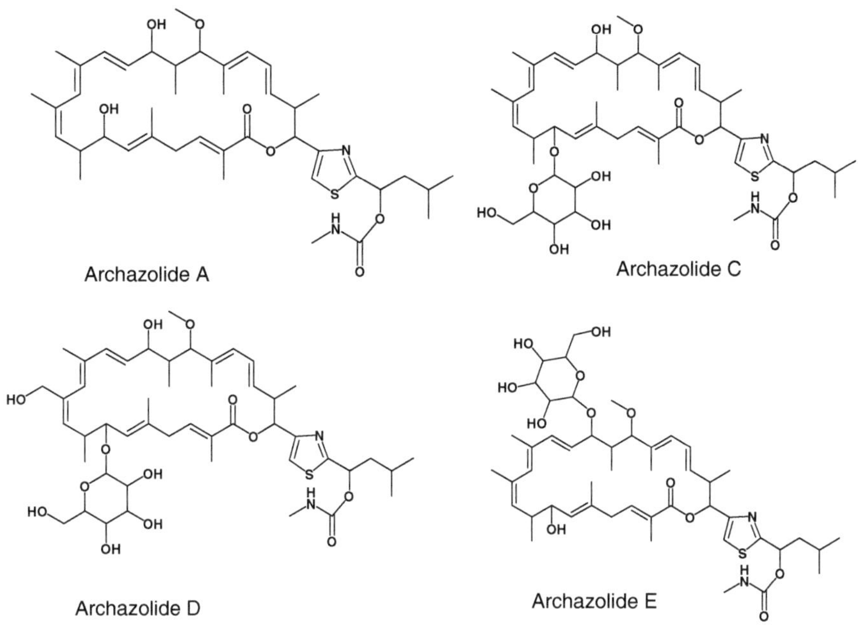

Fig. 3 Chemical structure of Archazolides

of polyketides and nonribosomal-made polypeptides. Unlike metabolites from Actinomycetes, Myxobacterial metabolites are not glycosylated, and their target areas are not the same as other microbial products [74]. They are found mostly in the soil as opposed to marine environments, and are prolific producers of secondary metabolites, which aid their role as predators. Also fascinating is the fact that these bacteria possess the ability to assault their prey in a "pack" or as a single bacterium with cell-to-cell contact [75]. The majority of myxobacteria have been isolated from the soil, a habitat rich in both organic matter and microbial life, including fungi and Actinomycetes. Compound production rates are typically highest during the exponential phase of growth. This behavior is unlike that of the Actinomycetes, in which secondary metabolism correlates with the onset of the stationary phase [76]. Recently, secondary metabolites from myxobacteria have been well reviewed by Weissman and coworkers [77].

Some of the recent scaffolds reported from the myxobacteria class of microbes are as follows.

Archazolides (Fig. 3)
Three compounds, Archazolide C (MW: 901.169), Archazolide D (MW: 917.168), and Archazolide E (901.169) have been reported from the *Cystobacter violaceus* Cb vi105 strain [78]. An amorphous solid, Archazolide D, was reported to have vacuolar-type H+–ATPase (V-ATPase) inhibitor activity, whereas the parent compound Archazolide A (MW: 739.027) was examined and confirmed for its V-ATPase inhibition, antifungal, and antineoplastic activity [79].

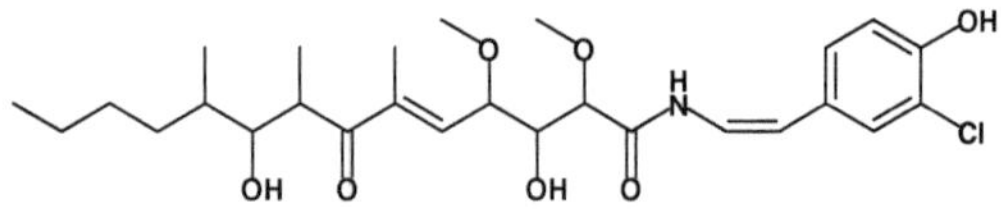

Chondrochloren A

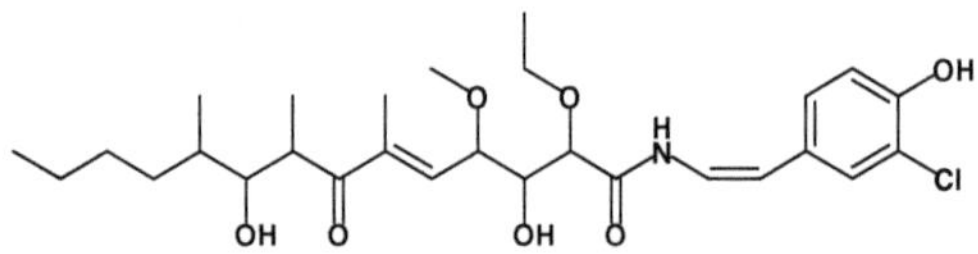

Chondrochloren B

Fig. 4 Chemical structure of Chondrochloren

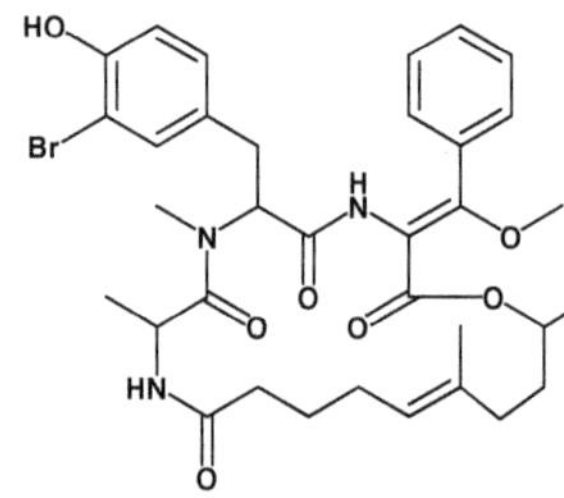

Fig. 5 Chemical structure of Miuraenamide A

Chondrochloren (Fig. 4)
Chondrochloren A (MW: 526.068), an antibacterial compound, was reported from *Chondromyces crocatus* strain Cm c5 in 2003. The attempts of unusual chemistry in the biosynthesis of the antibiotic Chondrochloren A and B had been well documented in 2009 [80].

Miuraenamide (Fig. 5)
A series of potent antifungal compounds, Miuraenamide A–F were reported from the slightly halophilic myxobacterium *Paraliomyxa miuraensis* strain SMH-27-4 [81, 82].

Pedein (Fig. 6)
Chondromyces pediculatus strain Cm p3 has been reported to produce the antifungal compounds Pedein A (MW: 925.390) and Pedein B (MW: 890.945) [83].

Myxobacteria, with their variety of secondary metabolites, unique structures, and new modes of action, are emerging as a highly valuable source of natural products. Myxobacteria are also known to produce different metabolite compounds from different structural classes. Steroid synthesis is extremely rare in bacteria, but both cholesterols and lanosterols have been isolated from myxobacterial extracts [84]. Iron transport metabolites, nannochelins, and myxochelins A and B are produced by myxobacteria. With genome sequencing and metabolic profiling of

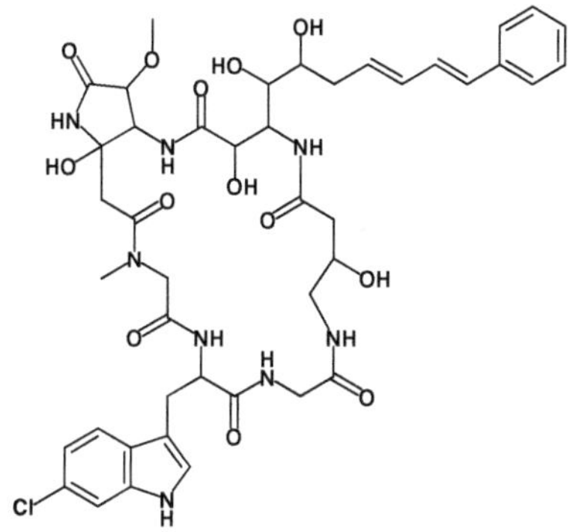

Fig. 6 Chemical structure of Pedein A

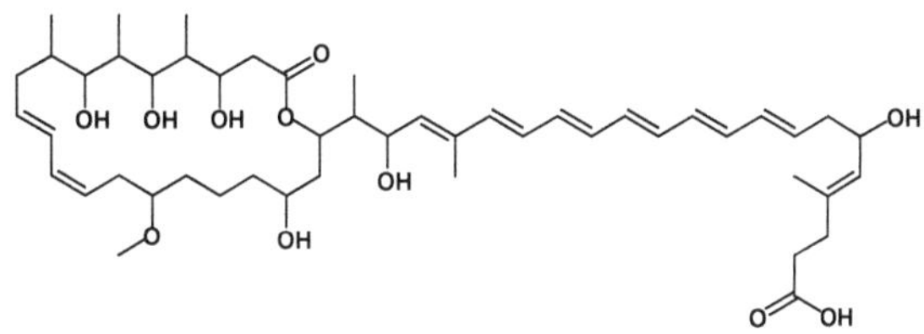

Fig. 7 Chemical structure of Etnangien

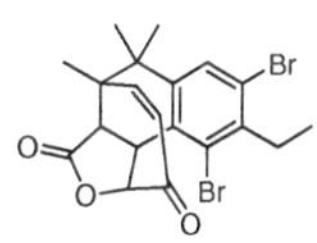

Fig. 8 Chemical structure of Salimabromide

myxobacteria, new strains may be unveiled leading to more promising metabolites with antibiotic potential [85].

Etnangien (Fig. 7)
It is a macrolide antibiotic isolated from the myxobacterium *Sorangium cellulosum*, strains So ce750 and So ce1045. Initial studies have indicated that bacterial and viral nucleic acid polymerases are inhibited by etnangien [86].

Recently, a novel analogue, comparable to that of etnangien has been obtained from the fermentation broths of *Sorangium cellulosum* [87].

Salimabromide (Fig. 8)
Salimabromide is the first natural product from the marine myxobacteria *Plesiocystis/Enhygromyxa* and has revealed antibiotic activity against *Arthrobacter cristallopoietes* [76].

4 Eubacteria-Derived Compounds

In this section we dwell on those prokaryotes under the subcategory of true bacteria or "Eubacteria". In the past decade, within the Eubacteria phyla, many microorganisms have been identified as a sources of bioactive compounds. For instance, marine bacteria have often been reported to produce antibacterial and anticancer compounds, allowing the (i) ecological steadiness of manifold marine ecosystems (ii) interrelations between epiphytic microorganism ambiences, and (iii) inhibition of rival organisms and pathogenic microbes [88]. The sharing or competition mechanisms that are known between these microorganisms are varied, such as antibiotic production, bacteriocins, siderophores, and even pH alteration through the production of organic acids [89]. In the past five years many more bioactive compounds have been reported, however, very few have progressed beyond the discovery or preclinical stage. Moreover, in recent years, research has to a large extent focused on altering existing, naturally occurring antibiotics. These have paved the way for a new class of antibiotics called lantibiotics.

Lantibiotics are peptides with lanthionine and/or methyllanthionine residues produced by Gram-positive bacteria. Modified amino acids such as dehydroalanine and dehydrobutyrine may be also components of the lantibiotics. More recently, they have been the focus of much attention as a consequence of the increasing understanding of their biosynthesis and mode of action, and their high specific activity against multidrug-resistant bacteria [90].

5 Fungal-Derived Compounds

The identification of antibiotics was heralded by the discovery of penicillin from a fungus, the *Penicillin notatum*. Since then, several genera of fungi have been extensively screened for bioactive compounds. However, publications and reviews until now attribute only 5 % of the fungi as producers [91, 92], and the rest await their turn to be tapped for human benefit. This indicates a huge cache of potentially useful fungi that can be tapped with modern techniques of cultivation and identification. Techniques used until now include media optimization, coculturing, chemical induction, epigenetic modulation, and metabolite remodeling, coupled with the fermentation technology for scale-up [93]. These techniques will thus enable their extensive cultivation for the mass production of natural products, both known and novel [93], along with bioprospecting of fungi from every possible source including extreme environments such as marine sediments, geothermal vents, cold deserts, and antarctic and arctic regions.

In recent times, endophytic fungi associated with plants have been viewed as a new source of these pharmacologically active natural products. It is evident that in some cases these associated fungi might be involved in the biosynthesis of compounds that had been previously isolated from plants and might by themselves be the producers of a multitude of new metabolites. However, it is only recently that

Table 4 List of compounds sourced from fungi

Antibiotic	Fungus	Potential use	Reference
FR (KARST)	*Ganoderma lucidum*	Antimicrobial	[94]
Ganodermycin	*Ganoderma applanatum*	Anti-inflammtory	[95]
Aspergiolide	*Aspergillus glaucus*	Antitumor	[96]
Bioxanthracenes	*Cordyceps pseudomilitaris*	Antimalarial	[97]
Chaetominine	*Chaetomium sp*	Anticancer	[98]
Communesins	*Penicillium expansum*	Cytotoxic	[99]
Dolastatin	*Marine mollusks*	Antineoplastic	[100]
Gliocladins	Gliocladium roseum	Antinematode	[101]
Spirolaxine	*Sporotrichum laxum*	Antiproliferative	[102]
Topopyrones A, B, C	*Phoma sp*	Antibacterial	[103]
Variecolorquinines	*Aspergillus variecolor*	Cytotoxic	[104]
Variecolortides	*Aspergillus variecolor*	Cytotoxic	[105]

their capacity for producing biologically active compounds has been explored. Examples are taxol from *Taxomyces andreanae*, podophyllum from *Phialocephala fortinii*, camptothecin from the endophytic fungus of *Camptotheca acuminata*, and hypericin from *Chaetomium globosum*.

Some recently derived compounds from fungi, in various stages of development, are tabulated in Table 4.

Several of these marine-fungal–derived compounds have been well reviewed by Abdessamad et al. [106].

5.1 Slime-Molds–Derived Compounds

Slime molds is a general term used to describe organisms that reproduce by spores. The Myxomycetes (true slime molds) are an unusual group of organisms that may be assigned to one of the lowest classes of eukaryotes. As their fruiting bodies are very small and it is very difficult to collect an adequate quantity of slime molds, few studies have been conducted on the chemistry of Myxomycetes. In a certain stage of their life cycle, they form jellylike plasmodia that feed on bacteria and are able to move by a synchronized perpendicular flow of their protoplasm. Later, the plasmodium transforms in a few hours into small fruiting bodies. These bodies (peridia) often exhibit delicate structures and colors. They release spores from which protozoa like amoeba originate that mate and finally aggregate again to the plasmodia stage. Initially classified under fungi, they are now a separate group as they are quite unrelated to fungi. Among fungi, the number of bioactive compounds reported from slime molds have been less compared to imperfect fungi, the Ascomycetes, and several other filamentous and endophytic fungal species [107]. Approximate 60 bioactive metabolites have been reported from slime molds [107]. The three main groups include *Physarum*, cellular slime molds, and *Labyrinthulomycota*. Of these,

Physarum gyrosum has been shown to express metabolites with antibacterial activity [108]. Masami reported new antimicrobial naphthoquinone pigments, tyrosine–kinase inhibitory bisindole alkaloids, a cytotoxic triterpenoid aldehyde lactone with a reversal effect of drug resistance, a cycloanthranilylproline with sensitizing effect of TRAIL-induced apoptosis through activation of COX2, a dibenzofuran glycoside, and, moreover, sterols with a 2,6-dioxabicyclo[2.2.2] octan-3-one ring system were also isolated from field-collected fruit bodies of Myxomycetes [109]. Secondary metabolites of slime molds were well reviewed in 2005 by Dembitskya et al. [110]. The review included several well-defined and characterized bioactive compounds. In the past few years there have been very sparse reports on significant bioactive compounds from this class of microbes. Nevertheless recently aquatic Myxomycetes have been thoroughly reviewed by Mitsunori and Harold in a 2013 review article [111].

6 Pipeline of Microbial Bioactive Compounds

Numerous companies worldwide are involved in bioprospecting, drug discovery, and drug development programs. However, the past 10–12 years have witnessed major progress in relying on innovation-driven natural products as the sole source of new compounds.

Recent advances in screening, analytical methods in isolating minor compounds, and genomic mining approaches have propelled natural products research to the next stage in the pharmaceutical business. Marine microbes, hitherto not readily accessible as compared to microbes from other sources, have been a source of unique compounds, leading to an increase in the number of drugs entering the drug development phase [112].

In the last decade, 13 new antibiotics have been approved by the FDA, of which just three—Linezolid, Daptomycin, and Retapamulin—have novel action mechanisms [113]. Recently, Fidaxomicin (Dificid, by Optimer Pharmaceuticals), a new scaffold from an Actinomycetes genera (*Dactylosporangium auranticum*) and an anti-*Clostridium difficile* antibiotic have been approved by the FDA and launched to the market in May 2011. Dificid (fidaxomicin) is a narrow-spectrum macrocyclic antibiotic. Dificid is specifically indicated in adults for treatment of *C. difficile*-associated diarrhea [114]. To reduce the development of drug-resistant bacteria and maintain the effectiveness of Dificid, it should be used only to treat infections that are proven or strongly suspected to be caused by *C. difficile*. Dificid is supplied as a tablet designed for oral administration. It is reported that Fidaxomicin is bactericidal against *C. difficile* in vitro, inhibiting RNA synthesis by RNA polymerases [114].

The pharmaceutical industry's main markets are under serious performance pressure. Higher R&D costs, a relatively dry pipeline for new drugs, the increasing demands from payers and providers for reduced healthcare costs, and a host of other factors are putting pressure on global pharmaceutical companies [115].

Cancer is the most important cause of global fatality, with 7.6 million deaths (around 13.6 % of all deaths) in 2008 [116]. Half of the deaths can be attributed to lung, stomach, liver, colorectal, and female breast cancers. About 47 % of cancer cases and 55 % of the cancer deaths occur in less-developed regions of the world. One of the recent reports predicts that the world market for anticancer agents will reach $116.5 billion in 2017, and expand further to 2023 [117]. Ten anticancer drugs have been approved by the FDA in 2013 [118], although none are from microbial resources. However, 86 anticancer compounds from natural products are reported to be under development, of which nine compounds are undergoing Phase III trials [119].

Despite a slowdown of the discovery programs of many pharmaceutical companies, at present there are numerous promising drug candidates in the current development pipeline. Interestingly, many of these promising candidates are of microbial origin. Scientific and practical shortcomings associated with microbial product research are being minimized, and better prospects are envisaged with the exploration of microbial compounds expressed by microbes in ecosystems that were not accessible before. Extrapolating the current situation, it will not be long until the second golden era of microbial compounds will be unveiled.

Acknowledgments The authors wish to thank the management of Piramal Enterprises Limited for their encouragement in the dissemination of science, which deals with the discovery of new drugs for alleviating disease on this planet. Very special thanks to Dr. P. D. Mishra, Head, Natural Product Chemistry Division-Piramal Enterprises Limited for his help in providing the necessary structures of requested compounds.

References

1. Newman DJ, Cragg GM (2009) Microbial anti tumor drugs: natural products of microbial origin as anticancer agents. Curr Opin Investig Drugs 10(12):1280–1296
2. Demain AL, Sanchez S (2009) Microbial drug discovery: 80 years of progress. J Antibiot 62:5–16
3. Alan HL, Qiong Y, Karen AL, Susan AF, William FB III, Roger JS, Anthony JT (2008) Structural diversity of organic chemistry. A scaffold analysis of the CAS registry. J Org Chem 73:444–4451
4. DeLong EF, Preston CM, Mincer T, Rich V, Hallam SJ, Frigaard NU, Martinez A, Sulli-van MB, Ed-wards R, Brito BR, Chisholm SW, Karl DM (2006) Community genomics among stratified microbial assemblages in the ocean's interior. Science 311:496–503
5. Simon C, Daniel R (2009) Achievements and new knowledge unravelled by metagenomic approaches. Appl Microbiol Biotechnol 85:265–276
6. Gullo VP, McAlpine J, Lam KS, Baker D, Petersen F (2006) Drug discovery from natural products. J Ind Microbiol Biotechnol 33:523–531
7. Cragg GM, Newman DJ (2013) Natural products: a continuous source of novel drug leads. Biochem Biophys Acta 1830(6):3670–3695
8. Molinari G (2009) Natural products in drug discovery: present status and perspectives. Adv Exp Med Biol 655:13–27
9. Raiijmakers JM, Mazzola M (2012) Diversity and natural functions of antibiotics produced by beneficial and plant pathogenic bacteria. Annu Rev Phytopathol 50:403–424

10. Jimeno J, Faircloth G, Fernández Sousa-Faro JM, Scheuer P, Rinehart K (2004) New marine derived anticancer therapeutics a journey from the sea to clinical trials. Mar Drugs 1:14–29
11. Donadio S, Brandi L, Monsiardini P, Sosio M, Gualerzi CO (2007) Novel assays and novel strains-promising routes to new antibiotics? Expert Opin Drug Discov 2(6):789–798
12. Dondero NC, Scotti T (1957) Excretion by streptomycetes of factors causing formation of aerial hyphae by old cultures. J Bacteriol 73:584–585
13. Waksman SA (1940) On the classification of Actinomycetes. J Bacteriol 39(5):549–558
14. Williams ST, Goodfellow M, Alderson G (1989) Genus Streptomyces Waksman and Henrici 1943, 339AL. In: Williams ST, Sharpe ME, Holt JG (eds) Bergey's manual of systematic bacteriology, vol 4. Williams and Wilkins, Baltimore, pp 2452–2492
15. Williams ST, Goodfellow M, Alderson G, Wellington EMH, Sneath PHA, Sackin MJ (1983) Numerical classification of Streptomyces and related genera. J Gen Microbiol 129:1743–1813
16. Buchanan RE, Gibbons NE (1974) Bergey's manual of determinative bacteriology, vol 8. The Williams and Wilkins Co., Baltimore, pp 747–842
17. Korn-Wendisch F, Kutzner HJ (1992) The family Streptomycetaceae. In: Balows A, Trooper HG, Dworkin M,Harder W, Schleifer KH (eds) The prokaryotes, a handbook on the biology of bacteria: ecophysiology, isolation, identification, application, vol 1, 2nd edn. Springer, New York, pp 921–995
18. http://taxonomicon.taxonomy.nl/TaxonTree.aspx. Accessed 13 July 2013
19. Mahajan GB, Balachandran L (2012) Antibacterial agents from actinomycetes—a review. Front Biosci (Elite Ed) 4:240–253
20. http://chem.sis.nlm.nih.gov/chemidplus/cas/4696-76-8. Accessed 13 July 2013
21. Ketaki B, Majumdar SK (1973) Utilization of carbon and nitrogen sources by Streptomyces kanamyceticus for kanamycin production. Antimicrob Agents Chemother 4(1):6–10
22. Gonzalez R, Isla L, Obregon AAM, Escalante L, Sanchez S (1995) Gentamicin formation in Micromonospora purpurea: stimulatory effect of ammonium. J Antibiot 48(6):479–839
23. http://www.drugs.com/mtm/neomycin.html. Accessed 25 Aug 2013
24. Kupferberg AB, Styles H, Singher HO, Selman AW (1950) The production of Streptocin by different strains of Streptomyces griseus. J Bacteriol 59(4):523–526
25. Waksman SA (1952) Streptomycin: background, isolation, properties and utilization, nobel lecture. Elsevier, Amsterdam, pp 370–388
26. Wongtavatchai J, McLean JG, Ramos F, Arnold D (2004) Chloramphenicol. whqldoc.who.int/publications/.../9241660538_chloramphenicol.pdf
27. Steffensky M, Mühlenweg A, Wang ZX, Li SM, Heide L (2000) Identification of the novobiocin biosynthetic gene cluster of Streptomyces spheroides NCIB 11891. Antimicrob Agents Chemother 44(5):1214–1222
28. From the Centers for Disease Control and Prevention (2001) Update on spectinomycin availability in the United States. JAMA 286(11):1308–1309
29. Donald MR, Selman AW (1948) Grisein, an antibiotic produced by certain strains of Streptomyces griseus. J Bacteriol 55(5):739–752
30. Bryan A, Dorothy CH, Viswamitra MA (1970) The structure of thiostrepton. Nature 225:233–235
31. Spizek J, Rezanka T (2004) Lincomycin, cultivation of producing strains and biosynthesis. Appl Microbiol Biotechnol 63(5):510–519
32. http://www.merck.com/mmpe/lexicomp/clindamycin.html. Accessed 25 Aug 2013
33. Jonker HRA, Baumann S, Wolf A, Schoof S, Hiller F, Schulte KW, Kirschner KN, Schwalbe H, Arndt HD (2011) NMR Structures of thiostrepton derivatives for characterization of the ribosomal binding site. Angew Chem Int Ed 50:3308–3312
34. Farver DK, Hedge DD, Lee SC (2005) Ramoplanin: a lipoglycodepsipeptide antibiotic. Ann Pharmacother 39(5):863–868
35. Levine D (2006) Vancomycin: a history. Clin Infect Dis 42:5–12
36. http://www.aic.cuhk.edu.hk/web8/glycopeptides.htm. Accessed 25 Aug 2013
37. Waksman SA (1943) Production and activity of streptothricin. J Bacteriol 46(3):299–310

38. Vivian M, Coëffet-LeGal Marie-Françoise, Paul B, Renee B, Julia P, Andrew W, Steven M, Robert F, Ian P, Mario B, Christopher JS, Stephen KW, Richard HB (2005) Daptomycin biosynthesis in Streptomyces roseosporus: cloning and analysis of the gene cluster revision of peptide stereochemistry. Microbiology 151:1507–1523
39. http://www.steadyhealth.com/encyclopedia/Carbomycin. Accessed 25 Aug 2013
40. http://www.rxlist.com/erythromycin-ethylsuccinate-drug.htm. Accessed 25 Aug 2013
41. Luis MQ, Jos AS (1995) Biosynthesis of macrolide oleandomycin by Streptomyces antibioticus: purification and kinetic characterisation of an oleandomycin glucosyltrans ferase. J Biol Chem 270:18234–18239
42. Anise L, Ahmed L, Choke B, Gerard L, Pierre G (1995) Glycerol effect on spiramycin production and valine catabolism in Streptomyces ambofaciens. Curr Microbiol 31(5): 304–311
43. Hafizur R, Brian A, Wilfrid JM, Peter CM, Derek JJ, David RA, Andrew MS, Michael S (2010) Novel anti-infective compounds from marine bacteria. Mar Drugs 8(3):498–518
44. Ogawara H, Maeda K, Umezawa H (1968) Biosynthesis of Pyridomycin I. Biochem 7(9):3296–3302
45. Mahajan GB, George SD, Ranadive P, Mishra PD, Sreekumar E, Panshikar RM, Sawant SN, Krishna S, Sivakumar M, Pari K, Thomas B, Patel Z, Vishwakarma RA, Naik CG, D'Souza L, Devi P (Piramal Life Sciences Limited, Mumbai and NIO, Goa) (2007) PM181104 and related antibacterial compounds, production, pharmaceutical compositions, and therapeutic use. PCT Int Appl WO 2007119201
46. Mahajan GB, Shanbhag P, Sivaramakrishnan H (2011) Poster entitled "Mode of action of antibiotic PM181104 on bacteria" at 1st Global Forum on Bacterial Infections: balancing treatment access and antibiotic resistance, organised by Center for Disease Dynamics, Economics and Policy, USA at New Delhi
47. http://www.merck.com/mmpe/print/sec14/ch170/ch170g.html. Accessed 25 Sept 2013
48. Waksman SA, Geiger WB, Bugie E (1947) Micromonosporin, an antibiotic from a little known group of microorganisms. J Bacteriol 53(3):355–357
49. Biffi G, Boretti G, Di Marco A, Pennella P (1954, September issue) Metabolic behaviour and chlortetracycline production by Streptomyces aureofaciens in liquid culture. Appl Microbiol 2:288–293
50. Zygmunt WA (1961) Oxytetracycline formation by Streptomyces rimosus in chemically defined media. Appl Microbiol 9(6):502–507
51. Hiroshi S, Hideo O, Toshiaki H, Ikutoshi M, Kunio A, Mikio S (1982) Thiolactomycin, a new antibiotic. J Antibiot 35(4):396–400
52. Schumacher RW, Talmage SC, Miller SA, Sarris KE, Davidson BS, Goldberg A (2003) Isolation and structure determination of an antimicrobial ester from a marine-derived bacterium. J Nat Prod 66:1291–1293
53. Lucas X, Senger C, Erxleben A, Grüning BA, Döring K, Mosch J, Flemming S, Günther S (2013) StreptomeDB: a resource for natural compounds isolated from Streptomyces species. Nucl Acids Res 41(Database Issue):D1130–D1136
54. Genilloud O, González I, Salazar O, Martín J, Tormo JR, Vincente F (2011) Current approaches to exploit actinomycetes as a source of novel natural products. J Ind Microbiol Biotechnol 38(3):375–389
55. Rahman H, Austin B, Mitchell WJ, Morris PC, Jamieson DJ, Adams DR, Spragg AM, Schweizer M (2010) Novel anti-infective compounds from marine bacteria. Mar Drugs 5(3):498–518
56. Hopp DC, Milanowski DJ, Rhea J, Jacobsen D, Rabenstein J, Smith C, Romari K, Clarke M, Francis L, Irigoyen M, Luche M, Carr GJ, Mocek U (2008) Citreamicins with potent gram-positive activity. J Nat Prod 71(12):2032–2035
57. Pei G, Dai H, Ren B, Liu X, Zhang L (2010) Exploiting bioactive Enediynes from marine microbe based on activity and gene screening. Wei Sheng Wu Xue Bao 50(4):472–477

58. Yunt Z, Reinhardt K, Li A, Engeser M, Dahse HM, Gütschow M, Bruhn T, Bringmann G, Piel J (2009) Cleavage of 4 carbon- carbon bonds during biosynthesis of griseorhod in a spiroketal pharmacophore. J Am Chem Soc 131(6):2297–2305
59. Singh SB, Zink DL, Dorso K, Motyl M, Salazar O, Basilio A, Vicente F, Byrne KM, Ha S, Genilloud O (2009) Isolation, structure and anti-bacterial activities of lucensimycins D-G, discovered from Streptomyces lucencis MA7349 using an antisense strategy. J Nat Prod 72(3):345–352
60. Cai P, Kong F, Fink P, Ruppen ME, Williamson RT, Keiko T (2007) Polyene antibiotics from Streptomyces mediocidicus. J Nat Prod 70(2):215–219
61. Nicolaou KC, Chen JS, Dalby SM (2009) From nature to the laboratory and into the clinic. Bioorg Med Chem 17(6):2290–2303
62. Xi Y, Chen R, Si S, Sun C, Xu H (2007) A new nucleosidyl peptide antibiotic, sansanmycin. J Antibiot (Tokyo) 60(2):158–161
63. Jose PA, Jebakumar SRD (2013) Non-streptomycete actinomycetes nourish the current antimicrobial discovery. Front Microbiol 4(240):1–3
64. Islam VI, Saravanan S, Ignacimuthu S (2014) Microbicidal and anti-inflammatory effects of Actinomadura spadix (EHA-2) active metabolites from Himalayan soils, India. World J Microbiol Biotechnol 30(1):9–18
65. Taurino C, Frattini L, Marcone GL, Gastaldo L, Marinelli F (2011) Actinoplanes teichomyceticus ATCC 31121 as a cell factory for producing teicoplanin. Microb Cell Fact 10:82
66. Jeong H, Sim YM, Kim HJ, Lee DW, Lim SK, Lee SJ (2013) Genome sequence of the vancomycin producing Amycolatopsis orientalis subsp. orientalis strain KCTC 9412T. Genome Announc 1(3):pii e00408–13
67. Hartkoorn RC, Sala C, Neres J, Pojer F, Magnet S, Mukherjee R, Uplekar S, Boy-Röttger S, Altmann KH, Cole ST (2012) Towards a new tuberculosis drug: pyridomycin—nature's isoniazid. EMBO Mol Med 4:1032–1042
68. Grappel SF, Giovenella AJ, Phillips L, Pitkin DH, Nisbet LJ (1985) Antimicrobial activity of aricidins, novel glycopeptides antibiotics with high and prolonged levels in blood. Antimicrob Agents Chemother 28(5):660–662
69. Evans PA, Huang MH, Lawler MJ, Maroto S (2012) Total synthesis of marinomycin A using salicylate as a molecular switch to mediate dimerization. Nat Chem 4:680–684
70. Cheng YB, Jensen PR, Fenical W (2013) Cytotoxic and antimicrobial napyradiomycins from two marine-derived streptomyces strains. Eur J Org Chem 18:3751–3757
71. Thorsten B, Doreen F, John V, Sonja V, Christine K, Sebastian T, Marc M, Brigitte K, Wagner-D Irene, Rolf D, Meinhard S (2012) Biogeography and phylogenetic diversity of a cluster of exclusively marine myxobacteria. ISME J 6:1260–1272
72. Wenzel SC, Muller R (2009) Myxobacteria-microbial factories for the production of bioactive secondary metabolites. Mol BioSyst 5:567–574
73. Reichenbach H, Gerth K, Irschik H, Brigitte K, Gerhard H (1988) Myxobacteria: a source of new antibiotics. Trends Biotechnol 6(6):115–121
74. Diez J, Martinez JP, Mestres J, Sasse F, Frank R, Meyerhans A (2012) Myxobacteria: natural pharmaceutical factories. Microb Cell Fact 11:52
75. Xiao Y, Wei X, Ebright R, Wall D (2011) Antibiotic production by myxobacteria plays a role in predation. J Bacteriol 193(18):4626–4683
76. Weissman KJ, Müller R (2009) A brief tour of myxobacterial secondary metabolism. Bioorg Med Chem 17:2121–2136
77. Irschik H, Schummer D, Höfle G, Reichenbach H, Steinmetz H, Jansen R (2007) Etnangien, a macrolide polyene antibiotic from Sorangium cellulosum that inhibits nucleic acid polymerases. J Nat Prod 70(6):1060–1063
78. Menche D, Arikan F, Perlova O, Horstmann N, Ahlbrecht W, Wenzel SC, Jansen R, Irschik H, Müller R (2008) Stereochemical determination and complex biosynthetic assembly of etnangien, a highly potent RNA polymerase inhibitor from the myxobacterium Sorangium cellulosum. J Am Chem Soc 130(43):14234–14243

79. Felder S, Dreisigacker S, Kehraus S, Neu E, Bierbaum G, Wright PR, Menche D, Schäberle TF, König GM (2013) Salimabromide: unexpected chemistry from the obligate marine Myxobacterium Enhygromxya salina. Chemistry 19(28):9319–9324
80. Iizuka T, Fudou R, Jojima Y, Ogawa S, Yamanaka S, Inukai Y, Ojika M (2006) Miuraenamides A and B, novel antimicrobial cyclic depsipeptides from a new slightly halophilic myxobacterium: taxonomy, production, and biological properties. J Antibiot (Tokyo) 59(7):385–391
81. Horstmann N, Essig S, Bockelmann S, Wieczorek H, Huss M, Sasse F, Menche D (2011) Arc-hazolid A-15-O-β-D-glucopyranoside and iso-archazolid B: potent V-ATPase inhibitory polyketides from the myxobacteria Cystobacter violaceus and Archangium gephyra. J Nat Prod 74(5):1100–1105
82. Menche D, Hassfeld J, Steinmetz H, Huss M, Wieczorek H, Sasse F (2007) The first hydrox-ylated archazolid from the myxobacterium Cystobacter violaceus: isolation, structural elucidation and V-ATPase inhibition. J Antibiot (Tokyo) 60(5):328–331
83. Kunze B, Böhlendorf B, Reichenbach H, Höfle G (2008) Pedein A and B: production, isolation, structure elucidation and biological properties of new antifungal cyclopeptides from Chondromyces pediculatus (Myxobacteria). J Antibiot (Tokyo) 61(1):18–26
84. Desmond E, Gribaldo S (2009) Phylogenomics of sterol synthesis: insights into the origin, evolution, and diversity of a key eukaryotic feature. Genome Biol Evol 1:364–381
85. Jansen R, Kunze B, Reichenbach H, Höfle G (2003) Chondrochloren A and B, new β-amino styrenes from Chondromyces crocatus (Myxobacteria). Eur J Org Chem 14:2684–2689
86. Rachid S, Scharfe M, Blöcker H, Weissman KJ, Müller R (2009) Unusual chemistry in the biosynthesis of the antibiotic chondrochlorens. Chem Biol 16(1):70–81
87. Iizuka T, Jojima Y, Fudou R, Tokura M, Hiraishi A, Yamanaka S (2003) Enhygromyxa salina gen. nov., sp. nov., a slightly halophilic myxobacterium isolated from the coastal areas of Japan. Syst Appl Microbiol 26(2):189–196
88. Irma ESM, Gómez LJV, Rivas GG, Sánchez NEA (2012) Bioactive compounds from bacteria associated to marine algae. In: Sammour R (ed) Biotechnology—molecular studies and novel applications for improved quality of human life, pp 25–44
89. Avendaño-Herrera R, Lody M, Riquelme CE (2005) Producción de substancias inhibitorias en-tre bacterias de biopelículas en substratos marinos. Revista Biología Marina y Oceanografía 40(2):117–125
90. Ojika M, Inukai Y, Kito Y, Hirata M, Iizuka T, Fudou R (2008) Miuraenamides: antimicrobial cyclic depsipeptides isolated from a rare and slightly halophilic myxobacterium. Chem Asian J 3(1):126–133
91. Hawksworth DL (1991) The fungal dimension of biodiversity: magnitude, significance and conservation. Mycol Res 95:641–655
92. Hawksworth DL (2001) The magnitude of fungal diversity: the 1.5 million species estimate re-visited. Mycol Res 105:1422–1431
93. http://www.springerplus.com/content/2/1/8. Accessed 25 Sept 2013
94. Cotter PD, Hill C, Ross RP (2005) Bacterial lantibiotics: strategies to improve the therapeutic potential. Curr Protein Pept Sci 6(1):61–75
95. Gray AI, Igoli JO, Edradebel R (2012) Natural products isolation in modern drug discovery programs. In: Sarkar SD, Nahar L (eds) Natural product isolation, pp 515–534
96. Yoon SY, Eo SK, Kim YS, Lee CK, Kan SS (1994) Antimicrobial activity of Ganoderma lucidum extract alone or in combination with some antibiotics. Arch Pharm Res 17(6): 438–442
97. Jung M, Liermann JC, OPatz T, Erkel G (2011) Ganodermycin, a novel inhibitor of CXCL 10 expression from Ganoderma applanatum. J Antibiot (Tokyo) 64(10):683–686
98. Sun X, Zhou X, Cai M, Zhou J, Zhang Y (2010) Significant stimulation of o-phthalic acid in biosynthesis of Aspergiolide A by a marine fungus Aspergillus glaucus. Bioresour Technol 101(10):3609–3616

99. Jaturapat A, Isaka M, Hywel-Jones NL, Lertwerawat Y, Kamchonwongpaisan S, Kirtikara K, Tan-ticharoen M, Thebtaranonth Y (2001) Bioxanthracenes from the insect pathogenic fungus. Cordyceps pseu-domilitaris BCC 1620. I: taxonomy, fermentation, isolation and antimalarial activity. J Antibiot (Tokyo) 54(1):29–35
100. Jiao RH, Xu S, Liu JY, Ge HM, Ding H, Xu C, Zhu HL, Tan RX (2006) Chaetominine, a cytotoxic alkaloid produced by endophytic Chaetomium sp IFB E015. Org Lett 8(25): 5709–5712
101. Kerzaon I, Pouchus YF, Monteau F, Le Bizec B, Nourrisson MR, Biard JF, Grovel O (2009) Structural investigation and elucidation of new communesins from a marine derived Penicillium expansum link by liquid chromatography/electrospray ionization mass spectroscopy. Rapid Commun Mass Spectrom 23(24):3928–3938
102. Piplani H, Rana C, Vaish V, Vaiphei K (1830) Sanyal SN (2013) Dolastatin, along with Celecoxib, stimulates apoptosis by a mechanism involving oxidative stress, membrane potential change and P13 K/AKT pathway down regulation. Biochim Biophys Acta 11:5142–5156
103. Dong JY, He HP, Shen YM, Zhang KQ (2005) Nematicidal epipolysulfanyldioxopiperazines from Gliocladium roseum. J Nat Prod 68(10):1510–1513
104. Nasini G, Bava A, Fronza G, Giannini G (2007) Microbial transformation of spirolaxine, a bioactive undecaketide fungal metabolite from the basidiomycete Sporotrichum laxum. Chem Biodivers 4(12):2772–2779
105. Kanai Y, Ishiyama D, Senda H, Iwatani W, Takahashi H, Konno H, Tokumasu S, Kanazawa S (2000) Novel human topoisomerase I inhibitors A, B, C and D. I: producing strain, fermentation, isolation, physico chemical properties and biological activity. J Antibiot (Tokyo) 53(9):863–872
106. Abdessamad D, Amal HA, Wen HL, Peter P (2010) Bioactive compounds from marine bacteria and fungi. Microb Biotechnol 3(5):544–563
107. Bérdy J (2005) Bioactive microbial metabolites. J Antibiot 58(1):26
108. Wang W, Zhu T, Tao H, Lu Z, Fang Y, Gu Q, Zhu W (2007) Two new cytotoxic quinone type compounds from the halotolerant fungus Aspergillus variecolor. J Antibiot. (Tokyo) 60(10):603–607
109. Ishibashi M (2007) Study on myxomycetes as a new source of bioactive natural products. Yakugaku Zasshi 127(9):1369–1381
110. Dembitsky VM, Rezanka T, Spízek J, Hanus LO (2005) Secondary metabolites of slime molds (myxomycetes). Phytochemistry 66(7):747–769
111. Mitsunori T, Harold WK (2013) Aquatic myxomycetes. FUNGI 6(3):18–25
112. Wang WL, Zhu TJ, Tao HW, Lu ZY, Fang YC, Gu QQ, Zhu WM (2007) Three novel, structurally unique spirocyclic alkaloids from the halotolerant B 17 fungal strain of Aspergillus variecolor. Chem Biodivers 4(12):2913–2919
113. Schroeder HR, Mallette MF (1973) Isolation and purification of antibiotic material from Physarum gyrosum. Antimicrob Agents Chemother 4(2):160–166
114. https://www.centerwatch.com/drug-information/fda-approved-drugs/drug/1152/dificid-fidaxomicin. Accessed 22 Dec 2013
115. Waters AL, Hill RT, Place AR, Hamann MT (2010) The expanding role of marine microbes in pharmaceutical development. Curr Opin Biotechnol 21(6):780–786
116. http://www.who.int/mediacentre/factsheets/fs297/en/. Accessed 22 Dec 2013
117. http://www.prnewswire.com/news-releases/leading-anti-cancer-drugs-and-associated-market-2013-2023-219612141.html. Accessed 22 Dec 2013
118. http://www.centerwatch.com/drug-information/fda-approvals/. Accessed 22 Dec 2013
119. Harvey AL (2008) Natural products in drug discovery. Drug Discov Today 13(19/20): 894–901

Adv Biochem Eng Biotechnol (2015) 147: 59–110
DOI: 10.1007/10_2014_273

Published Online: 8 July 2014

Medicinal Plants, Human Health and Biodiversity: A Broad Review

Tuhinadri Sen and Samir Kumar Samanta

Abstract Biodiversity contributes significantly towards human livelihood and development and thus plays a predominant role in the well being of the global population. According to WHO reports, around 80 % of the global population still relies on botanical drugs; today several medicines owe their origin to medicinal plants. Natural substances have long served as sources of therapeutic drugs, where drugs including digitalis (from foxglove), ergotamine (from contaminated rye), quinine (from cinchona), and salicylates (willow bark) can be cited as some classical examples.Drug discovery from natural sources involve a multifaceted approach combining botanical, phytochemical, biological, and molecular techniques. Accordingly, medicinal-plant–based drug discovery still remains an important area, hitherto unexplored, where a systematic search may definitely provide important leads against various pharmacological targets.Ironically, the potential benefits of plant-based medicines have led to unscientific exploitation of the natural resources, a phenomenon that is being observed globally. This decline in biodiversity is largely the result of the rise in the global population, rapid and sometimes unplanned industrialization, indiscriminate deforestation, overexploitation of natural resources, pollution, and finally global climate change.Therefore, it is of utmost importance that plant biodiversity be preserved, to provide future structural diversity and lead compounds for the sustainable development of human civilization at large. This becomes even more important for developing nations, where well-planned bioprospecting coupled with nondestructive commercialization could help in the conservation of biodiversity, ultimately benefiting mankind in the long run.Based on these findings, the present review is an attempt to update our knowledge about the diverse therapeutic application of different plant products against various pharmacological targets including cancer, human brain, cardiovascular function, microbial infection, inflammation, pain, and many more.

Keywords Biodiversity · CNS · Cardiovascular · Anticancer · Antimicrobial

T. Sen (✉)
Department of Pharmaceutical Technology and School of Natural Product Studies,
Jadavpur University, Kolkata 700032, India
e-mail: tssen@hotmail.com

S. K. Samanta
Calcutta Institute of Pharmaceutical Technology, Banitabla, Uluberia
Howrah 711316, India

Contents

1 Introduction 60
2 New Bioresources 62
2.1 Phytochemicals and the Human Brain 61
2.2 Phytochemicals as Potential Anti-inflammatory Agents 63
2.3 Phytochemicals and Cardiovascular Disease 72
2.4 Medicinal Plants Used as a Source of Anticancer Agents 74
2.5 Medicinal Plants Used as a Source of Antimicrobial Agents 79
2.6 Others 89
3 Pharmaceutical Bioprospecting and Bioconservation 90
3.1 Impacts on Biodiversity of Major Pressures and Associated Effects on Ecosystem Services and Human Well-Being 90
4 Conclusions 93
References 94

1 Introduction

"This curious world which we inhabit is more wonderful than it is convenient; more beautiful than it is useful; it is more to be admired and enjoyed than it is to be used" (Henry David Thoreau 1837).

For millennia humankind has relied heavily on plants for food and also for the alleviation of diseases. Natural products have always contributed extensively towards the development of modern medicine, and still continue to play a significant role in drug discovery. While taking a look at the history of modern medicine, we find the application of digitalis glycosides, during the eighteenth century, for the treatment of cardiovascular disorders, and willow bark was popularly used for the management of pain and fever-like conditions. The search for novel therapeutic leads from natural resources has been going on for ages and has resulted in several important discoveries that include antibiotics, anticancer agents, anti-inflammatory compounds, and analgesics. The vast genetic diversity available in plants, animals, and microorganisms presents a wealth of possibilities for the betterment of humankind in the production of food, materials, and medicine. Terrestrial plants offer a unique and renewable resource for the discovery of therapeutically active novel biomolecules owing to the structural and biological diversity of their constituents. However, only a small fraction of the plant kingdom has yet been analyzed for their possible medicinal uses. Newer techniques of combinatorial chemistry and high-throughput screening of plant products as well as de novo design is now a mainstay for new drug discovery. Thus the search for novel chemical entities (of natural origin) continues to serve as an important source of structural diversity. Genomic research continues to identify molecular targets for disease that can derive specific screening assays. All major drug companies screen plant extracts as well as synthetics. More often, natural products provide lead structures which are starting points for chemical modification to derive an optimal drug. In parallel, new techniques of combinatorial biosynthesis including the "omics" approach offer possibilities for identification of novel substances. According to WHO reports, treatments with herbal medicine or vegetable extracts are practiced by approximately 80 % of the worlds population [1]. Currently, phytotherapies represent an approximately $14 billion/year industry, which is about 5 % of the current $280 billion/year market. Here, it is pertinent to mention that significant regional differences exist between developed and developing countries, where herbal products represent 25 and 80 % of medications, respectively [2]. Among the 56 % of currently prescribed synthetic drugs, 24 % are derivatives from plant species, 9 % are synthetic products modeled from natural products, 6 % are extracted directly from the plant species, and 5 % are of animal origin [3]. However, the vast repertoire of natural products still remains to be tapped. The estimated total number of existing species is between 350,000 and 550,000, of which less than 20 % have been investigated for medicinal potential [4]. Brazil, for example, has around 10 % of the world's flora, where less than 1 % of its plant species have been investigated for chemical and/or pharmacological properties [5].

Certain groups of people have for many years immensely benefited from the conversion of natural ecosystems to human-dominated ecosystems and from the exploitation of biodiversity. Unfortunately, such gains have always been achieved at the cost of losses in biodiversity, degradation of many ecosystem services, and the exacerbation of poverty for other groups of people. Ironically, just as we have begun to recognize some of the potential benefits that might accrue from a systematic search of this vast storehouse, the plant kingdom, we have also started realizing that there is a simultaneous decline in the number of available species and which in turn may have catastrophic consequences [6]. This decline in biodiversity is largely the result of human activities such as drastic transformation of natural landscapes or deforestation. These phenomena pose a serious threat to sustainable development because the species diversity of our planet is one of the most important as well as irreplaceable resources we possess. Therefore, preservation of biodiversity has become a paramount issue for human civilization and thus a matter of utmost concern, one of which warrants urgent measures to prevent further diminution of potential medicinal and biological agents. In this chapter attention is focused on bioresources and their possible therapeutic targets including a discussion of conservation strategies.

In this review we focus on the importance of biodiversity with respect to modern therapeutic challenges. Chemical diversity and biodiversity are two sides of the same coin where plant products offer a vast repertoire of chemical diversity which in turn may provide an array of lead structures. Interestingly, most of the therapeutically active molecules are plant secondary metabolites, capable of interacting with a diverse range of macromolecules such as proteins, DNA, and the like, and thus exhibiting important biological functions that can be utilized to yield biomolecules of therapeutic importance.

2 New Bioresources

2.1 Phytochemicals and the Human Brain

Humans consume a wide range of plant-derived foods, drugs, and dietary supplements that modify the functioning of the central nervous systems (CNS). The psychoactive properties of these substances are attributable to the presence of plant secondary metabolites. The roles of secondary metabolites are relatively straightforward; for instance, they participate in general protective roles (e.g., as antioxidant, free radical-scavenging, UV light-absorbing, and antiproliferative agents) and protect the plant from herbivorous animals (grazing) including different pathogenic microorganisms such as bacteria, fungi, and viruses. They also manage interplant relationships, acting as allelopathic defenders of the plant's growing space against competitor plants [7, 8]. In many cases, the effects of these phytochemicals on the human CNS might be linked either to their ecological roles in the life of the plant or to molecular and biochemical similarities in the biology of plants and higher animals.

2.1.1 Biological Similarities Across Taxa

It is well established that groups of enzymes occur in all living organisms and are involved in the biosynthesis, detoxification, and metabolism of compounds [1]. Similarly, a raft of interrelated, ancestral, signaling molecules and pathways are preserved in both plants and animals [3]. For example, nitric oxide (NO) plays a key role in cellular signaling, both in plants as well as in animals [5]. Additionally, multiple aspects of cellular and redox signaling are conserved between the taxa [9, 10], including similar gene expression in response to cellular stressors, which are regulated by common transcription factors [10]. Plant signaling molecules such as fatty acid-derived, growth-regulating jasmonate (*cis*-jasmone, jasmonic acid, and methyl jasmonate) and many mammalian paracrine molecules, including prostaglandins and other eicosanoids, are synthesized via similar, genetically preserved pathways [11]. For instance, most "human" neurochemicals, such as neuropeptides [12], hormones [13], and neurotransmitters, including dopamine, serotonin, glutamate, and gamma-aminobutyric acid [14, 15], can also be found in insects. Even the uniquely nonvertebrate neurotransmitter/modulator octopamine is functionally and structurally analogous to noradrenaline [16]. These neurochemicals can play similar or at times different roles in both animals as well as in insects. Insects have also been used as models to study behavioral responses associated with diet, addictive drugs [15, 17], alcohol [18], sleep deprivation [19], and age-associated decline of cognitive functions [20] including behavioral effects of serotonergic [21], dopaminergic [22], glutamatergic [23], GABAergic, and cholinergic [24] pharmacological agents. It has also been observed that certain pharmacological agents known to upregulate the activity of the cholinergic system, may in turn improve memory processes in both mammals and insects, whereas downregulation of the same is known to produce the opposite effects [25].

2.1.2 Hypotheses: Why Secondary Metabolites Affect Human Brain Function

There are two hypotheses to explain the effects of secondary metabolites on human brain function. First, many molecular signaling pathways that are conserved between the taxa have been known to contribute towards production/synthesis of secondary metabolites [26]. Second, the effects are predicated on the similarities between the nervous systems of humans and herbivores. Therefore, in such situations, phytochemicals (whose synthesis has been retained by a process of natural selection), on the basis of their ability to interact with the CNS of herbivorous (sometimes in symbiotic insects), may also interact with the human nervous system, possibly via similar mechanisms, with either similar, or in some cases dissimilar, behavioral effects.

2.1.3 Current Status of Knowledge

A vast number of natural, plant-based extracts and chemicals are purported to have beneficial effects on human brain function. However, few plant-based products have been methodically assessed with particular reference to human trials. Presently, many plants and plant-derived substances, such as *Cannabis sativa* (marijuana), *Papaver somniferum* (morphine and heroin), *Coffea arabica* (caffeine), *Catha edulis* (cathinone), and *Withania somnifera* (withaferin and other withanolides) are widely used and abused throughout the world. In Table 1 we have tried to illustrate some of the CNS active biomolecules, displaying diversity of plant origin and biological functions.

According to Roth et al., a number of online resources (enethogen.com, erowid.org, botanical.com, maps.org, heffter.org, http://kidb.cwru.edu/) are currently available for obtaining diverse kinds of information (botanical and chemical information, molecular targets) on psychoactive botanicals [30].

2.2 *Phytochemicals as Potential Anti-inflammatory Agents*

Inflammation is a complex response to a tissue injury or an infection, often characterized by several characteristic features such as redness, heat, swelling, pain, and loss of function. *Acute inflammation* is characterized by a vascular response (increase of blood flow into the area) and recruitment of polymorphonuclear cells, typically neutrophils, followed by monocytes, which later differentiate into macrophages. The inflammatory response leads to a synchronized activation of various signaling pathways involved in the regulation and expression

Table 1 CNS active biomolecules of plant origin

Plant(s)	CNS active molecules	CNS pharmacology	References
Areca catechu L.	Arecoline	Stimulant (nicotinic receptor agonist causing a cortical arousal response)	[27]
Atropa belladonna L. *Brugmansia aurea* Lagerh. *Mandragora officinalis* L. *Datura stramonium* L.	Atropine	Depressant and euphoric (anticholinergic–muscarinic receptor antagonist)	[27, 28]
Paullinia cupana *Camellia sinensis* *Coffea arabica* L. E. *Coffea canephora* *Coffea liberica* *Ilex paraguariensis*	Caffeine (and other methylxanthines)	Stimulant (increases norepinephrine secretion via competitive antagonism at adenosine receptors)	[27, 29]
Catha edulis Forssk.	Cathinone	Stimulant (amphetamine-like adrenergic agonist, inhibits dopamine reuptake)	[27, 30]
Erythroxylum coca Lam.	Cocaine	Stimulant (euphoria primarily due to inhibition of catecholamine uptake)	[27]
Anadenanthera peregrina (L.) *Virola theiodora* (Spruce ex Benth.)	Dimethyltryptamine (DMT)	Hallucinogen (serotonin receptor agonist)	[27]
Ephedra nevadensis *Ephedra sinica*	Ephedrine	Stimulant (agonist activation of adrenergic receptors)	[27]
Ginkgo biloba L.	Ginkgolides, bilobalide	Nootropic	[31]
Banisteriopsis caapi *Banisteriopsis inebrians.* *Passiflora incarnata* L. NA *Peganum harmala* L. Asia	Harmine, harmaline	Hallucinogen (MAO_A inhibitor, sedative)	[28, 32]
Amanita muscaria L.	Ibotenic acid	Hallucinogen (activates glutamate receptors)	[27, 33]
Piper methysticum G.	Methysticin, dihydrometh ysticin, yangonin, desmethoxyyangonin, kavain, dihydrokavain	Anxiolytic (kavain inhibits reuptake of norepinephrine), desmethoxyyangonin is a reversible MAO_B inhibitor	[31, 34]
Amanita muscaria L.	Muscimol	Sedative (γ-aminobutyric acid receptor agonist)	[332]
Psilocybe cubensis.	Psilocin, psilocybin	Hallucinogenic ($5\text{-}HT_{1A}$ and $5\text{-}HT_{2A}$/2C agonist)	[28]

(continued)

Table 1 (continued)

Plant(s)	CNS active molecules	CNS pharmacology	References
Salvia divinorum Epling et Játiva NA	Salvinorin-A	Hallucinogen (κ-opioid receptor agonist)	[30, 35]
Cannabis sativa L. Asia		Stimulant, produce euphoria (agonist to cannabinoid receptors)	[27, 36]
Withania somnifera	Withanolide A	Reconstruct neuronal networks	[37]

of various inflammatory mediators including chemokines, cytokines, vasoactive amines, eicosanoids, and different proteolytic enzymes. In situations where the inflammatory condition persists over a period of time, it may lead to chronic inflammation, a condition known to be associated with several chronic diseases, including cancer, arthritis, inflammatory bowel disease, and several others.

During the last two decades, there have been remarkable advances in the field of immunology and molecular pharmacology, and today we have been able to identify a number of different molecular targets for effective management of acute and chronic inflammatory conditions, including (i) COX1/COX2 mediated production of arachidonic acid metabolites (prostaglandins, leukotrienes, PAF, lipoxins); (ii) NO; (iii) reactive oxygen species (ROS) (ii) cytokines (TNFα and interleukins IL-1, IL-6, IL-10), and TNFα-converting enzyme (TACE); (iv) interferons (IFNa2, b1, g); (v) G-protein coupled receptors; (vi) cell interaction molecules such as LFA (leukocyte function-associated antigen; (vi) cytotoxic T lymphocyte antigen-4 immunoglobulin; (viii) transcription factors including nuclear factor (NF)-kB, mitogen-activated protein kinases (MAPKs), c-Jun-N-terminal kinase (JNK), and p38 kinases; and (ix) adhesion molecules. Interestingly, (NF)-kB is known to regulate the transcription of a number of genes involved in the immune/inflammatory pathways, cellular stress, apoptosis, cell adhesion, and proliferation.

According to recent reports, pattern recognition receptors (PRRs) are also responsible for recognizing endogenous molecules liberated from damaged cells, and are referred to as damage-associated molecular patterns (DAMPs). Currently, four different classes of pattern recognition receptors have been identified [38, 39]. These PRRs include different transmembrane receptors such as the Toll-like receptors (TLRs) and C-type lectin receptors (CLRs), and cytoplasmic receptors such as retinoic acid-inducible gene (RIG)-I-like receptors (RLRs) and NOD-like receptors (NLRs). The PRRs are known to upregulate the transcription of genes involved in inflammatory responses. As already established, the inflammatory response is orchestrated by proinflammatory cytokines such as tumor necrosis factor (TNF) and interleukin (IL)-1, IL-1b, and IL-6. IL-1b maturation requires cleavage of pro-IL-1b by a protease, caspase-1, which is activated independently of TLR signaling. The complex that activates caspase-1 is referred to as inflammasome (namely NLRP1, NLRP3, IPAF, and AIM2). Activation of inflammasome complex formation leads to secretion of proinflammatory cytokines (IL-1b and IL-18) which may ultimately result in cell death. In some acquired diseases, NLRP3 inflammasome activity is dysregulated, and therefore effective management of these diseases can be achieved with IL-1b antagonists or with antagonists against the IL-1b receptor [40].

Similarly, there are a number of reports mentioning the role of purinergic receptors in inflammation and pain. It has been observed that P2X3R and the P2X2/3R are expressed selectively in the peripheral afferent fibers and are involved in pain response [1]. It has also been observed that stimulation of P2X7 by ATP causes secretion of different cytokines (IL-1β, IL-18, TNF-α) and NO. These purinergic receptors are known to be stimulated by ATP or glutamate, leading to neuronal excitation, along with pain and inflammation.

Due to the potential side effects and serious adverse effects, many synthetic drugs reported to be used for the treatment of inflammatory disorders are gradually losing their shine. As evident from the literature, there has been a gradual shift of scientific focus towards herbal medicines, as evident from their growing popularity in the management of different human ailments. Various natural products have been found to inhibit or suppress the inflammatory response by interfering with enzyme functions or suppression of the signaling cascades. Since ancient times, plant-based medicines (dispensed in the form of extracts, tinctures, powders, and poultices) have been used for the alleviation of pain and other inflammatory conditions (the bark of *Salix alba* is one of the earliest known remedies for pain and inflammation). Alkaloids have been used for their medicinal properties since the early nineteenth century. According to available reports, quinoline, isoquinoline, and indole alkaloids have been studied in detail for their anti-inflammatory properties. Isoquinoline, quinolone, and indole alkaloids were the most studied classes for anti-inflammatory activity. According to Koehn and Carter [41] about 49 % of new chemical entities introduced between 1981 and 2002 were found to be of natural origin. It is also interesting to note that different types of plant secondary metabolites displaying wide structural diversity coupled with biochemical specificity (alkaloids, steroids, terpenoids, polyphenolics, phenylpropanoids, fatty acids and lipids, etc.), make them effective lead compounds for the management of inflammatory conditions.

2.2.1 Phytoconstituents with Promising Anti-inflammatory Activity

According to the available reports, a large number of anti-inflammatory compounds have been isolated from herbal sources (Table 2). Alkaloids isolated from *Sophora subprostata, Alstonia scholaris, Isatis tinctoria*, and *Evodia rutaecarpa* have been found to display COX inhibitory properties, as evident from the reduced level of PGE2 production whereas alkaloids from *Berbaris crataegina, Hydrastis Canadensis, Caulerpa racemosa, Zanthoxylon ailanthoides, Chrysophyllus albidium, Alostonia scholaris*, and a number of others have been found to inhibit edema

Table 2 Some recent studies with natural products on various inflammation-related targets

Plant	Compound	Target	References
Alkaloids			
Sophora subprostrata	Matrine	COX-1 and COX-2	[60]
Alstonia scholaris	Picrinine	COX-1, COX-2, and 5-LOX	[61]
Berberis crataegina	Palmatine	Inhibition of COX-1, COX-2	[62]
Isatis tinctoria	Tryptanthrin	COX-2, LTB4, NO; also inhibitor of P-gp and MRP2	[63–65]
Evodiae Fructus	Dehydroevodiamine	COX-2, NF-kB, iNOS	[66]
Evodia Rutaecarpa	Rutaecarpine	iNOS, COX-1, COX-2, TNF-α, and IL-4	[67–69]
Evodia fruits	Evodiamine	PGE2, NF-kB,NO, and iNOS	[68, 70]
Sinomenium acutum	Sinomenine	PGE3 and LTC4, NO, and TNF-α	[71]
Stephania tetrandrae	Fangchinoline	IL-5	[72]
Stephania tetrandra	Tetrandrine	IL-5, iNOS, and COX-2, JNK, ERK, AP-1	[72–74]
Phellodendri cortex	Berberine	IL-6, 3T3-L1 NO, TNF-α	[75–77]
Coptidis rhizomaand			
Piper kadsura	Piperlactam S	TNF-α and IL-1β	[78]
Fatty acid			
Chromolaena odorata	*(s)-coriolic acid*	NO, NF-kB	[79]
	(s)-coriolic acid ester	NO, NF-kB	[79]
	Linoleamide	NO, NF-kB	[79]
Plantago major, Ziziphus jujuba	*Alpha-Linolenic acid*	COX-2	[80, 81]
Ziziphus jujuba	*Linoleic acid*	COX-2	[81]
Ziziphus jujuba	*Oleic acid*	COX-2	[81]
Hernandia ovigera	*(S)-coriolic acid*	COX-2	[82]
Steroids			
Antrodia salmonea,	*Methyl antcinate L*	NO	[83]
Antrodia camphorate	*Antcin*	NO	[83]
Antrodia cinnamomea	*Methyl antcinate K*	NO	[83]
Commiphora mukul	*Guggulsterol*	IFN-γ, IL-12, TNF-α, IL-1β, and NO	[84]
Commiphora mukul	*Guggulsterone*	COX-2, MMP-9, NF-kB	[85]
Nerium oleander	*Neridienone A*	ICAM 1	[86]
Flavonoids			

(continued)

Table 2 (continued)

Plant	Compound	Target	References
Terminalia chebula	Luteolin	TNF-α, MMP-2, NO, IL-4	[87–90]
Different fruits, vegetables, spices	Apigenin	MMP-2, NO, IL-4,TNF-α	[87]
Citrus peel	Tangeretin	IL-1β,COX-2, iNOS, MAPK, Akt	[91, 92]
	Nobiletin	IL-1b, COX-2, LPS/IFN-γ, MAPK, Akt	[86, 91]
	Naringenin	TNF-α, PGE2	[93, 94]
	Hesperetin	TNF-α PGE2	[93, 94]
Eriodictyon californicum	Eriodictyol	TNF-α PGE2	[93, 94]
Genista tinctoria	Genistein	TNF-α, PGE2, NO, IL-1b	[93–98]
Miscellaneous			
Scutellaria baicalensis	Baicalin	ROS, iNOS and TNF-α, IL-1β, IL-6	[95]
Scutellaria Baicalensis	Wogonin	iNOS, COX-2, IL-6 and -8	[88, 95]
Kaempferia pandurata,	Panduratin A	iNOS, PGE2, COX-2,	[92]
Hypericum geminiflorum	Gemichalcone A	Beta-glucuronidase and histamine, mast	[102]
	Gemichalcone B	β-glucuronidase and lysozyme, mast	
Camellia sinensis	Epigallocatechin-3-gallate	IL-1 β	[100]
Acacia species	Isoliquiritigenin	LPS	[101]
Echinochloa colona	Tricin	COX-2	[102]
Terpenoids			
Croton tonkinensis	Ent-akurane Diterpenoids	NFkB and NO	[103]
Isodon excises	Ent-akurane Diterpenoids	NFkB and NO	[104]
Pluchea sagittalis	Taraxasteryl acetate	ROS and RNS	[105, 106]
Vitex peduncularis	Agnuside	COX-2	[107, 108]
Fomitopsis pinicola fruits	Triterpenoids	COX-2, COX-1	[109]
Calocedrus formosana	Sugiol	IL-1β,TNF-α, reduces ROS	[110]
Tripterigium wilfordii	Triptolide	COX-2, iNOS, and IL-1b	[111, 112]
	Tripdiolide	COX-2, iNOS, and IL-1b	
	Celastrol	NF-kB	
Tanacetum parthenium	Parthenolide	NO, PAF1 and fMLP2-induced human neutrophils	[113, 114]
Magnolia grandiflora	Costunolide Parthenolide	NO,NF-kB,	[113, 115]

(continued)

Table 2 (continued)

Plant	Compound	Target	References
Magnolia grandiflora	7-hydroxycostunolide	NF-kB	[115]
Arnica Montana	Helenalin	NF-kB, 5-LOX1 and LTC4, Human platelets, NFkB	[116, 117]
Laurus nobilis	Terpenoids	NO	[118]
Elephantopus mollis	Molephantin	PAF1 and fMLP2-induced human neutrophils	[116]
Milleria quinqueflora	Terpenoids	PAF1 and fMLP2-induced human neutrophils	[119]
Chloranthus serratus	Terpenoids	NO	[125]
Ligustrum lucidum	Oleanolic acid	COX-2, NO	[120–123]
Plantago major	Ursolic acid	COX-2, free enzyme NO, mouse macrophages	[120, 124, 126]
Synthetic analogs	Oleanonic acid	NO	[122–126]
Lignans			
Coptis japonica	Woorenoside IV Second analogue	TNF-α and NO TNF-α	[127, 128]
Saururus chinensis	Sauchinone	iNOS, TNF-α, and COX-2	[129–131]
Haplophyllum Hispanicum	Diphyllin acetylapioside	LTB4, 5-HETE, and LT	[132]
S. chinensis	Manassantin A	NF-kβ, LTC 4,IL-6, TNF-α	[133–135]
S. chinensis	Manassantin B	NF-kβ of IL 1-B,IL 6,TNFα	[134–136]
Ocotea bullata	Sibyllenone	5-LOX	[137]
Arctium lappa	Arctigenin	ERK, p38, p13 K pathway, JNK, iNOS, TNF-a and NF-kB, COX, LOX	[129, 138–140]
Garcinia subelliptica	Garcinielliptone M	b-glucoronidase and histamine, mast cells, NO	[141]
Helichrysum italicum ssp. Microphyllum	Arzanol	IL-1b and TNF- α, IL-6, IL-8	[143]
Quinones			
H. perforatum	Hypericin	NF-kB, IL-12,	[144, 145]
Kniphofia foliosa	Knipholone	LT, 12(S) HETE	[146]
Maesa lanceolata	Maesanin	5-LOX	[143]
Nigella sativa	Thymohydroquinone	COX-1 and -2	[147]
Nigella sativa	Thymol	COX-1 and -2,MPO	[147]
Nigella sativa	Thymoquinone	COX-2, TNFa, NFkB, 5-LOX	[142, 147]
Aloe vera	Emodin	NF-kB and IkB, casein kinase II, HER2/neu, HIF-1α, AKT/mTOR, STAT3, CXCR4, topoisomerase II, p53, p21 involved in cancer, inhibit TLR-4	[148–151]
Phenylpropanoids Illicium species	Phenylpropanoids	TNF-α	[152]
Buddleja officinalis	Acteoside	iNOS and AP-1	[153, 154]

formation in rodents. The alkaloids from *Prunus persica* and *Cissampelos sympodialis* are known to alter NO production. The anti-inflammatory properties of *Evodia rutaecarpa* have been known for many years, and the alkaloids rutaecarpine and evodiamine have been found to inhibit PGE_2 in vitro. Moreover, these alkaloids are also known to suppress COX expression by virtue of NF-kB inhibition [42]. Ligunstrazine, an alkaloid isolated from *Ligusticum wallichii*, inhibits ATP-induced membrane depolarization through inhibition of P2X3R [1]. Verminoside (an iridoid glycoside), found abundantly in the dichloromethane extract of *Kigelia ricana* (Bignoniaceae), displays significant anti-inflammatory properties, attributed to inhibition of iNOS expression and subsequent NO release in the mouse J774.A1 macrophage cell line [43]. The lignans isolated from *Phyllanthus amarus* (Euphorbiaceae) include a group of phytoconstituents such as phyltetralin, nirtetralin, niranthin, and phyllanthin. Among these, only nirtetralin was found to inhibit IL1-β production in the infammatory tissues, whereas the whole extract was found to produce prominent reduction of paw edema induced with bradykinin, platelet activating factors (PAF), and endothelin-1 [44]. Neolignans, isolated from stem barks of several species of *Piper kadsura* (Piperaceae), are considered to be important for the treatment of several inflammatory disorders [45]. Anti-inflammatory phenylpropanoids, isolated from *Illicium* species (*Illicium tashiroi, Illicium anisotum*, and *Illicium arborescens*), were found to inhibit compound A23187-induced histamine release, when tested on rat basophilic RBL-2H3 leukemia cells [46]. A novel class of plant constituents (Fig. 1) such as naphthoquinone, isolated from plants belonging to the family of Iridaceae (*Eleutherine Americana*) produced potent anti-inflammatory properties [47].

Luteolin, a flavone isolated from the leaves of *Perilla nankinensis,* also occurring in several other plants, displays potential antioxidant, anti-inflammatory, and antiallergic properties as compared to other flavonoids [48]. It was found to suppress leukocyte infiltration and also reduced the level of 6-keto-PGF1α in the inflammatory exudates, through downregulation of COX2 [49].

Luteins with basic carotenoid skeleton (Fig. 2) are derived from *Tagetes erecta* (family Compositae), known widely as "marigold". They display potent anti-inflammatory properties, probably attributed to their ability to scavenge superoxide radicals [50].

Sesquiterpenes isolated from the leaves of the Yacon tree, *Smalianthus sonchifolia* (Asteraceae), are reported to produce significant anti-inflammatory activity (as evident from their actions on murine macrophage RAW264.7 cells), probably through inhibition of NO production [51]. Hinkitiol belongs to the class of tropolone derivatives, and is abundantly present in the heartwood of plants belonging to the family Cupressaceae. It produces promising anti-inflammatory activity in the LPS-induced macrophage cell line through inhibition of TNF-α [52]. Novel anti-inflammatory agents including Evodiamine, rutaecarpine, and goshuyuamide II, isolated from the fruits of *Evodia rutaecarpa* (Rutaceae) produced their activity through inhibitory action on PGE_2 generation [53]. Polyzellin and polysylvin (Fig. 3), possessing the stilbene skeleton, isolated from the fruiting bodies and leaves of *Polyozellus multiplex* (Thelephoraceae) and *Pinus densiflora*

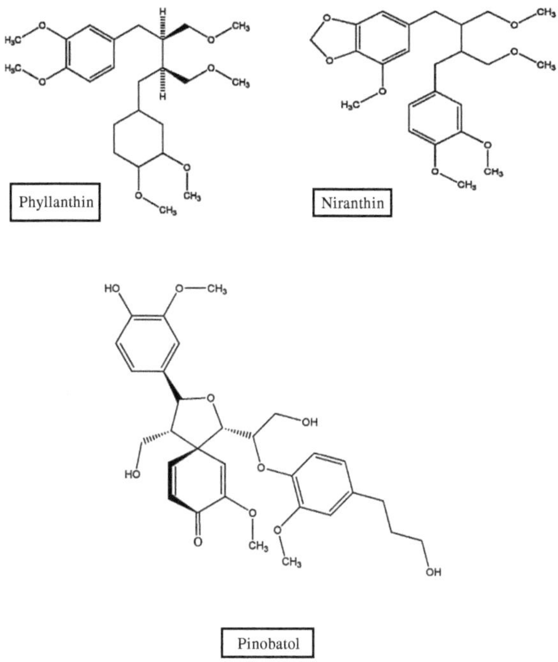

Fig. 1 Structure of some potent naphthoquinones from *Eleutherine Americana*

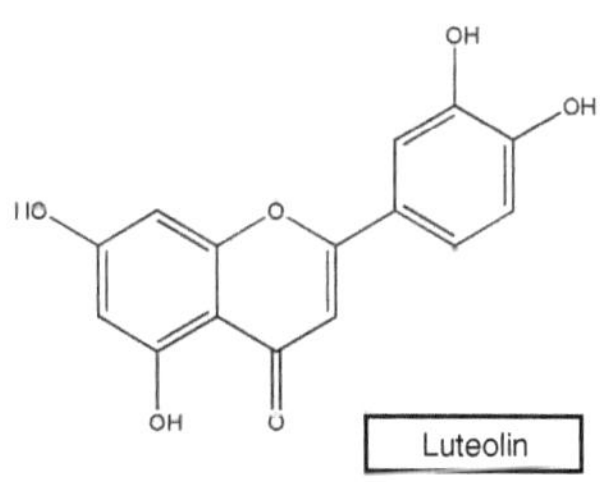

Fig. 2 Carotenoid from *Tagetes erecta*

(Pinaceae), respectively, have been found to downregulate NF-kB along with suppression of LPS-induced NO production [54, 55].

Prenylated derivatives of resveratrol, a newer class of anti-inflammatory agents, display anti-inflammatory activity probably through inhibition of the COX2 enzyme [56]. A newer group of plant secondary metabolites such as dithymoquinone, thymo hydroquinone, and thymoquinone has been isolated from the seeds of *Nigella sativa* (Ranunculaceae). These compounds have been found to inhibit both COX1 and

Fig. 3 Stilbene containing compounds isolated from *Polyozellus multiplex*

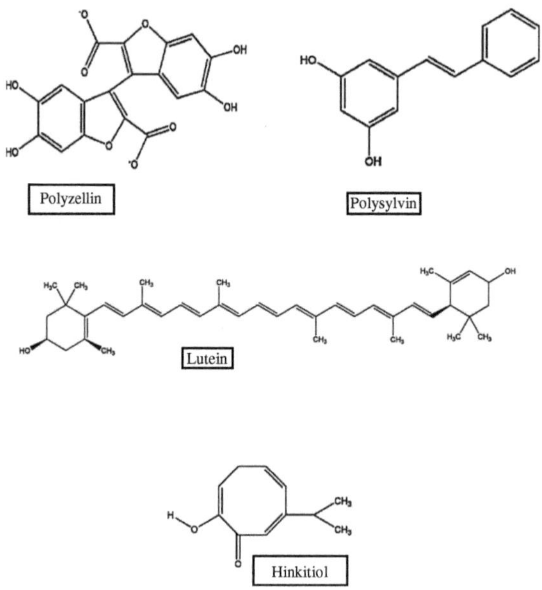

COX2 enzymes significantly [57]. Gigantol (Fig. 4), a chemical constituent isolated from the whole plant extract of *Cymbidium goeringii* (Orchidaceae), possesses significant inhibitory effect on the expression of both iNOS and COX2, as evident from the mRNA level in RAW 264.7 cell line [58].

There are a number of natural analgesics such as emodin, amentoflavone from *Rheedia longifólia* [59], ligunstrazine, and puerarin that have been found to act as antagonists of the purinergic receptors.The analgesic property of the anthraquinone glycoside, emodin (*Rheum officinale*), are also known to be mediated through antagonism of the P2X3R and P2X7R, expressed in the primary sensory neurons [1].

2.3 Phytochemicals and Cardiovascular Disease

Over the last several decades, cardiovascular diseases (CVDs) are considered to be one of the major causes of morbidity and mortality, both in developed as well as in the developing nations. Today, CVD is a major health burden across the globe, but the severity of the disease has a varied nature and is closely related to food intake (vegetables and fruits), hence evidence-based studies have shown variations in the nature and severity, when comparisons are drawn between different geographical regions [155].

CVDs cover a complex range of conditions that arise due to several factors. Although some of the conditions are a result of defects in the organ itself, others result due to problems related to the vascular system. The major risk factors for these disorders were recognized over many years, and they include high levels of

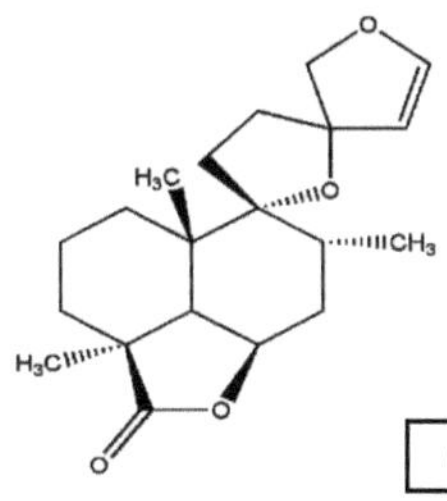

Fig. 4 Chemical constituent isolated from *Cymbidium goeringii*

low-density lipoprotein (LDL) cholesterol, psychosocial factors (stress), smoking, hypertension, renal disorders, diabetes, endocrine dysfunctions, obesity, excess consumption of alcohol, and lack of regular physical activity. Among the various CVDs, hypertensive disorder is considered to be the most prevalent and often predisposes myocardial infarction, stroke, and renal failure [156]. There has also been continued research to help define more precisely the cardiovascular risk of an individual with respect to genetic factors, more complex lipid traits, and inflammatory markers, but these issues need to be validated through extensive studies.

A large group of plants have compounds that have a direct effect on the heart and blood vessels and may cause severe adverse reactions in animals that consume them. The most recognized of these compounds are the cardiac glycosides, of which digoxin, found in foxglove (*Digitalis* spp.), is best known. The pharmacologic properties of digoxin have been known for a long time. Because of its effects on the heart at therapeutic levels, it is routinely used to treat congestive heart failure in humans and animals. Similarly, plants such as *Crataegus oxycantha* (cardiotonic, antianginal, antihypertensive, and lipid-lowering effects) and Terminalia *arjuna* (cardiotonic, coronary artery disease, heart failure, and hypolipidemic) have been widely exploited for their therapeutic benefits in cardiac diseases; a combination of *Inula racemosa*, and *Commiphora mukul* have been used in the Ayurveda for the management of anginal pain and in certain cases for controlling dypsnea associated with angina pectoris [157]. Several plants have been shown to contain active principles which have been identified over the years for the treatment of CVD. A search of the literature reveals a huge amount of information regarding properties of different plants and plant-derived compounds. In most cases, the plant extracts or the pure compounds were found to act on multiple targets [158], namely (i) NO generation, cGMP pathway; (ii) PGI2, cAMP pathway; (iii) potassium channel activators; (iv) inhibitors of voltage-gated calcium channels; (v) phosphodiesterase inhibitors (PDE5); (v) activation of endothelial transient receptor potential (TRP) channels; (vi) inhibitors of protein kinase C (PKC); and (vii) free radical scavenging. A number of plant-derived compounds, particularly alkaloids [99, 159], polyphenols [160], flavonoids [161,

162], saponins [163], proanthocyanindins [164], xanthones [165], and glycosides [166] have been found to exhibit cardioprotective properties.

In this section an attempt to summarize the drugs from plant sources that affect the cardiovascular system both in terms of efficacy and safety as available in scientific reports has been made. Table 3 describes the cardioactive principles from medicinal plants that have been identified in recent years. However, for many of these plants, the active principles are yet to be identified. Moreover, there is need for understanding the mechanism of action of many of these active compounds derived from plants. Thus, it is the need of the day to study these plants extensively that have been traditionally used for the management of CVS or have been screened in laboratories for cardiovascular effects. Some plant extracts (Table 4) produce promising cardiovascular activities which in turn may provide valuable leads for further research and development. Hence, there is a lot of scope for developing standardized herbal extracts or pure molecules that may work alone or as combinations for the management of some of the CVS disorders. This is particularly important in light of the fact that there are better chances of getting more effective drugs acting on multiple targets [167].

2.4 Medicinal Plants Used as a Source of Anticancer Agents

Today, cancer is a major health problem around the world. According to WHO, there were more than 14.1 million cases of cancer reported in the year 2012 (7.4 million males and 6.7 million females were affected). According to the World Cancer Research Fund International (http://www.wcrf.org), there would be about 24 million cancer cases by the year 2035. Survey reports indicate a prevalence of lung, colon, prostate, and breast cancer cases in the Western countries; cervical and cancers of the head and neck are common in India, whereas stomach cancer is found to be common among the Japanese population [214].

During the 1950s, the discovery of the vinca alkaloids (vinblastine and vincristine) and the isolation of podophyllotoxins paved the way for natural product scientists to explore plant biodiversity further for novel anticancer biomolecules. Today, apart from the vinca alkaloids, a number of naturally occurring anticancer molecules—namely taxanes (paclitaxel, docetaxel), podophyllotoxin (derivatives including etoposide, teniposide), camptothecin (topothecan and irinothecan), and anthracyclines (doxorubicin, daunorubicin, epirubicin, idarubicin)—have been proven to be clinically effective. Similarly, a number of dietary phytochemicals such as curcumin, genistein, resveratrol (red grapes), diallyl sulphide, allicin, and lycopene have been found to possess anticancer properties. The last few years have seen a globalized thrust in anticancer research and many researchers have started exploring the plant kingdom for lead compounds against cancer. In a report published on African medicinal plants, Hostettman et al. mention the anticancer potential of a large number of plant extracts derived from plants belonging to Guittiferae, Rubiaceae, Apocynaceae, Euphorbiaceae, and Solanaceae [215].

Table 3 Active principles from plants having activity on cardiovascular system

Scientific name of the plant	Parts used	Active constituents	Mechanism of action	References
Tabernaemontana dichotoma	Barks	Cathafoline	Inhibits receptor-operated calcium channels	[168]
Boophone disticha (Amaryllidaceae)	Bulbs	Alkaloid	Reduce acute mixing stress induced heart rate and BP	[169]
Xylopia aethiopica (Annonacea)	Dried fruit	Xylopic acid, kaurenoids	Hypotensive and coronary vasodialatory action by blocking voltage-gated calcium channel. Diuretic and natriuretic effect	[167]
Halenia elliptica	Whole plant	Xanthone	Vasodilator	[170]
Dioclea grandiflora (Leguminoseae)	Root	Dioclein	Vasorelaxant (inhibit phosphodestease type I)	[171]
Fragaria vesca (Rosaceae)	Herb	Polysaccharidephenolic Glyco-conjugate. Hexuronic acid	Anticoagulant due to hexuronic acids and phenolic glycoconjugates	[172]
Echinacea purpurea (Asteraceae)	Flower			
Gynostemma pentaphyllum (Cucurbitaceae)	Leaves	Gypenoside III and VIII	Protective effect against vasopressin-induced coronary spasm; antiarrhythmic due to inhibition of calcium overload	[173]
Stephania tetrandra (Menispermaceae)	Root	Tetrandrine	Calcium channel antagonist (block L type Ca^{2+} channels) regulate steroidogenesis in adrenal glomerulosa cells	[174]
Lingusticum wallichii (Apiaceae)	Root	Tetramethylpyrazine	Antagonist of calcium channel (selective) and α-adrenergic receptors (nonselective); antioxidant properties	[175]
Uncaria rhynchophylla (Rubiaceae)	Root	Rhynchophylline and hirsutine	Antihypertensive (block voltage-dependent calcium channel); antiplatelet aggregation (inhibits release of platelet factor 4)	[175]
Evodia rutaecarpa (Rutaceae)	Fruit	Rutaecarpine	Endothelium dependent vasorelaxant properties; prevent antiplatelet aggregation (blockade of phospholipase C)	[176]

(continued)

Table 3 (continued)

Scientific name of the plant	Parts used	Active constituents	Mechanism of action	References
Panex notoginseng (Araliaceae)	Root	Notoginsenoside-R1	Fibrinolytic effect by increasing the amount of tissue plasminogen activator (tPA)	[177]
Petasites formosanus	Root	Angelic ester of 2β-hydroxy-8βH-7(11)-eremophilene-12,8-olide	Block L type voltage-gated calcium channels	[178]
Aesculus hippocastanum (Sapindaceae)	Seeds	Ascin	Vein tonic and lymphagogue properties; Vasculotropic action due to blockade of cutaneous capillary hyperpermeability induced by histamine and serotonin; anti-inflammatory properties	[179]
Amomum subulatum	Seeds	Cardamonin	Block voltage operated calcium channel and ryanodine receptor; activate Ca^{2+} operated K^{+} channel	[180]
Scutellaria baicalensis	Leaf	Baicalin	Block voltage operated calcium channel	[181]
Erigeron Canadensis (Asteraceae)	Flower	Polysaccharide-polyphenolic macromolecules contains hexuronic acid	Anticoagulant activity	[182]
Chondrondendron platyphyllum	Root, bark	Curine (bisbenzylisoquinoline alkaloid)	Block voltage operated L- type calcium channels	[183]
Commiphora mukul (Burseraceae)	OleogumResin (known as guggul)	Guggulsterone	Hypolipidemic; antioxidant; anti-inflammatory activities; antagonist of Farnesoid X receptor	[184]
Terminalia arjuna (Combretaceae)	Stem bark	Various bioactive compounds, such as triterpinoids, tannins, flavonoids, and minerals, have been isolated	Positive inotropic; hypolipidemic; coronary vasodilator; antioxidant properties	[185, 186]
Inula recemosa (Asteraceae)	Root	Alantolactone	Cardioprotective activity in ischemic rats; prevents the myocardium of rat heart from oxidative damage	[187, 188]

Table 4 Plant extracts having promising cardiovascular activity

Plant/extract (family)	Pharmacological effect	References
Ethanolic extract of aerial parts of *Ocimum basilicum* (Lamiaceae)	*Ocimum basilicum* recovered the arterial pressure and improved the left ventricular performance along with a simultaneous reduction in the left ventricular end-diastolic pressure induced with Isoproterenol (ISO). Endothelium-dependant vasorelaxant and antiplatelet aggregation activities. Inhibits lipid peroxidation both in the serum and the myocardium	[189]
Extract of *Ginkgo biloba* (Ginkgoaceae) and *Ocimum sanctum* (Lamiaceae.)	Cardioprotective activity due to its antioxidant effects; marked myocardial protective activity in ISO-induced cardiac necrosis	[190]
Linum usitatissimum (Linaceae.)	Cardioprotective effect due to its antioxidant properties was established by hemodynamic, biochemical, and histopathological results	[191]
Ethanolic extract *of Hybanthus enneaspermus* (Violaceae)	Normalization of cardiac marker enzymes (CK, LDH, SGOT, SGPT) and Troponin I; reduction of tissue lipid peroxidation	[192]
Allium sativum oil (Amaryllidaceae)	Reduce ventricular tachycardia and fibrillation; reduced serum total cholesterol, LDL, platelet aggregation; antihypertensive; decreased lipid peroxidation and improved antioxidant status; inhibit angiotension-converting enzyme	[193]
Water extract of *Buchanania axillaries* (Anacardiaceae)	Cardioprotective effects due to an augmentation of the endogenous antioxidants; inhibition of lipid peroxidation	[194]
Cucumis trigonus Roxb (Cucurbitaceae) *Commiphora mukul*	ALT, AST, LDH, and CPK were decreased significantly due to its action on membrane integrity; heart rate, R-wave amplitude, and ST-segment elevation normalized. Commiphora mukul, protects the myocardial cellular membrane against oxidative damage by regulating the redox status of proteins; stabilizes myocardial membranes and prevents necrotic damage	[195, 196]
Water extract of *Withania somnifera* (Solanaceae)	Restored serum levels cardiac markers (CK-MB, LDH, SGOT, and SGPT); improvement of antioxidant status (superoxide dismutase, catalase, and glutathione peroxidase)	[197]

(continued)

Table 4 (continued)

Plant/extract (family)	Pharmacological effect	References
Ethanolic extract of *Premna serratifolia* (Verbenaceae)	Reduced elevation of ST segments in rat; decrease in heart tissue glycogen; cardiotonic; anticoagulant; antioxidant properties	[198]
Water extract of *Daucus carota* Linn. (Umbeliferae)	Inotropic (decrease of Na^{+}/K^{+} ATPase and Mg^{2+}ATPase and an increase in Ca^{2+}ATPase); antioxidant and antilipid peroxidative	[199]
Ethanolic extract of *Callistemon lanceolatus* (Myrtaceae)	Cardioprotective; improvement of biochemical marker enzymes (CK-MB, AST, ALT, and LDH); improvement of antioxidant enzyme status (superoxide dismutase and catalase)	[200]
Aqueous and alcoholic extracts *Curcuma longa* (Zingiberaceae)	Cardioprotective; antioxidant	[201]
Alcoholic extracts *Spathodea campanulata* (Bignoniaceae)	Cardioprotective; antioxidant; reduction of serum (CK-MB, LDH, SGOT, SGPT); restoration of HDL and LDL concentration	[202]
Whole plants of *Cyathula prostrata Linn* (Amaranthaceae)	Antioxidant; prevent myocardial necrosis	[203]
Ethanolic root extract of *Momordica cymbalaria* Fenzl (Cucurbitaceae)	Significant reduction of serum CK-MB, LDH, AST, ALT; improves antioxidant status (SOD, CAT); normalization of ECG; reduces blood pressure	[204]
Amaranthus spinosus (Amaranthaceae)	Decreases atherogenic index; increases serum HDL	[205]
Aqueous extract of *Terminalia Chebula* and Ethanolic extract of unripe pods and leaves of *Bauhinia purpurea*	Significant decrease in the levels of serum cholesterol, phospholipids, triglycerides, LDL, VLDL atherogenic index, also decrease in aortic plaque and fatty liver formation	[206, 207]
Methanol extract of pericarps of *Sapindus emarginatus* (Sapindaceae)	Improves serum lipid profile	[208]
Aqueous and ethanolic leaf extract of *Aegle marmelos* (L) Corr., (Rutaceae)	Improves serum lipid profile	[209]
Erythrina indica Lam, Ginkgo biloba L. Family Ginkgoaceae, Ethanol and water (1:1) Extract of leaves of *Carissa carandas* and chloroform extract of Mimosa pudica (Mimosaceae) leaves	Antioxidant; improves serum lipid profile	[210–213]

On the basis of published reports, it may be mentioned that the phytomolecules work on a range of targets, namely on the cell cycle, signaling pathway in apoptosis, PI3/Akt, p38/MAPK, nuclear factor-kappa B (NF-κB), cyclooxygenase enzyme. Apart from these, many plant-derived molecules (alkaloids, anthocyanidines, anthraquinones, chalcones, catechins, flavones, terpeinoids, and xanthenes) are also known to modulate epigenetic mechanisms, thereby displaying a promising potential against cancer [216].

It is also well known that traditional Chinese medicine, Kampo medicine, Ayurveda (the ancient Indian system of medicine) have been used for ages for the management of cancer, in different Asian countries. These traditional approaches are now becoming popular (as alternative therapies) in various other nations. Here, we have tried to summarize the current progress of plant-derived therapies for the management of cancer.

2.4.1 Patented Biomolecules with Anticancer Activity

In recent years there has been an increasing thrust for the identification and isolation of molecules from natural resources, displaying inhibitory properties towards tumor growth and metastasis. Accordingly, a number of compounds with promising anticancer activity have now been identified. In Table 5, we put forward a brief account of such isolated compounds that have been patented in the United States in recent years [217].

2.4.2 Herbal Extract with Anticancer Activity

Drug development (from bench to bedside) needs huge financial investments and intensive research including extended timeframes. In many cases such studies often lead to the development of molecules that may not have any distinct advantages over the existing ones. Hence in recent years, we observe a renewed interest in alternative therapies, particularly involving products of botanical origin. Table 6 describes the list of different plant extracts that have now been patented or are under application in the United States [217].

2.5 *Medicinal Plants Used as a Source of Antimicrobial Agents*

Studies of the potential antimicrobial activity of plant-derived compounds have been gaining importance due to a rapid increase in the incidences of antimicrobial drug resistance. On the basis of available scientific information on plant extracts and oils, and knowledge of traditional anti-infective therapies, the attention of the scientific community has now shifted towards natural products for isolation and identification of novel molecules with antimicrobial properties. As per the available reports, the antimicrobial pipelines, as compared to the 1970s and 1980s are gradually drying up due to reduced interest of the pharmaceutical industries [257, 258]. Hence, antimicrobial drug discovery from alternative sources is gradually

Table 5 Some patented biomolecules with anticancer activity

Source	Nature of chemical constituent(s)	Mechanism of anticancer action	Major claims (publication no.)	Assignee, country [references]
Swainsona canescens	Swainsonine	Golgi mannosidase II Inhibitors	A method for stimulating the immune system, treating proliferative disorders, or microbial or parasitic infections (US6395745 B1)	Glycodesign Inc., Canada [218]
Glycirrhiza glabra	1-Propanone-1-(2,4-dihydroxyphenyl)-3-hydroxy-3-(4′-hydroxyphenyl),	Induces apoptosis, inhibits tumor cell growth	A method of inhibiting the growth of tumor with the help of mentioned composition (US6498195 B2)	Rutgers, The State University of New Jersey, University of Medicine And Dentistry of New Jersey, New Jersey [219]
Saururus chinensis	Eepimanassantin A and manassantin	Apoptosis	Effective in liver cancer, prostate cancer, breast cancer, mammary gland cancer, uterus epidermal cancer, glioma, neuroblastoma, glioblastoma, stomach cancer, throat epidermal cancer, fibrous carcinoma or malignant melanosis (US20040024055 A1)	Jong-Cheon Hahm, Dae-Sang Lee, Jae-Pil Ko, In-Kyoung Lee, Lee Hyun-Woo, Jeong-Sook Park, Korea [220]
Euphorbia peplus, *Euphorbia hirta* and *Euphorbia drummondii*	Angeloyl substituted ingenane	Cytotoxic	Effective in colon cancer, lung cancer, prostate cancer, cervical cancer, breast cancer (US6787161B2)	Peplin Biotech Pty. Ltd., Australia [221]
Indigofera tinctoria	Isoindigo, I ndigo and indirubin derivatives	Inhibits cell growth Pathway	Treatment of colon cancer; hormone dependent or independent prostate cancer; breast cancer; lung cancer (US6933315 B2)	Longgui Wang, Xiaomei Liu, Ruihuan Chen,Wisconsin [222]
Streptomyces hygroscopius	39-desmethoxyrapamycin	Cytotoxic	A method of treating cancer or B-cell malignancies (US7183289 B2)	Biotica Technology Limited, UK [223]

(continued)

Table 5 (continued)

Source	Nature of chemical constituent(s)	Mechanism of anticancer action	Major claims (publication no.)	Assignee, country [references]
Everniastrum cirrhatum	Methyl beta orcinol carboxylate	Cytotoxic	Treatment of liver, colon, ovarian and mouth (oral) cancer (US20070099993 A1)	Council Of Scientific And Industrial Research, India [224]
Coptis teeta,	Berberine	Cytotoxic	Selectively inhibits growth of lung cancer cells (US20070298132 A1)	Pharmchem Inc., US [225]
Boswella sacra	Analogues from of resin	Induces apoptosis	(US20090298938 A1)	Council Of Scientific & Industrial Research, India [226]
Garcinia anburyi	C-2 Epimeric xanthones	Cytotoxic	A pharmaceutical composition, consist of a pharmaceutically acceptable carrier (US7592367 B2)	Hong Kong Jockey Club Institute Of Chinese Medicine Ltd., Hong Kong [227]
Xanthoceras sorbifolia	Saponins	Anticancer	A composition effective in inhibiting bladder, bone, leukocyte, liver, prostate, breast, or brain cancer (US7514412 B2)	Pacific Arrow Ltd., Hong Kong [228]
Curcuma longa	Curcumin and curcumin derivatives	Prevent tumor formation, tumor cell invasion, metastasis	A method for reducing cancer cell growth; decrease tumor size; prevent tumor formation; reduce tumor cell invasion (US20100197584 A1)	Research Foundations of the City University of New York, New York [229]
Scutellaria baicalensis	*Scutellaria* flavonoid	Cyclin-dependent Protein kinases (Cdks) Inhibitors	A method for prevention and treatment of cancer and AIDS (US20100197619 A1)	Research Foundations of the City University of New York, New York [230]
Triptolide derivatives		Modulation of apoptosis; immunosuppressant	Structural claim of the compound (US7662976 B2)	Pharmagenesis, Inc., California [231]

Table 6 Patented plant extracts

Extracts	Major claims (publication no.)	Assignee, country [references]
Tannins and procyanidin extracts from *Fagopyrum cymosum* (Trev.)	Used in lung cancer, gastric cancer, cervical cancer, sarcoma, and other neoplasms, relieving inflammation, and alleviating toxic and adverse effects associated with chemotherapy and radiotherapy US6451353B1	Han Pei, Guo Qiyu, Chen Bo, Zhu Hongwu, China [232]
Lipid–sterol extract of *Serenoa repens*	A method of treatment of prostate cancer by administering a lipid–sterol extract of *Serenoa Repens (*US6599540B1)	Pierre Fabre Medicament, France [233]
Banyan tree bark fraction	A process of isolation of a combination that inhibits insulin secretion in *β*TC-6 cells and HIT-T15 cells and noncytotoxic to *β*TC-6 cells, but cytotoxic to HIT-T15 cells (US6660309B2)	Biozak, Inc., California [234]
Euphorbia antiquorum extract	Herbal extract from *Euphorbia antiquorum* for inhibiting tumor and cancer growth (US20030165579A1)	Chih-Hui Lin, Wen-Ching Cheng, Taiwan [235]
Polysaccharide-based extract of Ganoderma genus	The oral pharmaceutical combination, providing immune-potentiating and antitumor effects (US6613754B1)	National Yang-Ming University, Taiwan [236]
Artemisolide compound from the aerial parts of *Artemisia sylvatica,*	A composition for treatment of leukemia and colon cancers (US6808724B2)	Korea Research Institute Of Bioscience And Biotechnology, Korea [237]
Sesquiterpenes isolated from *Resina ferulae*	A pharmaceutical composition for treating lung cancer, ovarian cancer, melanoma cancer, or colic cancer (US20040043083A1)	Shi-Yong Ryu, Chong-Ock Lee, Sang-Un Choi, Park Sung-Hee, Young-Sup Kim, Sung-Ki Kim, Sang-Keun Kim, Shin-Kwon Kang, Korea [238]
Herbal compositions from *Radix asparagi*	A pharmaceutical composition for treating or reducing the risk of prostate disorders (US6790464B2)	Healthaid Enterprise Pte. Ltd., Singapore [239]

(continued)

Table 6 (continued)

Extracts	Major claims (publication no.)	Assignee, country [references]
Extract from 20-year-old platycodon (*Platycodon grandiflorum*)	Anti-inflammatory agent in rheumatic arthritis; antihyperlipemia, antidiabetes agent, and anticancer agent (US6902748B1)	Jang Saeng Doraji Co., Ltd, Korea [240]
Triterpenoid and steroidal saponins from *Quillaja saponaria Molina* (soap tree)	The saponins can directly or indirectly inhibit cancer cell growth in vitro or in vivo (US20050175623A1)	Zheng-Pin Wang, US [241]
Leaf extract of *Melissa officinalis*	Inhibits angiogenesis (angiogenesis-associated diseases) and matrix metalloproteinase (US20040009244B2)	Min-Young Kim, Byung-Young Park, Chang-Hee Moon, Eun-Kyu Park, Kyoung-Mi Kim, Korea [242]
A mixture of *Tinospora cordifolia, Aloa vera, Curcuma longa, Withania somnifera, Achyranthus aspera, Ocimum sanctum,* and *Picorrhiza kurroa*	A pharmaceutical preparation for use in the treatment of cancer (hematological malignancies) (US6649185B2)	Sahajanand Biotech Private Limited, India [243]
An extract of the leaves and/or stems of plants belonging to *Panax* genus (*Panax ginseng* C. A. Mayer, *Panax quinquefolum, Panax notoginseng, Panax pseudoginseng, Panax japonicum, Panax vietnamensis)*	Anticancer drug, cancer metastasis inhibitor, hematopoiesis enhancer, radiation side-effect inhibitor, decrease side effects of anticancer drugs, auto-immune disease treatment (US20060034951B2)	Kwak Tae H, Shin Myoung S, Kim Ji Y, Jong-Kook Park, Korea [244]
Flavonoid glycosides (butrin and isobutrin) from flowers of *Butea monosperma*	Hepatocellular carcinoma (US20060280817A1)	Saxena Ajit K, Gupta Bishan D, Kapahi Bal K, Shanmugavel Muthiah, Mondhe Dilip M, India [245]
Extract of *Anoectochilus formosanus*	Chemoprevention, control of various human malignant diseases (US7033617B2)	Academia Sinica, Taiwan [246]
A water from the plants of *Solanum* (at least 60–90 % of solamargine and solasonine)	Inhibitory effect on tumor/cancer cells (liver, lung, and breast cancer cells (US7078063B2)	G & E Herbal Biotechnology Co., Ltd., Taiwan [247]
Purified extract of *Solanaceae Dulcamara* root (Amazonian variety)	A method of treating mammals with prostate cancer (US7250180B2)	Edwin Cevallos Arellano, Quito [248]

(continued)

Table 6 (continued)

Extracts	Major claims (publication no.)	Assignee, country [references]
Synergistic composition of lignans from *Cedrus deodra*	Anticancer activities against breast, cervix, neuroblastoma, colon, liver, lung, mouth, ovary, and prostate cancer (US7285571B2)	Council of Scientific And Industrial Research, India [249]
Extract of *Sphaeranthus indicus*	Anticancer (US20080199550A3)	Shanker Kumar Mitra, Ekta Saxena,Mallikarjun Narayan Dixit, Venkanna Babu Uddagiri, Venkata Ranganna Marikunte,Shivamurthaiah Arun Mathad, Sunil Vaikunth Shanbhag, India [250]
Formulation containing cucurbitacins (cucurbitacin B and cucurbitacin D)	Anticancer (antiproliferation and inducing cellular apoptosis) cells (US20080207578A1)	Chu Kee Hung, Hongtao Xing, U.K [251]
Extracts of *Gleditsia Sinensis*	Treatment of estrogen receptor (ER) negative breast cancer (US20090258096A1)	Bionovo, Inc., California [252]
Rhus verniciflua extracts	Anticancer (US7618661B2)	Azi Co., Ltd., Korea [253]
Extract obtained from *Anemarrhena asphodeloides* Bunge (containing Timosaponin A3 and Timosaponin B2)	Anticancer (US20100009017A1)	Bionovo, Inc, California [254]
Leaf extract of *Piper betle*	Treatment of chronic myeloid leukemia resistant to imatinib (US20100028472A1)	Piramal Life Sciences Limited, India [255]
Extract of *Rubus suavissimus*	Angiogensesis inhibition (US7709031B2)	[256] Board of Supervisors of Louisiana State University and Agricultural and Mechanical College, USA

gaining importance owing to a slump in corporate investments, particularly related to antimicrobial drug discovery. Secondary metabolites are a major reservoir of chemical diversity, therefore, they are considered a potential source of new drugs for combating the perils of drug resistance. The diversity of plants with respect to their potential to generate newer antibacterial agents has been reviewed by Shahid et al. [259]. The focus of this review is to provide recent insights towards an untapped source of antimicrobial chemotypes that are used in the traditional systems of medicine in different countries.

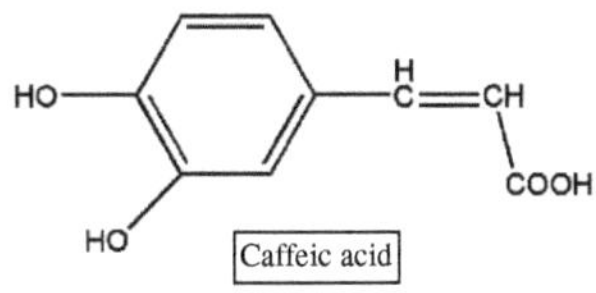

Fig. 5 Structures of caffeic acid

2.5.1 Recent Update of Antimicrobial Phytochemicals

Plant-derived antimicrobials are mostly secondary plant metabolites and have a wide range of activity, according to the species, topography, and climate of the country of origin, and may contain different categories of active principles [260, 261].

Simple Phenols and Phenolic Acid Polyphenols

Some of the simplest bioactive phytochemicals such as cinnamic and caffeic acids, consisting of a single substituted phenolic ring are common representatives of a wide group of phenyl propane-derived compounds (Fig. 5).

Caffeic acid obtained from the common herbs tarragon and thyme are known to be effective against viruses, bacteria, and fungi. Hydroxylated phenols including catechol and pyrogallol have been found to be toxic to the microorganisms [262].

Quinones

Quinones (Fig. 6) are a class of highly reactive organic compounds, usually acting as electron acceptors. These compounds are ubiquitous in nature and are known to be readily produced from phenols and catechols.

Stable free radicals produced by quinone produce an irreversible complex with nucleophilic amino acids present in microbial proteins, leading to the loss of their function [263]. Anthraquinones possess a wide spectrum of antibacterial activity (including their toxicity against the mycobacterium) due to their ability to inactivate bacterial proteins such as adhesins, cell wall polypeptides, or membrane-bound enzymes, consequently leading to the destruction of the pathogens [264].

Flavones, Flavonoids and Flavonols

Flavones are phenolic structures containing the carbonyl group and flavonoids are hydroxylated phenolic derivatives (Fig. 7). They are known to be synthesized by plants in response to microbial infection and are known to be present in different parts of the plants. The compounds are found to be effective against a wide range of microorganisms. Their activity is thought to be produced due to their ability to

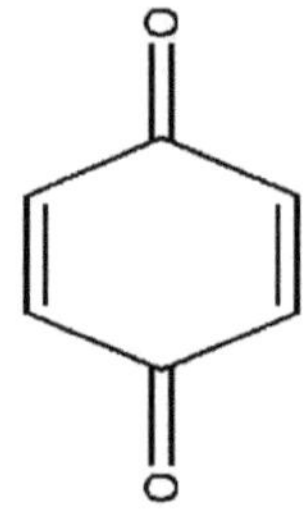

Fig. 6 Structure of quinone

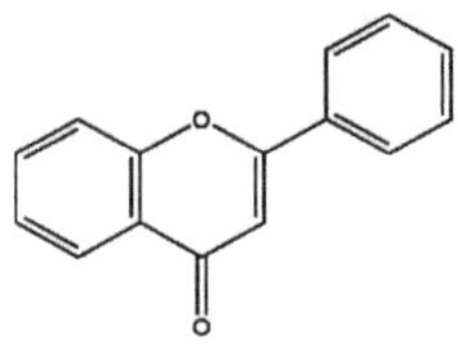

Fig. 7 General structures of flavones

form complexes with both extracellular and soluble proteins, as well as with bacterial membranes [265, 266].

Recent reports indicated that the antimicrobial activity of flavonoids (6-hydroxy-7-methoxyluteolin and the xanthone 8-carboxymethyl-1,5,6-trihydroxy-3-methoxyxanthone) isolated from the leaf extract of the *Leiothrix spiralis* (Eriocaulaceae) family, produced a promising antimicrobial activity [267]. Owing to the emergence of new cases and the increased incidence of multidrug-resistant strains of *Mycobacterium tuberculosis*, researchers around the globe are also exploring the natural resources for antitubercular leads, and some flavonoids have been found to demonstrate such antimycobacterial properties [268]. Flavonoid compounds also showed inhibitory effects against multiple viruses. Numerous studies have documented the effectiveness of flavonoids such as swertifrancheside, glycyrrhizin (from licorice), and chrysin [269] against HIV. Antifungal flavonoids isolated from mango (*Mangifera indica*) display promising activity on different species of fungi, including the Aspergillus sp. [270].

Tannins

Tannins are a group of polymeric phenolic substances known to demonstrate promising antimicrobial activity through inactivation of adhesins, cell envelope, enzymes, and different transport proteins. They can be divided into two groups, namely hydrolyzable and condensed tannins (Fig. 8). Both groups of tannins produce antimicrobial activity through antiperoxidative properties, inhibiting in

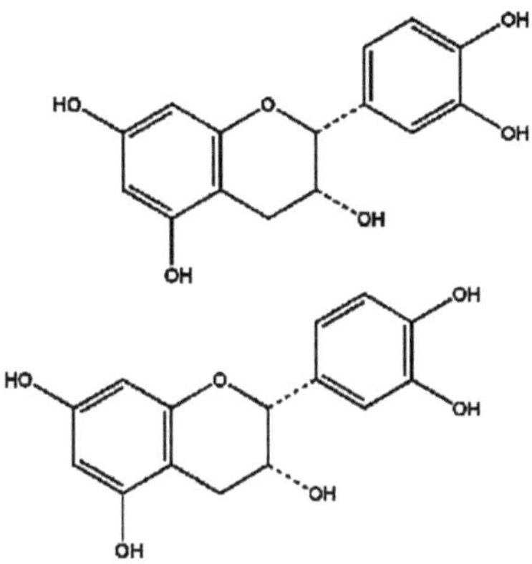

Fig. 8 Structure of tannin

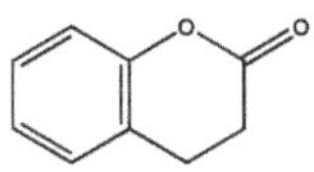

Fig. 9 Structure of coumarins

particular the growth of uropathogenic *E. coli*. According to recent reports, antimicrobial activity of gallotannin-rich plant extracts is attributed to the inactivation of membrane-bound proteins [271].

Coumarins

Coumarins are known to contain fused benzene and pyrone rings (Fig. 9), with a characteristic odor of hay. The antimicrobial activity of the coumarins has been reviewed and documented by Marjorie Murphy Cowan [272]. Scopoletin, a coumarin and two chalcones from *Fatuoa pilosa* have been found to display promising anti-tubercular properties against multidrug resistant mycobacteria [273], whereas hydroxylated derivatives of coumarins revealed potent antifungal activity [274].

Terpenoids and Essential Oils

The aroma of plants is related to the presence of essential oil, containing high concentrations of terpenes and terpenoids (Fig. 10). Essential oils derived from different families such as Pinaceae, Cupressaceae, Apiaceae, Burseraceae, Anacardiaceae, Palmaceae, Euphorbiaceae, Dracenaceae, and Fabaceae demonstrate antifungal, antibacterial, and antiprotozoal properties [275, 276]. Essential oils, in

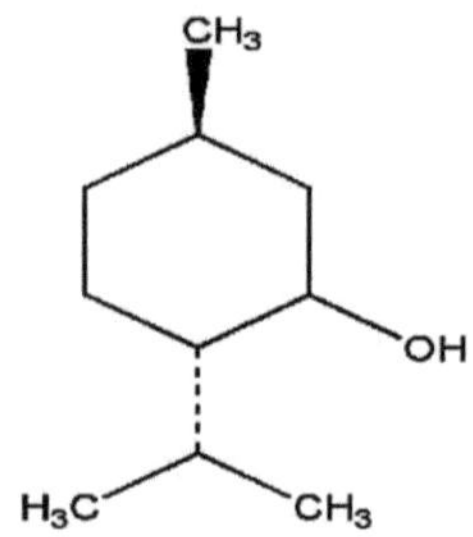

Fig. 10 Structure of menthol

particular from carrot (*Daucus carota*), have been found to be effective against both *Mycoplasma pneumoniae* and *Helicobacter pylori*, probably through their action on the intracellular cytoskeleton [277]. Recent studies report the antifungal properties of essential oils obtained from different medicinal plants [278–287]. Pulegone and piperitone oxide, present in the essential oils derived from *Mentha suaveolens*, are particularly effective against vaginal candidiasis [288]. Interestingly, eugenol and cinnamaldehyde, found in essential oils, showed promising activity against *C. albicans* biofilm formation and these compounds also displayed synergy with fluconazole in vitro [289].

Alkaloids

Alkaloids are a diverse group of organic nitrogen-containing compounds, displaying potent pharmacological properties and are often toxic in nature. Since ancient times, alkaloid-rich plants have been exploited for their antimicrobial properties [290]. Diterpenoid alkaloids, isolated from the plants belonging to the family Ranunculaceae, have been found to possess antimicrobial properties [291]. An isoquinoline alkaloid, berberine (Berberis species), is a hydrophobic cation widely used in traditional medicine for the management of infections associated with bacteria, fungi, viruses, and protozoa [292]. As per available reports, this compound on accumulation inside the cells, intercalates with the DNA [293] and also interferes with the activities of RNA polymerase, gyrase, and topoisomerase IV [294] thereby leading suppression of cellular growth and proliferation; it is an excellent DNA intercalator.

Polypeptides

Apart from the plant secondary metabolites, antimicrobial peptides, produced constitutively or in response to microbial infections, happen to be a part of innate immunity, known to work as a first line of defense against different classes of pathogens. These peptides display wide molecular diversity and are now classified

Table 7 Plant-derived proteins/peptides possessing antimicrobial properties

Plant	Molecular weight	References
Euonymus europaeus	45 a.a. residue	[297]
Triticum aestivum	23 kDa	[298]
Andean cropoca	18 kDa	[299]
Vigna angularis	8 kDa	[300]
Solanumtuberosum	7025 Da	[301]
Oryctes rhinoceros	4080 Da	[302]
Fagopyrumesculenium	3879 Da	[303]
Tulia gesneriana	5006 Da	[304]
Nicotianatabacum	26 kDa	[305]
TriticumKiharae	4844 Da	[306]
Allium sativum	13 kDa	[307]
Cocos nucifera	10 kDa	[308]
Solanum tuberosum	5.6 kDa	[310]
Phaseolus lunatu	7 kDa	[309]
Araucaria angustifolia	8 kDa	[310]
Withania somnifera	28 kDa	[311]
Cucurbita moschata	30665 Da	[312]
Capsicum annuum	10 kDa	[313]
Glycine max		[314]
Arabidopsis thaliana		[315]
Baby lima beans	6.5 kDa	[316]
Amaranthus tricolor	27 kDa	[317]
Pouteria torta	14 kDa	[318]
Ginkgo biloba	134-a.a	[319]
Capsicum annuum L.	6.5 kDa	[320]

according to their tertiary structures, the most common being thionins, defensin, and lipid transfer proteins [295]. Even though the antimicrobial peptides show wide structural diversity, however, most of these are short in length, hydrophobic, and with a net positive charge [296]. As they are cationic, these peptides tend to interact with negatively charged microbial cell membranes, ultimately leading to pore formation along with altered membrane permeability. Interestingly, as these peptides interact with the membranes, the chances of resistance development is comparatively reduced [296], thereby making them an attractive alternative to other synthetic antimicrobials. Information related to an increasing number of such naturally occurring proteins/peptides, possessing antimicrobial properties, in vitro, are now available in the scientific literature. An attempt has been made here to summarize the antimicrobial peptides/proteins derived from plant sources (Table 7).

2.6 Others

Medicinal plants have also played a significant role in the management of hepatic disorders [321–324], gastrointestinal (GI) ulceration [325–331], and also for controlling blood glucose levels in diabetes mellitus [332–335]. The use of traditional systems of medicine has always been popular in the developing and underdeveloped countries for the management of GI and hepatic disorders. A number of medicinal plants including *Phyllanthus niruri*, *Phyllanthus emebellica, Phyllanthus amaraus, Tinospora cordifolia, Silybum adans, Silybum marianum, Androgaphis paniculata,* and *Eclipta alba*, among others, have been popularly used for the treatment of hepatic disorders and related GI disturbances. Similarly, plants such as *Momordica charantia* (containing charantin, vicine, polypeptide-p, alkaloids), *Coptis chinensis* (rich source of berberine), *Gymnema sylvestre*, fenugreek leaves, and cinnamon have been found to be effective in Type II diabetes. Similarly, a number of Chinese herbal formulations (Jiangtangkeli, Yerbe Mate-Guarana-Damianaa, and Byakko-ka-ninjin-to) have also been used popularly for lowering blood glucose levels [336]. These antidiabetic plants and the herbal formulations have been found to act by (i) increasing insulin secretion, (ii) improving glucose uptake by adipose and muscle tissues, (iii) lowering glucose absorption, and (iv) reducing glucose production in heptocytes [336, 337].

3 Pharmaceutical Bioprospecting and Bioconservation

3.1 Impacts on Biodiversity of Major Pressures and Associated Effects on Ecosystem Services and Human Well-Being

As the basis for all ecosystem services, and the foundation for truly sustainable development, biodiversity plays fundamental roles in maintaining and enhancing the well-being of the global population, which includes rich and poor, rural and urban alike. Biodiversity comprises much of the renewable natural capital and therefore it is intricately linked to both livelihood and development. However, ongoing, and in many cases, accelerated, losses in biodiversity over the past 20 years have decreased the capacity of many ecosystems to provide services, and have had a profound negative impact on sustainable development (Fig. 11). These impacts are particularly pronounced in the developing world, largely related to the patterns of consumption and trade in the industrialized world.

Biodiversity contributes significantly towards livelihood and human development and thus plays a predominant role towards the well-being of the global population. The impending danger posed due a meteoric rise in the global population, rapid and sometimes unplanned industrialization, alongside indiscriminate deforestation, overexploitation of natural resources, pollution, and finally global climate change have now brought us to a crossroad where we need to act rapidly and decisively to conserve the biodiversity of our planet and for the sustainable development of human civilization at large. The United Nations Environment Programme (UNEP) has identified several risk [338–341] factors that may have a profound negative influence on biodiversity. These can be summarized as follows:

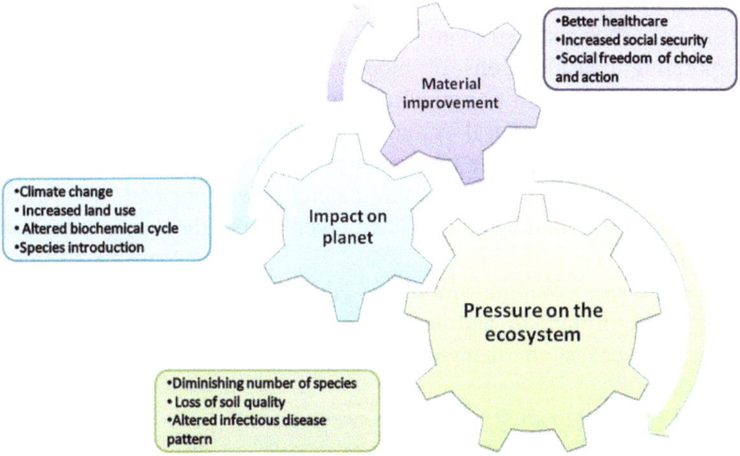

Fig. 11 Interrelation of human pressure on the global environment and ecosystem

(i) several million hectares of forest were lost annually during 1990–1997 and as a result it has led to a decrease in natural habitat as well as homogenization of species. (ii) Similarly, overexploitation of wild varieties, particularly in countries such as India (for meeting the demands of industry), have led to extinction and introduction of alien species along with changes in the functioning of the ecosystem. (iii) Climate change (especially due to overuse of fossil fuel) has resulted in the shrinkage of temperate rain forests, changing the species range and behavior and increasing the risks of invasive alien species. (iv) Pollution is another major threat and it often results in nutrient alteration, acidification, and accumulation of heavy metals and pesticides. The UNEP report on biodiversity mentions that loss of biodiversity may be considerably minimized by curbing production subsidies (sometimes politically motivated), undervaluation of biological resources, internalization of environmental costs into prices, and finally implementation of the conservation policies at local, national, and global levels.

The Millennium Ecosystem Assessment (MEA), an initiative of the United Nations, constituted of multistakeholders, was formed in the year 2001. MEA commenced the assessment of the effect of humans on the Earth's ecosystems and it was observed that during the past 50 years, human activities have changed the ecosystems more rapidly and extensively than in any other comparable period in the history of mankind (MEA [342]). MEA recommended some actions that have been (at least partly) successful in reducing biodiversity loss and therefore the same can be further strengthened in the future:

- Protected areas.
- Species protection and recovery measures for threatened species.
- Ex situ and in situ conservation of genetic diversity (e.g., genebanks).
- Ecosystem restoration.

- Payments and markets for biodiversity and ecosystem services (e.g., for eco-tourism or carbon sequestration).
- Incorporating considerations of biodiversity conservation into management practices in sectors such as agriculture, forestry, and fisheries.
- Capture of benefits by local communities (i.e., ensuring local people benefit from the conservation of the biodiversity around them).
- Increased co-ordination among multilateral environmental agreements and between environmental agreements and other international economic and social institutions.
- Public awareness, communication, and education.
- Enhancement of human and institutional capacity for assessing the consequences of ecosystem change for human well-being and acting on such assessments.
- Increased integration of sectoral responses.
- Elimination of subsidies that promote excessive use of ecosystem services.
- Sustainable intensification of agriculture.
- Slowing and adapting to climate change.
- Addressing unsustainable consumption patterns.
- Slowing the global growth in nutrient loading.
- Correction of market failures and internalization of environmental externalities that lead to the degradation of ecosystem services. (Because many ecosystem services are not formally traded, markets fail to provide appropriate signals that might otherwise contribute to their efficient allocation and sustainable use. In addition, many of the harmful trade-offs and costs associated with the management of one ecosystem service are borne by others and so are not weighed in sectoral decisions regarding the management of that service).
- Integration of biodiversity conservation and development planning.
- Increased transparency and accountability of government and private-sector performance in decisions that affect ecosystems, including through greater involvement of concerned stakeholders in decision making.
- Scientific findings and data need to be made available to all of society.

To further substantiate the global initiatives towards conservation of biodiversity, it may be relevant to mention the role of The National Botanic Garden of Wales (NBGW), which is working for the research and conservation of biodiversity and its sustainable utilization (http://www.gardenofwales.org.uk/). Presently, the NBGW contributes towards conservation of biodiversity in the following ways: (i) conducting a research program on the ecology, taxonomy, and conservation of the Welsh flora; (ii) plant collection and systematic gardening; (iii) in situ conservation (ii) and also provide scope for students to work on conservation; (iv) use of DNA barcoding technology for the first time in the world for the flowering plants and conifers, which could have a huge impact on biodiversity conservation; and (v) conservation of endangered plants species of the Wales region.

It may also be relevant to mention that WHO recommends the creation of a national or regional inventory of medicinal plants, so as to (i) facilitate the identification of medicinal plants used by communities (including endangered species), and (ii) outline their distribution and assess their abundance. Moreover, WHO recommends enhanced research activities to improve the agronomy of cultivated medicinal plants, encourage the exchange of information related to agricultural production, and investigate the social and environmental impact of medicinal plant cultivation and collection [WHO guidelines on good agricultural and collection practices (GACP) for medicinal plants World Health Organization, Geneva, 2003]. Thus, natural loss due to species extinction could be of paramount importance to society for a number of reasons, including the maintenance of ecosystem function and for ethical reasons. But species extinction is specifically important to pharmaceutical companies in their search for novel bioactive molecules. Among the different species that are being lost, there may be some with tremendous therapeutic potential that have remained undiscovered thus far. Having said this, there is a variety of biodiversity, conservation, and sampling issues associated with examining different types of biota for the production of novel natural products of pharmaceutical value.

There may be a perception among conservationists and the public that "large quantities" of material are being collected from the bush or the oceans for screening for novel natural products. Although small amounts of material may be used for the initial stages of the drug discovery process, there is a clear desire within the pharmaceutical industry to conserve the world's biota so that more species can be examined for novel chemical molecules, and that compounds of interest are produced via routes that do not involve the destructive and costly harvesting of samples.

4 Conclusions

Considering the unplanned utilization of natural resources coupled with the rapid decline of biodiversity, it may be predicted that by the turn of this century, some species of plants may cease to exist. Once depleted, species regeneration, if at all possible, might take 5–10 million years. This loss would be phenomenal and thus may have profound negative effects on the inhabitants of the earth. In medical sciences, the loss of species could adversely affect the process of drug discovery and finally disease management. Therefore, it is of utmost importance that plant biodiversity be preserved, so that it may continue to provide structural diversity in the form of novel lead compounds, for the already existing as well as emerging therapeutic targets, in the coming years. It may also be important to mention that countries (both developed and developing) with rich biodiversity should take necessary measures for harnessing such treasure troves in a scientific manner, safeguarding the biodiversity but at the same time making the herbal medicines more accessible to the common mass.

Acknowledgments The authors are thankful to Ratul Sarkar, Kirendra Kr. Yadav, D. Sinha, Md. Firdaus, Pritam Saha, Amrita Sharma, Prof. Samantak Das, and Aritro Bhattacharjee for their help during the preparation of the manuscript.

References

1. Bezerra RJS, Calheiros AS, Ferreira NCS, Frutuoso VS, Alves LA (2013) Natural products as a source for new anti-inflammatory and analgesic compounds through the inhibition of purinergic P2X receptors. Pharmaceuticals (Basel) 6(5):650–658
2. Scott JG, Liu N, Wen Z (1998) Insect cytochromes P450: diversity, insecticide resistance and tolerance to plant toxins. Comp Biochem Physiol C: Pharmacol Toxicol Endocrinol 121:147–155
3. Kushiro T, Nambara E, McCourt P (2003) Hormone evolution: the key to signalling. Nature 422:122
4. Wink M (2000) Interference of alkaloids with neuroreceptors and ion channels. Stud Nat Prod Chem 21:3–122
5. Palavan-Unsal N, Arisan D (2009) Nitric oxide signalling in plants. Bot Rev 75:203–229
6. Chivian E, McCally M, Hu H, Haines A (eds) (1993) Critical condition: human health and the environment. MIT Press, Cambridge, MA
7. Tahara S (2007) A journey of twenty-five years through the ecological biochemistry of flavonoids. Biosci Biotechnol Biochem 71:1387–1404
8. Wink M (2003) Evolution of secondary metabolites from an ecological and molecular phylogenetic perspective. Phytochemistry 64:3–19
9. Dalle-Donne I, Rossi R, Colombo G, Giustarini D, Milzani A (2009) Protein S-glutathionylation: a regulatory device from bacteria to humans. Trends Biochem Sci 34:85–96
10. Scandalios JG (2005) Oxidative stress: molecular perception and transduction of signals triggering antioxidant gene defenses. Braz J Med Biol Res 38:995–1014
11. Lee D-S, Nioche P, Hamberg M, Raman CS (2008) Structural insights into the evolutionary paths of oxylipin biosynthetic enzymes. Nature 455:363–368
12. Nässel DR, Winther AM (2010) Drosophila neuropeptides in regulation of physiology and behavior. Prog Neurobiol 92:42–104
13. Klowden MJ (2007) Physiological systems in insects. Academic, London
14. Marder E (2007) Searching for insight. In: North G, Greenspan RJ (eds) Invertebrate neurobiology. Cold Spring Harbor Laboratory, Cold Spring Harbor, New York
15. Wolf FW, Heberlein U (2003) Invertebrate models of drug abuse. J Neurobiol 54:161–178
16. Farooqui T (2007) Octopamine-mediated neuromodulation of insect senses. Neurochem Res 32:1511–1529
17. Giurfa M (2007) Invertebrate cognition: nonelemental learning beyond simple conditioning. In: North G, Greenspan RJ (eds) Invertebrate neurobiology. Cold Spring Harbor Laboratory, Cold Spring Harbor, New York
18. Mixson TA, Abramson CI, Bozi J (2010) The behavior and social communication of honey bees (*Apis mellifera carnica poll*). Under the influence of alcohol 1, 2. Psychol Rep 106:701–717
19. Ho KS, Sehgal A (2005) Drosophila melanogaster: an insect model for fundamental studies of sleep. Methods Enzymol 393:772–793
20. Horiuchi J, Saitoe M (2005) Can flies shed light on our own age-related memory impairment? Ageing Res Rev 4:83–101
21. Thamm M, Balfanz S, Scheiner R, Baumann A, Blenau W (2010) Characterization of the 5-HT 1A receptor of the honeybee (*Apis mellifera*) and involvement of serotonin in phototactic behavior. Cell Mol Life Sci 67:2467–2479

22. Mustard JA, Pham PM, Smith BH (2010) Modulation of motor behavior by dopamine and the D1-like dopamine receptor AmDOP2 in the honey bee. J Insect Physiol 56:422–430
23. Hassani AK, Dupuis JP, Gauthier M, Armengaud C (2009) Glutamatergic and GABAergic effects of fipronil on olfactory learning and memory in the honeybee. Invert Neurosci 9:91–100
24. Ismail N, Christine S, Robinson GE, Fahrbach SE (2008) Pilocarpine improves recognition of nestmates in young honey bees. Neurosci Lett 439:178–181
25. Guez D, Zhu H, Zhang S, Srinivasan M (2010) Enhanced cholinergic transmission promotes recall in honeybees. J Insect Physiol 56:1341–1348
26. Schultz JC (2002) Shared signals and the potential for phylogenetic espionage between plants and animals. Integr Comp Biol 42:454–462
27. Brunton L, Lazo J, Parker K (2005) Goodman & Gilman's the pharmacological basis of therapeutics, vol 11. McGraw-Hill, New York
28. Halpern JH (2004) Hallucinogens and dissociative agents naturally growing in the United States. Pharmacol Ther 102:131–138
29. Kitagawa M, Houzen H, Tashiro K (2007) Effects of caffeine on the freezing of gait in Parkinson's disease. Mov Disord 22(5):710–712
30. Roth BL, Lopez E, Beischel S, Westkaemper RB, Evans JM (2004) Screening the receptorome to discover themolecular targets for plant-derived psychoactive compounds: a novel approach for CNS drug discovery. Pharmacol Ther 102:99–110
31. Kumar V (2006) Potential medicinal plants for CNS disorders: an overview. Phytotherapy Research 20(12):1023–1035
32. Carlini EA (2003) Plants and the central nervous system. Pharmacol Biochem Behav 75:501–512
33. Halpern R, Sewell A (2005) Hallucinogenic botanicals of America: a growing need for focused drug education and research. Life Sci 78:519–526
34. Singh YN, Singh NN (2002) Therapeutic potential of kava in the treatment of anxiety disorders. CNS Drugs 16:731–743
35. Willmore-Fordham CB, Krall DM, McCurdy CR, Kinder DH (2007) The hallucinogen derived from *Salvia divinorum*, salvinorin A, has κ-opioid agonist discriminative stimulus effects in rats. Neuropharmacology 53:481–486
36. Calixto JB, Beirith A, Ferreira J, Santos ARS, Filho VC, Yunes RA (2000) Naturally occurring antinociceptive substances from plants. Phytother Res 14:401–418
37. Chen LX, He H, Qiu F (2011) Natural withanolides: an overview. Nat Prod Rep 28:705–740
38. Akira S, Uematsu S, Takeuchi O (2006) Pathogen recognition and innate immunity. Cell 124:783–801
39. Takeuchi O, Akira S (2010) Pattern Recognition Receptors and Inflammation. Cell 140:805–820
40. Schroder K, Tschopp J (2010) The inflammosomes. Cell 140:821–832
41. Koehn FE, Carter GT (2005) The evolving role of natural products in drug discovery. Nature 4:206–220
42. Gautam R, Jachak SM (2009) Recent developments in anti-inflammatory natural products. Med Res Rev 29:767–820
43. Picerno P, Autore G, Marzocco S, Meloni M, Sanogo R, Aquino RP (2005) Anti-inflammatory activity of verminoside from *Kigelia africana* and evaluation of cutaneous irritation in cell cultures and reconstituted human epidermis. J Nat Prod 68:1610–1614
44. Kassuya CA, Leite DF, de Melo LV, Rohder VL, Calixto JB (2005) Anti-inflammatory properties of extracts, fractions and lignans isolated from *Phyllanthus amarus*. Planta Med 71:721–726
45. Lin LC, Shen CC, Shen YC, Tsai TH (2006) Antiinflammatory neolignans from *Piper kadsura*. J Nat Prod 69:842–844
46. Ito C, Itoigawa M, Okada T, Furukawa H (2007) Anti-inflammatory activity of phenylpropanoids and phytoquinoids from *Illicium species* in RBL-2H3 cells. Planta Med 73:662–665

47. Min HY, Nam JW, Lee NY, Wiryawan A, Suprapto W et al (2008) Identification of a new naphthalene and its derivatives from the bulb of *Eleutherine Americana* with inhibitory activity on lipopolysaccharide induced nitric oxide production. Chem Pharm Bull 56:1314–1316
48. Seelinger G, Merfort I, Schempp CM (2008) Antioxidant, antiinflammatory and antiallergic activities of luteolin. Planta Med 74:1667–1677
49. Yongmei Z, Nan Z, Ning T, Baolin L (2007) Evaluation of the anti-inflammatory activity of luteolin in experimental animal models. Planta Med 73:221–226
50. Sindhu ER, Kuttan R (2007) Antioxidant and anti-inflammatory activity of lutein and its ester. Amala Res Bull 27:261–270
51. Lee SA, Han XH, Lee MH, Hwang S, Park JS et al (2008) Melampolides from the leaves of *Smallanthus sonchifolius* and their inhibitory activity of LPS-induced nitric oxide production. Chem Pharm Bull 56:199–202
52. Byeon SE, Lee YG, Kim JC, Han JG, Lee HY, Cho JY (2008) Hinokitiol: a natural tropolone derivative, inhibits TNF-alpha production in LPS-activated macrophages via suppression of NF-KB. Planta Med 74:828–833
53. Shin EM, Kim YS, Cai XF, Lee JJ, Kim HP (2006) Anti-inflammatory principles from the fruits of *Evodia rutaecarpa* and their cellular action mechanisms. Arch Pharmacal Res 29:293–297
54. Lee SH, Kim JY, Zhao YZ, Park EJ, Lee BS et al (2006) Polyozellin inhibits nitric oxide production by down-regulating LPS-induced activity of NF-kB and SAPK/JNK in RAW 264.7 cells. Planta Med 72:857–859
55. Lee J, Jung E, Lim J, Lee J, Hur S, Kim SS et al (2006) Involvement of nuclear factor-kB in the inhibition of pro-inflammatory mediators by pinosylvin. Planta Med 72:801–806
56. Patel B, Patel S, Hoffman R (2005) Inhibition of cyclo-oxygenase-2 expression in mouse macrophages by 4-(3-methyl-but-1-enyl)-3,5,3',4'-tetrahydroxystilbene: a resveratrol derivative from peanuts. Phytother Res 19:552–555
57. Marsik P, Kokoska L, Landa P, Nepovim A, Soudek P, Vanek T (2005) In vitro inhibitory effects of thymol and quinones of *Nigella sativa* seeds on cyclooxygenase-1-and-2-catalyzed prostaglandin E2 biosyntheses. Planta Med 71:739–742
58. Won JH, Kim JY, Yun KJ, Lee JH, Back NI, Chung HG et al (2006) Gigantol isolated from the whole plants of *Cymbidium goeringii* inhibits the LPS-induced INOS and COX-2 Expression via NF-kB inactivation in RAW 264.7 macrophages cells. Planta Med 72:1181–1187
59. Alves LA, Bezerra RJ, Faria RX, Ferreira LG, da Silva Frutuoso V (2013) Physiological roles and potential therapeutic applications of the P2X7 receptor in inflammation and pain. Molecules 18:10953–10972
60. Ao CW, Araki N, Tawata S (2009) Cyclooxygenase inhibitory compounds with antioxidant activities from *Sophora subprostrata*. Asian J Chem 21:745–754
61. Shang JH, Cai XH, Feng T, Zhao YL, Wang JK, Zhang LY, Yan M, Luo XD (2010) Pharmacological evaluation of *Alstonia scholaris*: anti-inflammatory and analgesic effects. J Ethnopharmacol 129(174):181
62. Kupeli E, Kosar M, Yesilada E, Baser KHC, Baser CA (2002) Comparative study on the anti-inflammatory, antinociceptive and antipyretic effects of isoquinoline alkaloids from the rootsof turkish berberis species. Life Sci 72:645–657
63. Danz H, Stoyanova S, Wippich P, Brattstrom A, Hamburger M (2001) Identification and isolation of the cyclooxygenase-2 inhibitory principle in *Isatis tinctoria*. Planta Med 67:411–416
64. Danz H, Stoyanova S, Thomet OAR, Simon HU, Dannhardt G, Ulbrich H, Hamburger M (2002) Inhibitory activity of tryptanthrin on prostaglandin and leukotriene synthesis. Planta Med 68:875–880
65. Ishihara T, Kohno K, Ushio S, Iwaki K, Ikeda M, Kurimoto M (2000) Tryptanthrin inhibits nitric oxide and prostaglandin E2 synthesis by murine macrophages. Eur J Pharmacol 407:197–204

66. Noh EJ, Ahn KS, Shin EM, Jung SH, Kim YS (2006) Inhibition of lipopolysaccharide-induced iNOS and COX-2 expression by dehydroevodiamine through suppression of NF-kB activation in RAW 264.7 macrophages. Life Sci 79:695–701
67. Woo HG, Lee CH, Noh M-S, Lee JJ, Jung Y-S, Baik EJ, Moon C-H, Lee SH (2001) Rutaecarpine, a quinazolinocarboline alkaloid, inhibits prostaglandin production in RAW264.7 macrophages. Planta Med 67:505–509
68. Ko HC, Wang YH, Liou KT, Chen CM, Chen CH, Wang WY, Chang S, Hou YC, Chen KT, Chen CF, Shen YC (2007) Anti-inflammatory effects and mechanisms of the ethanol extract of *Evodia rutaecarpa* and its bioactive components on neutrophils and microglial cells. Eur J Pharmacol 555:211–217
69. Moon TC, Murakami M, Kudo I, Son KH, Kim HP, Kang SS, Chang HW (1999) A new class of COX-2 inhibitor, rutaecarpine from *Evodia rutaecarpa*. Inflamm Res 48:621–625
70. Choi YH, Shin EM, Kim YS, Cai XF, Lee JJ, Kim HP (2006) Anti-inflammatory principles from the fruits of *Evodia rutaecarpa* and their cellular action mechanisms. Arch Pharm Res 29:293–297
71. Wang Y, Fang Y, Huang W, Zhou X, Wang M, Zhong B, Peng D (2005) Effect of sinomenine on cytokine expression of macrophages and synoviocytes in adjuvant arthritis rats. J Ethnopharmacol 98:37–43
72. Choi HS, Kim HS, Min KR, Kim Y, Lim HK, Chang YK, Chung MW (2000) Anti-inflammatory effects of fangchinoline and tetrandrine. J Ethnopharmacol 69:173–179
73. Shen YC, Chou CJ, Chiou WF, Chen C (2001) Anti-inflammatory effects of the partially purified extract of radix *Stephaniae tetrandrae*: comparative studies of its active principles tetrandrine and fangchinoline on human polymorphonuclear leukocyte functions. Mol Pharmacol 60(5):1083–1090
74. Wu SJ, Ng LT (2007) Tetrandrine inhibits proinflammatory cytokines, iNOS and COX-2 expression in human monocytic *cells*. Biol Pharm Bull 30:59–62
75. Choi BH, Ahn IS, Kim YH, Park JW, Lee SY, Hyun CK, Do MS (2006) Berberine reduces the expression of adipogenic enzymes and inflammatory molecules of 3T3-L1 adipocyte. Exp Mol Med 38:599–605
76. Lee CH, Chen JC, Hsiang CY, Wu SL, Wu HC, Ho TY (2007) Berberine suppresses inflammatory agents-induced interleukin-1b and tumor necrosis factor-a productions via the inhibition of IkB degradation in human lung cells. Pharmacol Res 56:193–201
77. Hu J-P, Nishishita K, Sakai E, Yoshida H, Kato Y, Tsukuba T, Okamoto K (2008) Berberine inhibits RANKL-induced osteoclast formation and survival through suppressing the NF-kB and Akt pathways. Eur J Pharmacol 580:70–79
78. Chiou WF, Peng CH, Chen CF, Chou CJ (2003) Anti-inflammatory properties of piperlactam S: modulation of complement 5a-induced chemotaxis and inflammatory cytokines production in macrophages. Planta Med 69:9–14
79. Hanh TTH, Hang DTT, van Minh C, Dat NT (2011) Anti-inflammatory effects of fatty acids isolated from *Chromolaena odorata*. Asian Pac J Trop Med 4(10):760–763
80. Ringbom T, Huss U, Stenholm A, Flock S, Skattebol L, Perera P, Bohlin L (2001) COX-2 inhibitory effects of naturally occurring and modified fatty acids. J Nat Prod 64:745–749
81. Su B-N, Cuendet M, Farnsworth NR, Fong HHS, Pezzuto JM, Kinghorn AD (2002) Activity-guided fractionation of the seeds of *Ziziphus jujuba* using a cyclooxygenase-2 inhibitor assay. Planta Med 68:1125–1128
82. Jang DS, Cuendet M, Su B-N, Totura S, Riswan S, Fong HHS, Pezzuto JM, Kinghorn AD (2004) Constituents of the seeds of *Hernandia ovigera* with inhibitory activity against cyclooxygenase-2. Planta Med 70:893–896
83. Shen CC, Wang YH, Chang TT, Lin LC, Don MJ, Hou YC, Lio KT, Chang S, Wang WY, Ko HC, Shen YC (2007) Anti-inflammatory ergostanes from the basidiomata of *Antrodia salmonea*. Planta Med 73:1208–1213

84. Manjula N, Gayathri B, Vinaykumar KS, Shankernarayanan NP, Vishwakarma RA, Balakrishnan A (2006) Inhibition of MAP kinases by crude extract and pure compound isolated from *Commiphora mukul* leads to down regulation of TNF-a, IL-1b and IL-2. Int Immunopharmacol 6:122–132
85. Khanna D, Sethi G, Ahn KS, Pandey MK, Kunnumakara AB, Sung B, Aggarwal A, Aggarwal BB (2007) Natural products as a gold mine for arthritis treatment. Curr Opin Pharmacol 7:344–351
86. Bai L, Wang L, Zhao M, Toki A, Hasegawa T, Ogura H, Kataoka T, Hiroe K, Sakai J, Bai J, Ando M (2007) Bioactive pregnanes from Nerium oleander. J Nat Prod 70:14–18
87. Ende C, Gebhardt R (2004) Inhibition of matrix metalloproteinase-2 and -9 activities by selected flavonoids. Planta Med 70:1006–1008
88. Matsuda H, Morikawa T, Ueda K, Managi H, Yoshikawa M (2002) Structural requirements of flavonoids for inhibition of antigen-induced degranulation, TNF-a and IL-4 production from RBL-2H3 cells. Bioorg Med Chem 10:3123–3128
89. Kim HK, Cheon BS, Kim YH, Kim SY, Kim HP (1999) Effects of naturally occurring flavonoids on nitric oxide production in the macrophage cell line RAW 264.7 and their structure-activity relationships. Biochem Pharmacol 58:759–765
90. Sadik CD, Sies H, Schewe T (2003) Inhibition of 15-lipoxygenases by flavonoids: Structure-activity relations and mode of action. Biochem Pharmacol 65:773–781
91. Venables BJ, Waggoner CA, Chapman KD (2005) N-acylethanolamines in seeds of selected legumes. Phytochemistry 66:1913–1918
92. Khanna D, Sethi G, Ahn KS, Pandey MK, Kunnumakara AB, Sung B, Aggarwal A, Aggarwal BB (2007) Natural products as a gold mine for arthritis treatment. Curr Opin Pharmacol 7:344–351
93. Paul AT, Gohil VM, Bhutani KK (2006) Modulating TNF-a signaling with natural products. Drug Discov Today 11:725–732
94. Ishikawa YT, Goto M, Yamaki K (2006) Structure–activity relations of inhibitory effects of various flavonoids on lipopolysaccharide-induced prostaglandin E2 production in rat peritoneal macrophages: comparison between subclasses of flavonoids. Phytomedicine 13:310–317
95. Calixto JB, Campos MM, Otuki MF, Santos ARS (2004) Anti-inflammatory compounds of plant origin. Part II. Modulation of pro-inflammatory cytokines, chemokines and adhesion molecules. Planta Med 70:93–103
96. Yamaki K, Kim DH, Ryu N, Kim YP, Shin KH, Ohuchi K (2002) Effects of naturally occurring isoflavones on prostaglandin E2 production. Planta Med 68:97–100
97. Laupattarakasem P, Houghton PJ, Hoult JR (2004) Anti-inflammatory isoflavonoids from the stems of *Derris scandens*. Planta Med 70:496–501
98. Matsuda H, Morikawa T, Ando S, Toguchida I, Yoshikawa M (2003) Structural requirements of flavonoids for nitric oxide production inhibitory activity and mechanism of action. Bioorg Med Chem 11:1995–2000
99. Zeng Y, Gu B, Ji X, Ding X, Song C, Wu F (2007) Sinomenine, an antirheumatic alkaloid, ameliorates clinical signs of disease in the lewis rat model of acute experimental autoimmune encephalolmyelitis. Biol Pharm Bull 30:1438–1444
100. Jachak SM, Saklani A (2007) Challenges and opportunities in drug discovery from plants. Curr Sci 92:1251–1257
101. Ho L-H, Juan T-Y, Chao P, Wu W-L, Chang D-M, Chang S-Y, Lai J-H (2004) Plant alkaloid tetrandrine downregulates IkBa kinases-IkBa-NF-kB signaling pathway in human peripheral blood T cell. Br J Pharmacol 143:919–927
102. Jeon TW, Jin CH, Lee SK, Jun IH, Kim GH, Lee DJ, Jeong HG, Lee K-B, Jahng Y, Jeong TC (2006) Immunosuppressive effects of rutaecarpine in female BALB/c mice. Toxicol Lett 164:155–166
103. Giang PM, Jin HZ, Son PT, Lee JH, Hong YS, Lee JJ (2003) ent-Akurane diterpenoids from Croton tonkinensis inhibit LPS-induced NF-kB activation and NO production. J Nat Prod 66:1217–1220

104. Hong SS, Lee SA, Han XH, Jin HZ, Lee JH, Lee D, Lee JJ, Hong JT, Kim Y, Ro JS, Hwang BY (2007) Akurane diterpenoids from *Isodon excisus* inhibit LPS-induced NF-kB activation and NO production in macrophage RAW264.7 cells. J Nat Prod 70:632–636
105. Perez-Garcia F, Martin E, Parella T, Adzet T, Canigueral S (2005) Activity of taraxasteryl acetate on inflammation and heat shock protein synthesis. Phytomedicine 12:278–284
106. Figueredo SM, do Nascimento FP, Freitas CS, Baggio CH, Soldi C, Pizzolatti MG, de Ibarrola Mdel C, de Arrua RL, Santos AR (2011) Antinociceptive and gastroprotective actions of ethanolic extract from *Pluchea sagittalis* (Lam.) Cabrera. J Ethnopharmacol 135(3):603-609
107. Pandey A, Bani S, Satti NK, Gupta BD, Suri KA (2012) Anti-arthritic activity of agnuside mediated through the down-regulation of inflammatory mediators and cytokines. Inflamm Res 61(4):293–304
108. Suksamrarn A, Kumpun S, Kirtikara K, Yingyongnarongkul, Suksamrarn S (2002) Iridoids with antiinflamatory activity from *Vitex peduncularis*. Planta Med 68:72–73
109. Yoshikawa K, Inoue M, Mastumoto Y, Sakkakibara C, Miyataka H, Mastumoto H, Arihara S (2005) Lanostane triterpenoids and triterpene glycosides from the fruit body of *Fomitopsis pinicola* and their inhibitory activity against COX-1 and COX-2. J Nat Prod 68(1):69–73
110. Chao KP, Hua KF, Hsu HY, Su YC, Chang ST (2005) Anti-inflammatory activity of Sugiol, a diterpene isolated from *Calocedrus formosana* bark. Planta Med 71:300–305
111. Ma J, Dey M, Yang H, Poulev A, Pouleva R, Dorn R, Lipsky PE, Kennelly EJ, Raskin I (2007) Antiinflammatory and immunosuppressive compounds from *Tripterygium wilfordii*. Phytochemistry 68:1172–1178
112. Sethi G, Ahn KS, Pandey MK, Aggarwal BB (2007) Celastrol, a novel triterpene, potentiates TNF-induced apoptosis and suppresses invasion of tumor cells by inhibiting NF-kB-regulated gene products and TAK1-mediated NF-kB activation. Blood 109:2727–2735
113. Koo TH, Lee J-H, Park YJ, Hong Y-S, Kim HS, Kim K-W, Lee JJ (2001) A sesquiterpene lactone, costunolide, from *Magnolia grandiflora* inhibits NF-kB by targeting IkB phosphorylation. Planta Med 67:103–107
114. Siedle B, Gustavsson L, Johansson S, Murillo R, Castro V, Bohlin L, Merfort I (2003) The effect of sesquiterpene lactones on the human neutrophil elastase. Biochem Pharmacol 65:897–903
115. Castro V, Murillo R, Klaas CA, Meunier C, Mora G, Pahl HL, Merfort I (2000) Inhibition of the transcription factor NF-kB by sesquiterpene lactones from *Podachaenium eminens*. Planta Med 66:591–595
116. LyX G, Schmidt TJ, Merfort I, Pahl HL (1997) Helenalin, an anti-inflammatory sesquiterpene lactone from Arnica, selectively inhibits transcription factor NF-kB. Biol Chem 378:951–961
117. LyX G, Knorre A, Schmidt TJ, Merfort I, Pahl HL (1998) The anti-inflammatory sesquiterpene lactone helenalin inhibits the transcription factor NF-kB by directly targeting p65. J Biol Chem 273:33508–33516
118. de Marino S, Borbone N, Zollo F, Lanaro A, di Meglio P, Iorizzi M (2005) New sesquiterpene lactones from *Laurus nobilis* leaves as inhibitors of nitric oxide production. Planta Med 71: 706–710
119. Rungeler P, Castro V, Mora G, Goren N, Vichnewski W, Pahl HL, Merfort I, Schmidt TJ (1999) Inhibition of transcription factor NF-kB by sesquiterpene lactones: a proposed molecular mechanism of action. Bioorg Med Chem 7:2343–2352
120. Honda T, Finlay HJ, Gribble GW, Suh N, Sporn MB (1997) New enone derivatives of oleanolic acid and ursolic acid as inhibitors of nitric oxide production in mouse macrophages. Bioorg Med Chem Lett 7:1623–1628
121. Honda T, Rounds BV, Gribble GW, Suh N, Wang Y, Sporn MB (1998) Design and synthesis of 2- cyano-3,12-dioxooolean-1,9-dien-28-oic acid, a novel and highly active inhibitors of nitric oxide production in mouse macrophages. Bioorg Med Chem Lett 8:2711–2724

122. Honda T, Rounds BV, Bore L Jr, Favaloro FG, Gribble GW, Suh N, Wang Y, Sporn MB (1999) Novel synthetic oleanane triterpenoids: a series of highly active inhibitors of nitric oxide production in mouse macrophages. Bioorg Med Chem Lett 9:3429–3434
123. Honda T, Gribble GW, Suh N, Finlay HJ, Rounds BV, Bore L Jr, Favaloro FG, Wang Y, Sporn MB (2000) Novel synthetic oleanane and ursane triterpenoids with various enone functionalities in ring A as inhibitors of nitric oxide production in mouse macrophages. J Med Chem 43:1866–1877
124. Honda T, Rounds BV, Bore L, Finlay HJ Jr, Favaloro FG, Suh N, Wang Y, Sporn MB, Gribble GW (2000) Synthetic oleanane and ursane triterpenoids with modified ring A and C: a series of highly active inhibitors of nitric oxide production in mouse macrophages. J Med Chem 43:4233–4246
125. Zhang M, Munekazu Iinuma, Jun-Song Wang Masayoshi Oyama, Tetsuro Ito, and Ling-Yi Kong (2012) Terpenoids from *Chloranthus serratus* and their anti-inflammatory activities. J Nat Prod 75(4):694–698
126. Honda T, Honda Y Jr, Favaloro FG, Gribble GW, Suh N, Place AE, Rendi MH, Sporn MB (2002) A novel dicyanotriterpenoid, 2-Cyano-3,12-dioxooleana-1,9(11)-dien-28-onitrile, active at picomolar concentrations for inhibition of nitric oxide production. Bioorg Med Chem Lett 12:1027–1030
127. Cho JY, Baik KU, Yoo ES, Yoshikawa K, Park MH (2000) In vitro anti-inflammatory effects of neolignan woorenosides from the rhizomes of *Coptis japonica.* J Nat Prod 63:1205–1209
128. Cho JY, Kim AR, Park MH (2001) Lignans from the rhizomes of *Coptis japonica* differentially act as anti-inflammatory principles. Planta Med 67:312–316
129. Saleem M, Kim HJ, Ali MS, Lee YS (2001) An update on bioactive plant lignans. Nat Prod Rep 22:696–716
130. Bin L, Lee YJ, Kim YC, Yoon JJ, Lee SM, Lee YP, Kang DG, Lee HS (2013) Sauchinone from *Saururus chinensis* protects vascular inflammation by heme oxygenase-1 induction in human umbilical vein endothelial cells. Phytomedicine 21(2):101–108
131. Han HJ, Li M, Son JK, Seo CS, Song SW, Kwak SH, Bae HB (2013) Sauchinone, a lignan from *Saururus chinensis,* attenuates neutrophil pro-inflammatory activity and acute lung injury. Int Immunopharmacol 17(2):471–477
132. Prieto JM, Giner RM, Recio MC, Schinella G, Manez S, Rios JL (2002) Diphyllin acetylapioside, a 5-lipoxygenase inhibitor from *Halophyllum hispanicum.* Planta Med 68:359–360
133. Kim SJ, Lu Y, Kwon O, Hwangbo K, Seo CS, Lee SH, Kim CH, Chang YC, Son JK, Chang HW (2011) Manassantin A isolated from *Saururus chinensis* inhibits 5-lipoxygenase-dependent leukotriene C4 generation by blocking mitogen-activated protein kinase activation in mast cells. Biol Pharm Bull 34(11):1769–1772
134. Kwon OE, Lee HS, Lee SW, Chung MY, Bae KHB, Rho MC, Kim YK (2005) Manassantin A and B isolated from *Saururus chinensis* inhibit TNF-induced cell adhesion molecule expression of human umbilical vein endothelial cells. Arch Pharm Res 28(1):55–60
135. Hwang BY, Lee J-H, Nam JB, Hong Y-S, Lee JJ (2003) Lignans from *Saururus chinensis* inhibiting the transcription factor NF-kB. Phytochemistry 64:765–771
136. Park HC, Bae HB, Jeong CW, Lee SH, Jeung HJ, Kwak SH (2012) Effect of manassantin B, a lignan isolated from *Saururus chinensis,* on lipopolysaccharide-induced interleukin-1β in RAW 264.7 cells. Korean J Anesthesiol 62(2):161–165
137. Zschocke S, van Staden J, Paulus K, Bauer R, Horn MM, Munro OQ, Brown NJ, Drewes SE (2000) Stereostructure and anti-inflammatory activity of three diastereomers of ocobullenone from *Ocotea bullata*. Phytochemistry 54:591–595
138. Zhao F, Wang L, Liu K (2009) In vitro anti-inflammatory effects of arctigenin, a lignan from *Arctium lappa* L., through inhibition on iNOS pathway. J Ethnopharmacol 122(3):457–462

139. Hyam SR, Lee IA, Gu W, Kim KA, Jeong JJ, Jang SE, Han MJ, Kim DH (2013) Arctigenin ameliorates inflammation in vitro and in vivo by inhibiting the PI3 K/AKT pathway and polarizing M1 macrophages to M2-like macrophages. Eur J Pharmacol 708(1–3):21–29
140. Lee JY, Kim CJ (2010) Arctigenin, a phenylpropanoid dibenzylbutyrolactone lignan inhibits type I-IV allergic inflammation and pro-inflammatory enzymes. Arch Pharm Res 6(33):947–957
141. Weng JR, Tsao LT, Wang JP, Wu RR, Lin CN (2004) Lin CN anti-inflammatory phloroglucinols and terpenoids from *Garcinia subelliptica*. J Nat Prod 67(11):1796–1799
142. Chehl N, Chipitsyna G, Gong Q, Yeo CJ, Arafat HA (2009) Anti-inflammatory effects of the *Nigella sativa* seed extract, thymoquinone, in pancreatic cancer cells. HPB (Oxford) 11(5):373–381
143. Abourashed EA, Muhammad I, Resch M, Bauer R, El-Feraly FS, Hufford CD (2001) Inhibitory effects of maesanin and analogues on arachidonic acid metabolizing enzymes. Planta Med 67:360–361
144. Calixto JB, Campos MM, Otuki MF, Santos ARS (2004) Anti-inflammatory compounds of plant origin. Part II. Modulation of pro-inflammatory cytokines, chemokines and adhesion molecules. Planta Med 70:93–103
145. Kang BK, Chung SW, Kim TS (2001) Inhibition of interleukin-12 production in lipopolysaccharide activated mouse macrophages by hypericin, an active component of *Hypericum perforatum*. Planta Med 67:364–366
146. Wube AA, Bucar F, Asres K, Gibbons S, Adams M, Streit B, Bodensieck A, Bauer R (2006) Knipholone, a selective inhibitor of leukotriene metabolism. Phytomedicine 13:452–456
147. Marsik P, Kokoska L, Landa P, Nepovim A, Soudek P, Vanek T (2005) In vitro effects of thymol and quinones of *Nigella sativa* seeds on cyclooxygenase-1 and -2 catalysed prostaglandin E2 biosynthesis. Planta Med 71:739–742
148. Meng G, Liu Y, Lou C, Yang H (2010) Emodin suppresses lipopolysaccharide-induced pro-inflammatory responses and NF-κB activation by disrupting lipid rafts in CD14-negative endothelial cells. Br J Pharmacol 161(7):1628–1644
149. Aggarwal BB, Shishodia S (2006) Molecular targets of dietary agents for prevention and therapy of cancer. Biochem Pharmacol 71:1397–1421
150. Folmer F, Jaspars M, Dicato M, Diederich M (2008) Marine natural products as targeted modulators of the transcription factor NF-kB. Biochem Pharmacol 75:603–617
151. Huang Q, Shen H-M, Ong C-N (2004) Inhibitory effect of emodin on tumor invasion through suppression of activator protein-1 and nuclear factor-kB. Biochem Pharmacol 68:361–371
152. Matsui T, Ito C, Itoigawa M, Okada T, Furukawa H (2007) Anti-inflammatory activity of phenylpropanoids and plastoquinoids from *Illicium species* in RBL-2H3 cells. Planta Med 73:662–665
153. Lee JY, Woo E-R, Kang KW (2005) Inhibition of lipopolysaccharide-inducible nitric oxide synthase expression by acteoside through blocking of AP-1 activation. J Ethnopharmacol 97:561–566
154. Shrimali D, Shanmugam MK, Kumar AP, Zhang J, Tan BK, Ahn KS, Sethi G (2013) Targeted abrogation of diverse signal transduction cascades by emodin for the treatment of inflammatory disorders and cancer. Cancer Lett 341(2):139–149. doi:10.1016/j.canlet.2013.08.023(Epub ahead of print)
155. Hartley L, Igbinedion E, Holmes J, Flowers N, Thorogood M, Clarke A, Stranges S, Hooper L, Rees K (2013) Increased consumption of fruit and vegetables for the primary prevention of cardiovascular diseases. Cochrane Database Syst Rev. doi:10.1002/14651858.CD009874.pub2
156. Nguelefack TB, Dongmo AB, Dimo T, Kamanyi A (2007) Phytopharmacology of some medicinal plants used in cameroonian traditional medicine to handle cardiovascular diseases. In: Capasso A (ed) Recent developments in medicinal plants research. Research Signpost, Trivandrum, Kerala, India, ISBN:978-81-308-0160-5

157. Miller AL (1998) Botanical influences on cardiovascular disease. Altern Med Rev 3(6):422–431
158. Luna-Vázquez FJ, Ibarra-Alvarado C, Rojas-Molina A, Rojas-Molina I, Zavala-Sánchez MA (2013) Vasodilator compounds derived from plants and their mechanisms of action. Molecules 18(5):5814–5857
159. Chiou WF, Chou CJ, Liao JF, Sham AY, Chen CF (1994) The mechanism of the vasodilator effect of rutaecarpine, an alkaloid isolated from Evodia rutaecarpa. Eur J Pharmacol 257:59–66
160. Oh KS, Ryu SY, Kim YS, Lee BH (2007) Large conductance Ca2+-activated K+(BKCa) channels are involved in the vascular relaxations elicited by piceatannol isolated from Rheum undulatum rhizome. Planta Med 73:1441–1446
161. Alvarez E, Campos-Toimil M, Justiniano-Basaran H, Lugnier C, Orallo F (2006) Study of the mechanisms involved in the vasorelaxation induced by (−)-epigallocatechin-3-gallate in rat aorta. Br J Pharmacol 147:269–280
162. de Sánchez Rojas VR, Somoza B, Ortega T, Villar AM (1996) Different mechanisms involved in the vasorelaxant effect of flavonoids isolated from Satureja obovata. Planta Med 62:554–556
163. Ojewole JA (2008) Cardiovascular effects of mollic acid glucoside, a 1alpha-hydroxycycloartenoid saponin extractive from Combretum molle R Br ex G Don (Combretaceae) leaf. Cardiovasc J Afr 19:128–134
164. Kawakami K, Aketa S, Sakai H, Watanabe Y, Nishida H, Hirayama M (2011) Antyhipertensive and vasorelaxant effects of water-soluble proanthocyanidins from persimmon leaf tea in spontaneously hypertensive rats. Biosci Biotechnol Biochem 75:1435–1439
165. Wang Y, Shi JG, Wang MZ, Che CT, Yeung JH (2009) Vasodilatory actions of xanthones isolated from a Tibetan herb Halenia elliptica. Phytomedicine 16:1144–1150
166. Luna-Vázquez FJ, Ibarra-Alvarado C, Rojas-Molina A, Rojas-Molina I, Zavala-Sánchez MA (2013) Vasodilator compounds derived from plants and their mechanisms of action. Molecules 18(5):5814–5857
167. Somova LI, Shode FO, Moodley K, Govender Y (2001) Cardiovascular and diuretic activity of kaurene derivatives of *Xylopia aethiopica* and *Alepidea amatymbica*. J Ethnopharmacol 77(2–3):165–174
168. Zaima K, Koga I, Iwasawa N, Hosoya T, Hirasawa Y, Kaneda T, Ismail IS, Lajis NH, Morita H (2013) Vasorelaxant activity of indole alkaloids from Tabernaemontana dichotoma. J Nat Med 67:9–16
169. Pote W, Tagwireyi D, Chinyanga HM, Musara C, Nyandoro G, Chifamba J, Nkomozepi P (2013) Cardiovascular effects of *Boophone disticha* aqueous ethanolic extract on early maternally separated BALB/C mice. J Ethnopharmacol 148(2):379–385
170. Wang Y, Shi JG, Wang MZ, Che CT, Yeung JH (2009) Vasodilatory actions of xanthones isolated from a Tibetan herb Halenia elliptica. Phytomedicine 16:1144–1150
171. Gonçalves RL, Lugnier C, Keravis T, Lopes MJ, Fantini FA, Schmitt M, Cortes SF, Lemos VS (2009) The flavonoid dioclein is a selective inhibitor of cyclic nucleotide phosphodiesterase type 1 (PDE1) and a cGMP-dependent protein kinase (PKG) vasorelaxant in human vascular tissue. Eur J Pharmacol 620:78–83
172. Pawlaczy I, Czerchawski L, Pilecki W, Zarawska EL, Gancarz R (2009) Polyphenolic-polysaccharide compounds from selected medicinal plants of Asteraceae and Rosaceae families: chemical characterization and blood anticoagulant activity. Carbohydr Polym 77:568–575
173. Circosta C, Pasquale RD, Occhiuto F (2005) Cardiovascular effects of the aqueous extract of *Gynostemma pentaphyllum Makino*. Phytomedicine 12:638–643
174. Rossler MF, Pyton CP, Capponi AM et al (1993) Blocking T type calcium channels with tetrandrine inhibits steroidgenesis in bovine adrenal glomerulosa cells. Endocrinology 132:1035–1043

175. Jing-Shan SHI, Jun-Xian YU, Xiu-Ping CHEN, Rui-Xia XU (2003) Pharmacological actions of *Uncaria* alkaloids, rhynchophylline and isorhynchophylline. Acta Pharmacol Sin 24(2):97–101
176. Jayakumar T, Sheu JR (2011) Cardiovascular pharmacological actions of rutaecarpine, a quinazolinocarboline alkaloid isolated from Evodia rutaecarpa. J Exp Clin Med 3(2):63e69
177. Zhang W, Wojta J, Binder BR (1994) Effect of notoginsenoside R1 on the synthesis of tissue-type plasminogen activator and plasminogen activator inhibitor-1 in cultured human umbilical vein endothelial cells. Arterioscler Thromb Vasc Biol 14:1040–1046
178. Sheykhzade M, Smajilovic S, Issa A, Haunso S, Christensen SB, Tfelt-Hansen J (2008) S-262 and butterbur lactones dilate vessels through blockage of voltage gated calcium channels and block DNA synthesis. Eur J Pharmacol 593:79–86
179. Guillaume M, Padioleall F (1994) Veinotoinc effect, vascular protection, antiinflammatory and free radical scavenging properties of horse chestnut extract. Arzneimitt Elforschung 44:25–35
180. Fusi F, Cavalli M, Mulholland D, Crouch N, Coombes P, Dawson G, Bova S, Sgaragli G, Saponara S (2010) Cardamonin is a bifunctional vasodilator that inhibits Ca(v)1.2 current and stimulates K(Ca)1.1 current in rat tail artery myocytes. J Pharmacol Exp Ther 332:531–540
181. Lin YL, Dai ZK, Lin RJ, Chu KS, Chen IJ, Wu JR, Wu BN (2010) Baicalin, a flavonoid from Scutellaria baicalensis Georgi, activates large-conductance Ca2+-activated K+channels via cyclic nucleotide-dependent protein kinases in mesenteric artery. Phytomedicine 17:760–770
182. Lzabela P, Czerchawski, Leszek, Kulliezkowski et al (2011) Anticoagulent and antiplatelet activity of polyphenolic polysaccharide preparation isolated from themedicinal plant *Erigeron Canadensis* L.Throm Res 127(4):328–340
183. Medeiros MA, Pinho JF, De-Lira DP, Barbosa-Filho JM, Araújo DA, Cortes SF, Lemos VS, Cruz JS (2011) Curine, a bisbenzylisoquinoline alkaloid, blocks L-type Ca2+ channels and decreases intracellular Ca2+ transients in A7r5 cells. Eur J Pharmacol 669:100–107
184. Deng R (2007) Therapeutic effects of guggul and its constituent guggulsterone: cardiovascular benefits. Cardiovasc Drug Rev 25(4):375–390
185. Maulik SK, Katiyar CK (2010) *Terminalia arjuna* in cardiovascular diseases: making the transition from traditional to modern medicine in India. Curr Pharm Biotechnol 11(8): 855–860
186. Dwivedi S (2007) *Terminalia arjuna* Wight & Arn.—a useful drug for cardiovascular disorders. J Ethnopharmacol 114(2):114–129
187. Ojha S, Nandave M, Kumara S, Dharmavhir S (2010) Cardioprotective activity of *Inula racemosa* hook in experimental model of myocardial ischemia reperfusion injury. Ind J Exp Biol 48:918–924
188. Chabukswar AR, Kuchekar BS, Jagdale SC, Lokhande PD, Raut CG (2010) Cardio protective activity of *Inula racemosa*. Int J Chem Sci 8(3):1545–1552
189. Fathiazad F, Matlobi A, Khorrami A, Hamedeyazda S, Soraya H, Hammami M, Maleki-Dizaji N, Garjani A (2012) Phytochemical screening and evaluation ofcardioprotective activity of ethanolic extract of *Ocimum basilicum* L. (basil) against isoproterenol induced myocardial infarction in rats. DARU J Pharm Sci 20:87
190. Panda VS, Naik SR (2009) Evaluation of cardioprotective activity of *Ginkgo biloba* and *Ocimum sanctum* in rodents. Altern Med Rev 14(2):161–171
191. Zanwar AA, Hegde MV, Bodhankar SL (2011) Cardioprotective activity of flax lignin concentrate extracted from seeds of *Linum usitatissimum* in isoprenalin induced myocardial necrosis in rats. Interdiscip Toxicol 4(2):90–97
192. Rahman K, Lowe GM (2006) Garlic and cardiovascular disease: a critical review. J Nutr 136:736S–740S
193. Senthilkumar GP, Moses MF, Sengottuvenlu S, Rajarajan T, Sarvanan G (2010) Cardioprotective activity of garlic *(Allium sativum)* in isoproterenol induced rat myocardial necrosis: a biochemical and histoarchitectural evaluation. Int J Pharm Sci Nanotech 2(4):779–784

194. Sakthivel K, Palani S,Kalash RS, Devi K, Kumar BS (2010) Phytoconstituents analysis by GC-MS, cardioprotective and antioxidant activity of *Buchanania Axillaris* against doxorubicin-induced cardio toxicity in albino rats. Int J Pharm Stud Res 1:34–48
195. Thippeswamy BS, Thakker SP, Tubachi S, Kalyani GA, Netra MK, Patil U, Desai S, Gavimath CC, Veerapur VP (2009) Cardioprotective effect of Cucumis trigonus Roxb on isoproterenol-induced myocardial infarction in rat. Am J Pharmacol Toxicol 4(2):29–37
196. Ojha S, Jagriti B, Sachin A, Golechha M, Kumari S (2011) Cardioprotective effects of *Commiphora mukul* againstisoprenaline-induced cardiotoxicity: a biochemical and histopathological evaluation. J Environ Biol 32:731–738
197. Hina S, Rehman KU, Dogar ZUH, Jahan N, Hameed M, Khan ZI, Ahmad K, Mukhtar K, Valeem EE (2010) Cardioprotective effect of gemmotherapeuticallytreated *withania somnifera* against chemically induced myocardial injury. Pak J Bot 42(2):1487–1499
198. Rajendran R, Basha NS (2008) Cardioprotective effect of ethanol extract of stem-bark and stem-wood of *Premna serratifolia* Lin., (Verbenaceae) Res J Pharm Tech 1(4):487–491
199. Muralidharan P, Balamurugan G, Kumar P (2008) Inotropic and cardioprotective effects of *Daucus carota* Linn. On isoproterenol-induced myocardial infarction. Bangladesh J Pharmacol 3:74–79
200. Firoz M, Bharatesh K, Nilesh P, Vijay G, Tabassum S, Nilofar N (2011) Cardioprotective activity of ethanolic extract of *Callistemon lanceolatus* leaves on doxorubicin-induced cardiomyopathy in rats. Bangladesh J Pharmacol 6:38–45
201. Eman M El-Sayed1, Amal S Abd El-azeem1, Abeer A Afify1, Manal H Shabana, Ahmed HH (2011) Cardioprotective effects of *Curcuma longa L.* extracts against doxorubicin-induced cardiotoxicity in rats. J Med Plants Res 5(17):4049–4058
202. Siddiq A, Shanmukha I, Jyoti TM, Gupt K (2012) Cardioprotective effect of *Spathodea campanulata* bark on isoproterenol-induced myocardial infarction in rats. Asian Pac J Trop Dis 2:S1–S5
203. Ramadoss S, Kannan K, Balamurugan K, Jeganathan NS, Manavalan R (2012) Cardioprotective effect of *Cyathula prostrata* Linn on isoproterenol induced myocardial infarction in rat. Int J Res Pharm Biomed Sci 3(2):551–555
204. Raju K, Balaraman R, Hariprasad A, Kumar MV, Ali A (2008) Cardioprotective effect of *Momordica Cymbalaria* fenzl in rats with isoproterenol-induced myocardial injury. J Clin Diagn Res 2(1):699–705
205. Girija K, Lakshman K (2011) Anti-hyperlipidemic activity of methanol extracts of three plants of Amaranthus in triton-WR 1339 induced hyperlipidemic rats. Asian Pac J Trop Biomed 1(1):S62–S65
206. Israni DA, Patel KV, Gandhi TR (2010) Anti-hyperlipidemic activity of aqueous extract of *Terminalia Chebula* & Gaumutra in high cholesterol diet fed rats. Int J Pharm Sci 1(1):48–59
207. Lakshmi BVS, Neelima N, Kasthuri N, Umarani V, Sudhakar M (2011) Antihyperlipidemic activity of *Bauhinia purpurea* extracts in hypercholesterolemic albino rats. Int J Pharm Tech Res 3(3):1265–1272
208. Jeyabalan S, Palayan M (2009) Antihyperlipidemic activity of *Sapindus emarginatus* in Triton WR-1339 induced albino rats. Res J Pharm Tech 2(2):319–323
209. Bhuvaneswari R, Sasikumar K (2013) Antihyperlipidemic activity of *Aegle marmelos* (l) corr., leaf extract in triton wr- 1339 induced hyperlipidemic rats. Pharmacie Globale (IJCP) 4(3):1
210. Kamalraj R (2011) Antihyperlipidemic studies on leaf extract of *Erythrina Indica* Lam IJRPC 1(3):488–490
211. Dubey AK, Devi A, Kutty G, Shankar RP (2005) Hypolipidemic activity of *Ginkgo biloba* extract, EGb 761 in hypercholesterolemic wistar rats. IJPT 4:9–12
212. Sumbul S, Ahmed SI (2012) Anti-hyperlipidemic activity of *Carissa carandas* (Auct.) leaves extract in egg yolk induced hyperlipidemic rats J Basic Appl Sci 8:124–134
213. Rajendran R, Krishnakumar E (2010) Hypolipidemic activity of chloroform extract of *Mimosa pudica* leaves. Avicenna J Med Biotech 2(4):215–221

214. Khuda-Bukhsh AR, Das S, Saha SK (2014) Molecular approaches toward targeted cancer prevention with some food plants and their products: inflammatory and other signal pathways. Nutr Cancer 66(2):194–205
215. Hostettmann K, Marston A, Ndjoko K, Wolfender JL (2000) The potential of African plants as source of drugs. Curr Org Chem 4:973–1010
216. Schnekenburger M et al (2014) Plant-derived epigenetic modulators for cancer treatment and prevention. Biotechnol Adv 29. doi:10.1016/j.biotechadv.2014.03.009. (Epub ahead of print)
217. Feng Yibin, Wang Ning, Zhu Meifen, Feng Yigang, Li Hongyun, Tsao Saiwah (2011) Recent progress on anticancer candidates in patents of herbal medicinal products. Recent patents on food. Nutr Agric 3:30–48
218. Dennis JW, Shah RN, Ziser L (2002) Alkaloid halide salts of swainsonine and methods of use. US6395745
219. Rosen, RTHo, Dipaola RS, Rafi MM, Vastano BC,Ghai G (2002) Use of 1- propanone-1-(2,4-dihydroxyphenyl)-3-hydroxy- 3-(4'-hydroxyphenyl) as an anti- carcinogenic agent. US6498195
220. Hahm JC, Lee DS, Ko JP, Lee IK, Lee HW, Park JS (2004) Pharmaceutical composition containing the extract of *Saururus chinensis* baill useful as an anticancer agent and a process for the preparation thereof. US20040024055
221. Aylward JH (2004) Anti-cancer compounds. US6787161
222. Wang L, Liu X, Chen R (2005) Derivatives of isoindigo, indigo andindirubin and methods of treating cancer. US6933315
223. Zhang MQ, Sheridan RM (2007) 39-desmethoxyrapamycin, compositions and methods of use thereof. US7183289
224. Khanuja SPS, Tiruppadiripuliyur RSK, Gupta VK, Chand P,Garg A, Srivastava SK, Verma SC, Saikia D, Darokar MP, Shasany AK, Pal A (2007) Antimicrobial an anticancer properties of methyl-beta-orcinolcarboxylate from lichen. US20070099993
225. Maung TW (2007) Berberine as a selective lung cancer agent. US20070298132
226. Qazi GN, Taneja SC, Singh J, Saxena AK, Sethi VK, Shah BA, Kapahi BK, Andotra SS, Kumar A, Bhushan S, Malik F, Mondhe DM, Muthiah S, Singh S, Verma M, Singh SK (2009) Use of semi synthetic analogues of boswellic acids for anticancer activity. US20090298938
227. Han QB, Song JZ, Qiao CF, Yang L, Xu HX (2009) Compounds from *Garcinia hanburyi*, their use in treating cancer and method of separating epimers thereof. US7592367
228. Chan PK, Mak MS, Wang Y (2009) Anticancer biangeloyl saponins. US7514412
229. Banerjee P, Raja KS (2010) Use of curcumin to block brain tumor formation in mice. US20100197584
230. Zhang S, Bao Y, Sun Y, Li K, Zou L, Ma J, Sun X, Shang H, Li J (2010) Cyclin-dependent protein kinases inhibitor of *Scutellaria* flavonoid organic amine derivatives, synthesis and use there of. US20100197619
231. Dai D, Musser JH, Lennox ES (2007) Triptolide derivatives for modulation of apoptosis and immunosuppression. US7662976
232. Pei H, Guo Q, Chen B, Zhu H (2002) *Fagopyrum cymosum* (Trev.) Meisn composition, method to prepare and analyze the same and uses thereof.US6451353
233. Fabre P, Raynaud JP, Cousse H (2003) Use of a *Serenoa repens* extract for the production of a medicament to treat prostate cancer. US6599540
234. Bassa BV (2003) Antitumor agent. US6660309
235. Lin CH (2003) *Euphorbia antiquorum* extract, a pharmaceutical composition containing the same and methods for treatment of cancers. US20030165579
236. Wu RT (2003) Polysaccharide-based extract from *Ganoderma*, pharmaceutical use thereof, and process for preparing the same. US6613754
237. Kwon BM, Son KH, Han DC, Kim JH, Kang HM, Jeon SB (2004) Artemisolide compound isolated from the aerial parts of *Artemisia sylvatica*, isolation method, and use thereof. US6808724

238. Ryu SY, Lee CO, Choi SU, Park SH, Kim YS, Kim SK, Kim SK, Kang SK (2004) Anticancer composition comprising sesquiterpenes isolated from *Resina ferulae*. US20040043083
239. Kuok KY, Ly H (2004) Herbal compositions for prostate conditions. US6790464
240. Lee SH, Lee YC, Jun YC, Seo JK, Noh JS (2005) Medicines manufactured from old *platycodon* extracts. US6902748
241. Wang ZP (2005) Saponins as anticancer agent. US20050175623
242. Kim MY (2004) Composition comprising Melissa leaf extract for antiangiogenic and matrix metalloproteinase inhibitory activity. US20040009244
243. Page TK (2003) Herbal formulation. US6649185
244. Kwak TH, Shin MS, Kim JY, Park JK (2006) Active fraction having anti-cancer and anti-metastasis isolated from leaves and stems of ginseng. US20060034951
245. Saxena AK, Gupta BD, Kapahi BK, Muthiah S, Mondhe DM, Baleshwar N, Qazi GN, Kumar V, Mathan G (2006) Pharmaceutical composition useful for the treatment of hepatocellular carcinoma. US20060280817
246. Shyur LF, Yang NS, Kang PL, Sun SJ, Wang SY (2006) Use of *Anoectochilus formosanus* plant extracts and their derived fractions as herbal medicines or nutraceutical supplements for chemoprevention or treatment of human malignancies. US7033617
247. Kuo KW (2006) Water soluble extract from plant of *Solanum* genus and the preparation process thereof, and pharmaceutical composition containing the water soluble extract. US7078063
248. Arellano EC (2007) Anti-prostate cancer composition and therapeutic uses therefor. US7250180
249. Rao JM, Srinivas PV, Yadav JS, Raghavan KV, Saxena AK, Shamugavel H, Qazi GN (2007) Herbal chemical composition for the treatment of cancer. US7285571
250. Mitra SK, Saxena E, Dixit MN, Uddagiri VB, Marikunte VR, Mathad SA, Shanbhag SV (2008) Novel anticancer agent, methods for obtaining the same and pharmaceutical compositions hereof. US20080199550
251. Chu KH (2008) Method of inducing apoptosis in cancer treatment by using Cucubritacins. US20080207578
252. Cohen I (2009) Anticancer methods employing extracts of *Gleditsia sinensis* lam. US20090258096
253. Choi WC, Park SJ, Kwon SP (2009) Process for preparation of *Rhus verniciflua* extracts having excellent anti-cancer activity and anticancer pharmaceutical composition containing the same. US7618661
254. Cohen I (2010) Anticancer methods using extracts of *Anemarrhena asphodeloides* bunge. US20100009017
255. Joshi KS, Wagh V, Sharma S (2010) Herbal composition and process for its preparation. US20100028472
256. Greenway FL (2010) Angiogenic agents from plant extracts, gallic acid, and derivatives. US7709031
257. Osbourn AE (1996) Preformed antimicrobial compounds and plant defense against fungal attack. Plant Cell 8:1821–1831
258. Sultana N (2011) Clinically useful anticancer, antitumor, and antiwrinkle agent, ursolic acid and related derivatives as medicinally important natural product. J Enzyme Inhib Med Chem 26:616–642
259. Shahid M, Shahzad A, Sobia F et al (2009) Plant natural products as a potential source for antibacterial agents: recent trends. AntiInfective Agents Med Chem 8:211–225
260. Assob JC, Kamga HL, Nsagha DS (2011) Antimicrobial and toxicological activities of five medicinal plant species from Cameroon Traditional Medicine. BMC Complement Altern Med 11:70–81
261. Arruda AL, Vieira CJ, Sousa DG, Oliveira RF, Castilho RO (2011) *Jacaranda cuspidifolia* Mart. (Bignoniaceae) as antibacterial agent. J Med Food 14:1604–1608

262. Ciocan ID, Băra II (2007) Plant products as antimicrobial agents. Analele Ştiinţifice ale Universităţii „Alexandru Ioan Cuza", Secţiunea Genetică şi Biologie Moleculară, TOM VIII
263. Saleem M, Nazir M, Ali MS et al (2010) Antimicrobial natural products: an update on future antibiotic drug candidates. Nat Prod Rep 27:238–254
264. Kurek A, Grudniak AM, Kraczkiewicz-Dowjat A, Wolska KI (2011) New antibacterial therapeutics and strategies. Pol J Microbiol 60:3–12
265. Cowan MM (1999) Plant product as antimicrobial agents. Clin Microbiol Rev 12:564–582
266. Fowler ZL, Baron CM, Panepinto JC, Koffas MA (2011) Melanization of flavonoids by fungal and bacterial laccases. Yeast 28:181–188
267. Araujo MG, Hilario F, Nogueira LG, Vilegas W, Santos LC, Bauab TM (2011) Chemical constituents of the methanolic extract of leaves of *Leiothrix spiralis* Ruhland and their antimicrobial activity. Molecules 16:10479–10490
268. Garcia A, Bocanegra-Garcia V, Palma-Nicolas JP, Rivera G (2012) Recent advances in antitubercular natural products. Eur J Med Chem 49:1–23
269. Critchfield JW, Butera ST, Folks TM (1996) AIDS Res Hum Retroviruses 12:39
270. Johann S, Sà NP, Lima LA et al (2010) Antifungal activity of schinol and a new biphenyl compound isolated from *Schinus terebinthifolius* against the pathogenic fungus *Paracoccidioides brasiliensis*. Ann Clin Microbiol Antimicrob 9:30
271. Engels C, Schieber A, Gänzle MG (2011) Inhibitory spectra and modes of antimicrobial action of gallotannins from mango kernels (*Mangifera indica* L.). Appl Environ Microbiol 77:2215–2223
272. Cowan MM (1999) Plant products as antimicrobial agents. Clin Microbiol Rev 12:564–582
273. Garcia A, Bocanegra-Garcia V, Palma-Nicolas JP, Rivera G (2012) Recent advances in antitubercular natural products. Eur J Med Chem 49:1–23
274. Tian J, Ban X, Zeng H et al (2012) The mechanism of antifungal action of essential oil from dill. (*Anethum graveolens*) on *Aspergillus flavus*. PLoS ONE 7:e30147
275. Termentzi A, Fokialakis N, Skaltsounis AL (2011) Natural resins and bioactive natural products thereof as potential antimicrobial agents. Curr Pharm Des 17:1267–1290
276. Paraschos S, Mitakou S, Skaltsounis AL (2012) Chios gum mastic: a review of its biological activities. Curr Med Chem 19:2292–2302
277. Wangchuk P, Keller PA, Stephen GP et al (2011) Evaluation of an ethnopharmacologically selected Bhutanese medicinal plants for their major classes of phytochemicals and biological activities. J Etnopharmacol 137:730–742
278. Yigit D, Yigit N, Ozgen U (2009) An investigation on the anticandidal activity of some traditional medicinal plants in Turkey. Mycoses 52:135–140
279. Agarwal V, Lal P, Pruthi V (2008) Prevention of Candida albicans biofilm by plant oils. Mycopathologia 165:13–19
280. Pietrella D, Angiolella L, Vavala E et al (2011) Beneficial effect of *Mentha suaveolens* essential oil in the treatment of vaginal candidiasis assessed by real-time monitoring of infection. BMC Complement Altern Med 11:18
281. Tomczychowa M, Leszczynsk K, Tomczyk M, Tryniszewska E, Kelemba D (2011) Composition of the essential oil of *Bidens tripartita* L. roots and its antibacterial and antifungal activities. J Med Food 14:428–433
282. Silva F, Ferreira S, Duarte A, Mendonca DI, Domingues FC (2011) Antifungal activity of *Coriandrum sativum* essential oil, its mode of action against Candida species and potential synergism with amphotericin A. Phytomedicine 19:42–47
283. Oliva Mde L, Carezzano ME, Gallucci MN, Demo MS (2011) Antimycotic effect of the essential oil of *Aloysia triphylla* against Candida species obtained from human pathologies. Nat Prod Commun 6:1039–1043
284. Shafagha A, Shafaghatlonbar M (2011) Antimicrobial activity and chemical constituents of the essential oils from flower, leaf and stem of *Gypsophila bicolor* from Iran. Nat Prod Commun 6:275–276

285. Zuzarte M, Goncalves MJ, Cavaleiro C et al (2011) Chemical composition and antifungal activity of the essential oils of *Lavandula viridis* L'Her. J Med Microbiol 60:612–618
286. Nazaruk J, Karna E, Wieczorek P, Sacha P, Tryniszewska E (2010) In vitro antiproliferative and antifungal activity of essential oils from *Erigeron acris* L. and E. annuus (L). Pers Z Naturforsch 65:642–646
287. Huang Y, Zhao J, Zhou L et al (2010) Antifungal activity of the essential oil of *Illicium verum* fruit and its main component trans-anethole. Molecules 15:7558–7569
288. Pietrella D, Angiolella L, Vavala E et al (2011) Beneficial effect of Mentha suaveolens essential oil in the treatment of vaginal candidiasis assessed by real-time monitoring of infection. BMC Complement Altern Med 11:18
289. Khan MS, Ahmad I (2012) Antibiofilm activity of certain phytocompounds and their synergy with fluconazole against Candida albicans biofilms. J Antimicrob Chemother 67:618–621
290. Ghoshal S, Prasad BNK, Lakshmi V (1996) Antiamoebicactivity of *Piper longum* fruits against Entamoeba histolytica in vitro and in vivo. J Ethnopharmacol 50:167–170
291. Omulokoli E, Khan B, Chhabra SC (1997) Antiplasmodial activity of four Kenyan medicinal plants. J Ethnopharmacol 56:133
292. Kim SH, Lee SJ, Lee JH, Sun WS, Kim JH (2002) Antimicrobial activity of 9-O-acyl and 9-O-alkylberberrubine derivatives. Planta Med 68:277–281
293. Iwasa K, Moriyasu M, Yamori T, Turuo T, Lee D, Wiegrebe V (2001) In vitro cytotoxicity of the protoberberine-type alkaloids. J Nat Prod 64:896–898
294. Yi ZB, Yu Y, Liang YZ, Zeng B (2007) Evaluation of the antimicrobial mode of berberine by LC/ESI-MS combined with principal component analysis. J Pharm Biomed Anal 44:301–304
295. Stotz HU, Waller F, Wang K (2013) Innate immunity in plants: the role of antimicrobial peptides. In: Hiemstra PS, Zaat SAJ (eds) Antimicrobial peptides and innate immunity, Springer, Basel
296. Zeitler B, Herrera Diaz A, Dangel A, Thellmann M, Meyer H et al (2013) De Novo Design of Antimicrobial Peptides for Plant Protection. PLoS ONE 8(8):e71687
297. Van KP, Proost P, Damme J, Coosemans J, Damine EJ, Peumans WJ (2002) Five disulfide bridges stabilize a hevein type antimicrobial peptide from the bark of spindle tree (*Euonymus europaeus* I). FEBS Lett 530:181–185
298. Koike M, Okamoto T, Tsuda S, Imai R (2002) A novel plant defensin like gene of winter wheat is specifically induced during cold acclimation. Biochem Biophys Res Commun 298:46–53
299. Flores T, Giron A, Diaz M, Flores HE (2002) Ocatin. A novel tuber storage protein from the Andean tuber crop oca with antibacterial and antifungal activities. Plant Physiol 128:1291–1302
300. Ye X, Ng TB (2002) Purification of angularin, a novel antifungal peptide from adzuki beans. J Pept Sci 8:101–106
301. Berrocal ML, Segura A, Moreno M, López G, García-Olmedo F, Molina A (2002) Snakin-2, an antimicrobial peptide from potato whose gene is locally induced by wounding and responds to pathogen infection. Plant Physiol 128(3):951–961
302. Tomie T, Ishibashi J, Furukawa S, Kobayashi S, Sawahata R, Asaoka A, Michito T, Yamakawab M (2003) Scarabaecin, a novel cysteine-containing antifungal peptide from the rhinoceros beetle, *Oryctes rhinoceros*. Biochem Biophys Res Commun 307:261–266
303. Fujimura M, Minami Y, Watanabe K, Tadera K (2003) Purification, characterization, and sequencing of a novel type of antimicrobial peptides, Fa-AMP1 and Fa-AMP2, from seeds of buckwheat *(Fagopyrum esculentum Moench.)*. Biosci Biotechnol Biochem 67(8): 1636–1642
304. Fujimura M, Ideguchi M, Minami Y, Watanabe K, Tadera K (2004) Purification, characterization, and sequencing of novel antimicrobial peptides, Tu-AMP 1 and Tu-AMP 2, from bulbs of tulip (*Tulipa gesneriana* L.). Biosci Biotechnol Biochem 68(3):571–577

305. Sharma N, Gruszewski HA, Park SW, Holm DG, Vivanco JM (2004) Purification and characterization of an isoform of patatin from late blight-resistant potatoes with antimicrobial activity against *Phytophthora infestans*. Plant Physiol Biochem 42:647–655
306. Egorov TA, Odintsova TI, Pukhalsky VA, Grishin EV (2005) Diversity of wheat antimicrobial peptides. Peptides 26:2064–2073
307. Xia L, NG T (2005) Isolation of alliumin, a novel protein with antimicrobial and antiproliferative activities from multiple-cloved garlic bulbs. Peptides 26(2):177–183
308. Wong JH, Ng TB (2005) Vulgarinin, a broad-spectrum antifungal peptide from haricot beans (*Phaseolus vulgaris*). Int J Biochem Cell Biol 37(8):1626–1632
309. Kim JY, Park SC, Kim MH, Lim HT, Park Y, Hahm KS (2005) Antimicrobial activity studies on a trypsin-chymotrypsin protease inhibitor obtained from potato. Biochem Biophys Res Commun 330:921–927
310. Santi-Gadelha T, Gadelha CAA, Aragão KS, Oliveira CC, Mota MRL, Gomes RC, Pires AF, Toyama MH, Toyama DO, Alencar NMN, Criddle DN, Assreuy AMS, Cavada BS (2006) Purification and biological effects of *Araucaria angustifolia* (Araucariaceae) seed lectin. Biochem Biophys Res Commun 350:1050–1055
311. Girish KS, Machiah KD, Ushanandini S, Harish Kumar K, Nagaraju S, Govindappa M et al (2006) Antimicrobial properties of a non-toxic glycoprotein (WSG) from *Withania somnifera* (Ashwagandha). J Basic Microbiol 46:365–374
312. Barbieri L, Polito L, Bolognesi A, Ciani M, Pelosi E, Farini V, Jha AK, Sharma N, Vivanco JM, Chambery A et al (2006) Ribosome-inactivating proteins in edible plants and purification and characterization of a new ribosome-inactivating protein from *Cucurbita moschata*. Biochim Biophys Acta 1760:783–792
313. Diz MS, Carvalho AO, Rodrigues R, Neves-Ferreira AG, da Cunha M et al (2006) Antimicrobial peptides from chili pepper seeds causes yeast plasma membrane permeabilization and inhibits the acidification of the medium by yeast cells. Biochim Biophys Acta 1760:1323–1332
314. Maruyama NIC, Mun M, Tatsuhara M, Sawada M, Ishimoto S, Utsumi (2006). Multiple vacuolar sorting determinants exist in soybean IIS globulin. Plant Cell 18:1253–1273
315. Oard VS, Enright FM (2006) Expression of the antimicrobial peptides in plants to control phytopathogenic bacteria and fungi. Plant Cell Rep 25:561–572
316. Wong JH, Ng TB (2006) An antifungal peptide from baby lima bean. Appl Microbiol Biotechnol 73(3):576–581
317. Roy S, Sadhana P, Begum M, Kumar S, Lodha ML, Kapoor HC (2006) Purification, characterization and cloning of antiviral/ribosome inactivating protein from Amaranthus tricolor leaves. Phytochemistry 67(17):1865–1873
318. Boleti AP, Freire MG, Coelho MB, Silva W, Baldasso PA, Gomes VM, Marangoni S, Novello JC, Macedo ML (2007) Insecticidal and antifungal activity of a protein from *Pouteria torta* seeds with lectin-like properties. J Agric Food Chem 55:2653–2658
319. Sawano Y, Miyakawa T, Yamazaki II, Tanokura M, Hatano K (2007) Purification, characterization, and molecular cloning of an antifungal protein gene from *Ginkgo biloba* seeds. Biol Chem 388:273–280
320. Ribeiro SFF, Carvalho AO, Cunha MD, Rodrigues R, Cruz LP, Melo VMM, Carvalho AO, Cunha MD, Rodrigues R, Cruz LP, Melo VMM, Vasconcelos IM, Melo EJT, Gomes VM (2007) Isolation and characterization of novel peptides from chilli pepper seeds: antimicrobial activities against pathogenic yeasts. Toxicon 50:600–611
321. Nema AK, Agarwal A (2011) Hepatoprotective prospective of herbal drugs and their vesicular carriers—a review. Int J Res Pharm Biomed Sci 2(2):360–374
322. Dixit PK, Mittal S (2013) A comprehensive review on Hepatoprotective herbal agents. Int J Pharm Sci Rev Res 20(1):165–170
323. Krishnaveni M, Mirunalini S (2010) Therapeutic potential of *Phyllanthus emblica* (amla): the ayurvedic wonder. J Basic Clin Physiol Pharmacol 21(1):93–105
324. Okamoto T, Hino O (2000) Drug development with hints from traditional Indian ayurveda medicine: hepatitis and rheumatoid as an example. Int J Mol Med 6(6):613–615

325. Kumar KTMS, Puia Z, Samanta SK, Barik R, Dutta A, Gorain B, Roy DK, Adhikari D, Karmakar S, Sen T (2012) The gastroprotective role of *Acanthus ilicifolius*—a study to unravel the underlying mechanism of anti-ulcer activity. Sci Pharm 80(3):701–717
326. Sen T, Ghosh TK, Chaudhuri AK (1993) Studies on the mechanism of anti-inflammatory and anti-ulcer activity of *Pluchea indica*–probable involvement of 5-lipooxygenase pathway. Life Sci 52(8):737–743
327. Sen T, Basu A, Chaudhuri AK (1998) Studies on the possible mechanism of the gastric mucosal protection by *Calotropis procera*–involvement of 5-lipoxygenase pathway. Fundam Clin Pharmacol 12(1):82–87
328. Hussain H, Al-Harrasi A, Abbas G, Rehman NU, Mabood F, Ahmed I, Saleem M, van Ree T, Green IR, Anwar S, Badshah A, Shah A, Ali I (2013) The genus *Pluchea*: phytochemistry, traditional uses, and biological activities. Chem Biodivers 10(11):1944–1971
329. Rozza AL, Pellizzon CH (2013) Essential oils from medicinal and aromatic plants: a review of the gastroprotective and ulcer-healing activities. Fundam Clin Pharmacol 27(1):51–63
330. Gadekar R, Singour PK, Chaurasiya PK, Pawar RS, Patil UK (2010) A potential of some medicinal plants as an antiulcer agents. Pharmacogn Rev 4(8):136–146
331. Schmeda-Hirschmann G, Yesilada E (2005) Traditional medicine and gastroprotective crude drugs. J Ethnopharmacol 100(1-2):61–66
332. Salimifar M, Fatehi-Hassanabad Z, Fatehi M (2013) A review on natural products for controlling type 2 diabetes with an emphasis on their mechanisms of actions. Curr Diabetes Rev 9(5):402–411
333. Li GQ, Kam A, Wong KH, Zhou X, Omar EA, Alqahtani A, Li KM, Razmovski-Naumovski V, Chan K (2012) Herbal medicines for the management of diabetes. Adv Exp Med Biol 771:396–413
334. DiNardo MM, Gibson JM, Siminerio L, Morell AR, Lee ES (2012) Complementary and alternative medicine in diabetes care. Curr Diab Rep 12(6):749–761
335. Hayasaka S, Kodama T, Ohira A (2012) Traditional Japanese herbal (kampo) medicines and treatment of ocular diseases: a review. Am J Chin Med 40(5):887–904
336. Hui H, Tang G, Go VLW (2009) Hypoglycemic herbs and their action mechanisms. Chin Med 4:11
337. Salimifar M, Fatehi-Hassanabad Z, Fatehi M (2013) A review on natural products for controlling type 2 diabetes with an emphasis on their mechanisms of actions. Curr Diabetes Rev 9(5):402–411
338. Achard F, Eva HD, Stibig HJ, Mayaux P, Gallego J, Richards T, Malingreau JP (2002) Determination of deforestation rates of the world's humid tropical forests. Science 297(5583):999–1002
339. Dold AP, Cocks ML (2002) The trade in medicinal plants in the Eastern Cape Province, South Africa. S Afr J Sci 98:11–12
340. Reid WV (1992) Conserving life's diversity. Environ Sci Technol 26:1090–1095
341. de Jonge VN, Elliot M, Orive E (2002) Causes, historical development, effects and future challenges of a common environmental problem: eutrophication. Hydrobiologia 475(476): 1–19
342. Reid WV et al (2005) Millennium ecosystem assessment. Ecosystems and human well-being: biodiversity synthesis. World Resources Institute, Washington, DC

Adv Biochem Eng Biotechnol (2015) 147: 111–135
DOI: 10.1007/10_2014_270

Published Online: 11 May 2014

Eco-Taxonomic Insights into Actinomycete Symbionts of Termites for Discovery of Novel Bioactive Compounds

D. İpek Kurtböke, John R. J. French, R. Andrew Hayes and Ronald J. Quinn

Abstract Termites play a major role in foraging and degradation of plant biomass as well as cultivating bioactive microorganisms for their defense. Current advances in "omics" sciences are revealing insights into function-related presence of these symbionts, and their related biosynthetic activities and genes identified in gut symbiotic bacteria might offer a significant potential for biotechnology and biodiscovery. Actinomycetes have been the major producers of bioactive compounds with an extraordinary range of biological activities. These metabolites have been in use as anticancer agents, immune suppressants, and most notably, as antibiotics. Insect-associated actinomycetes have also been reported to produce a range of antibiotics such as dentigerumycin and mycangimycin. Advances in genomics targeting a single species of the unculturable microbial members are currently aiding an improved understanding of the symbiotic interrelationships among the gut microorganisms as well as revealing the taxonomical identity and functions of the complex multilayered symbiotic actinofloral layers. If combined with target-directed approaches, these molecular advances can provide guidance towards the design of highly selective culturing methods to generate further information related to the physiology and growth requirements of these bioactive actinomycetes associated with the termite guts. This chapter provides an overview on the termite gut symbiotic actinoflora in the light of current advances in the

D. İ. Kurtböke (✉) · J. R. J. French
Faculty of Science, Health, Education and Engineering, University of the Sunshine Coast, Maroochydore DC QLD, 4558, Australia
e-mail: IKurtbok@usc.edu.au

R. A. Hayes
Department of Agriculture, Fisheries and Forestry, Ecosciences Precinct, Dutton Park, Brisbane QLD, 4102, Australia

R. J. Quinn
Eskitis Institute for Drug Discovery, Griffith University, Nathan QLD, 4111, Australia

"omics" science, with examples of their detection and selective isolation from the guts of the Sunshine Coast regional termite *Coptotermes lacteus* in Queensland, Australia.

Keywords Actinomycetes · Biodiscovery · Eco-taxonomy · Symbiosis · Termites

Contents

1 Introduction 112
2 Functional Basis of Symbiosis 114
2.1 Functional Chemistry 114
2.2 Actinomycete Symbionts of Termites 115
2.3 Actinomycete-Derived Chemicals in Termites 117
3 Diversity of Termite Microbial Symbionts and Actinoflora 122
3.1 Molecular Analysis of Uncultured Microflora and Metagenomic Approaches 122
3.2 Culturing the Representatives of Actinoflora 124
4 Future Prospects 128
References 129

1 Introduction

Specialized mutualistic relationships between hosts and microorganisms have taken place during co-evolution where microorganisms have been actively cultured in exchange for producing bioactive small molecules [96]. Since then symbionts have been proven to aid the physiological capabilities of their host by synthesizing essential metabolites promoting survival and, by harnessing the chemical potential of these microorganisms, hosts have been able to mediate lifestyle transitions, alter gene transcription, and combat pathogens and competitors [82, 90, 96, 134].

Symbionts often enhance their host's ability to acquire nutrients from the environment or provide the pathways for synthesis of essential organic compounds or for catabolism of molecules available in the environment [87]. One important example is the termite gut symbiosis with the first examples dating back to 20-million-year-old termite fossils preserved in amber [55, 132]. The symbionts in termite guts originate from all three Domains of life, namely the Eukarya (protists, yeasts, and other fungi), the Archaea, and Bacteria [7, 55, 64, 65] and contribute towards host defense and nutrition [55].

Termites feed predominantly on wood and litter tissue, which is rich in difficult-to-digest lignocelluloses, but deficient in vitamins and essential components for protein and fat synthesis. As a result termites are dependent on a beneficial symbiosis with a dense and diverse flora of microorganisms in the hindguts of the workers to digest lignocellulose and to acquire supplemental nutrition [1, 8, 55]. Wood-feeding termites live in a nest or colony with many thousands of individuals. The colony is akin to an extended family. Within this family, various groups of individuals have different functional roles according to a "caste system". The worker caste is the largest group (ca. 80–90 %), and workers are about the size of a grain of rice (ca. 3–4 mm long). They are generally white in color, have soft bodies, are wingless, and without composite eyes or ocelli. Workers perform cleaning, maintenance, and repair of the nest; gather food (cellulosic materials) and water; care for the queen; tend the eggs and subsequent young nymphal forms; and construct new tunnels and galleries. They ensure that all members of the nest have microbial exo- and endo symbionts that have a role in finding food sources and nestmate recognition [63].

The cockroach like termite ancestors most likely acquired symbionts during evolution by feeding on dead plant material colonized by microbes [38, 55, 93] and maintained their existence via coprophagy and trophallaxis among relatives [55, 93]. Symbionts then entered into a mutualistic beneficial relationship, supplementing nutrients and energy of their host, and in return, gaining a steady food supply and protection in the constant environment of the gut [55]. The mutualistic beneficial relationship of termites with intestinal symbionts has been suggested as one of the fundamental factors predisposing termites to a social lifestyle [55, 126]. Each worker termite must acquire an initial inoculum of symbionts from parents or nestmates after hatching, and again after each molt. Dependence on the symbionts therefore, requires extended parental care, group living, and overlapping generations setting the stage for eusocial behavior [55]. Increasing interdependency between hosts and symbionts, which were acquired via vertical transmission during co-evolution, resulted in continuity in the identity of the microorganisms transferred between generations [31, 55, 93].

One of the important classes of termite symbiotic microorganisms, the Actinobacteria, in particular the members of the order Actinomycetales in this class has recently become a focus to generate information on the rationale of functional chemistry during symbiosis. A collaborative effort to systematically explore such functional chemistry including the characterization of antimicrobial and volatile compounds from actinobacterial symbionts of termites has been established by the authors of this chapter. Their wealth of expertise was utilized in a complementary fashion (Actinomycetology, Termite Biology, Chemical Ecology, and Natural Product Chemistry) with a long-term objective to generate sound scientific understanding on the basis of termite and actinomycete symbiosis in Australian environments resulting in the production of a range of symbiont-derived and function-related bioactive compounds. This chapter is designed to provide an overview and insights into the rationale of the approach taken by the collaborators, in particular into the eco-taxonomical aspects of actinomycete symbionts not only aiding host vital biological and environmental functions but also becoming an important source for the discovery of novel bioactive compounds.

2 Functional Basis of Symbiosis

Insect endosymbionts have been placed into two groups: primary or secondary. The primary ones reside in specialized host cells called the bacteriome, and the associations are due to obligate metabolic needs between primary endosymbionts and their hosts [28, 31]. Secondary endosymbionts, on the other hand, are reported to be often facultative with a shorter co-evolutionary history with a single host species [31] and their occurrence may be sporadic [31]. These symbionts may not necessarily reside in specialized host tissues but may occur extracellularly in the hemocoel or in other body tissues including fat bodies [18], muscle, nervous tissue, or gut. Moreover, their occurrence can be in lower titers in comparison with primary endosymbionts [31].

Many traits of the holobiont mediated by symbiotic microbiota have thus far been identified revealing the roles of primary endosymbionts. These roles include aiding in digestion; other ecologically important traits such as increased host fitness, are frequently conferred by facultative secondary endosymbionts [31]. Husseneder et al. [56] detected a tight co-evolutionary link between termites and their gut flora that maintains a certain association of species and functional groups.

2.1 Functional Chemistry

Insect societies have collective defense mechanisms as a prominent characteristic, manifested by defendable nests and defensive adaptations of inhabitants [20, 66]. Termites display a range of anatomical and behavioral adaptations for mechanical defense. They were, however, also reported to possess efficient chemical weaponry, especially for the families Rhinotermitidae, Serritermitidae, and Termitidae [66]. Current termite taxonomy includes 2 suborders, 7 families, 21 subfamilies, and 16 tribes. The accepted families are: Termopsidae, Hodotermitidae, Mastotermitidae, Kalotermitidae, Heterotermitidae, Rhinotermitidae, and Termitidae [67]. In the past, termite taxonomy was based on morphological features. However, with the development of electron microscopy and DNA fingerprinting, the world of termites has been tremendously opened with molecular taxonomy and phylogenetic relationships inferred from mitochondrial COII and 16S sequences using polymerase chain reaction (PCR). Currently only the head of a termite worker or soldier is used as source tissue for identification purposes [67].

Soldiers are usually larger in size than the workers, also wingless, blind, and with powerful mandibles. They guard the nest site and protect foraging workers outside the nest from ants or other predatory insects. They are unable to feed themselves so are fed and groomed by workers. Like the workers, their life span is short in terms of years. They are also able to distinguish colony members and recognize termites from other nest colonies [63]. Chemical defense in soldiers is ensured by the exocrine glands, that is, the labial glands, the labral gland, and the frontal gland [122]. The labial gland evolved as a modification of glandular structures existing in the basic anatomical plan of insects, whereas the other two represent novel secretory organs, exclusive to termites [122]. A fascinating diversity of defensive chemicals produced by the frontal gland include alcohols, mono-, sesqui-, di-terpenoid hydrocarbons, ketoaldehydes, fatty acids, macrocyclic lactones, and heterocyclic and aromatic compounds. These defense chemicals are used as irritants, repellents, glues, antihealants, and contact poisons by the termites [66, 102, 103, 122]. In addition to their defensive function, volatiles from the frontal gland have also been suggested to be involved in signaling alarm by fighting or irritated soldiers [66, 122].

Defensive secretions of the frontal gland from termite soldiers were found to be a mixture of monoterpenes, sesquiterpenes, and diterpenes, with the latter being the most representative [22]. Analyses conducted on the dichloromethane extract from soldiers of the Brazilian termite, *Nasutitermes macrocephalus* (Blattaria: Nasutitermitinae), identified the presence of two monoterpenes (alpha-pinene and limonene) and two sesquiterpenes (beta-trans-caryophyllene and gamma-selinene), and the isolation of one rippertane and six trinervitane diterpenes with activity against antibiotic-resistant bacteria. Monoterpenes alpha-pinene and limonene have been reported to inhibit fungal growth by *Nasutitermes* soldiers [34, 122].

The functional chemistry of the frontal gland secretion was suggested to be co-evolved with structural aspects of the insect such as the anatomy of the glandular reservoir, the frontal pore, and the shape of the head and mandibles [66, 104, 122]. The related behavior used in association with defense chemicals exhibits a multitude of defensive strategies, ranging from contact discharge combined with mandibular biting, to non contact delivery by spraying [66, 122]. Thus, the chemistry and anatomy of the frontal gland provide information on the evolutionary history of defensive strategies in particular lineages. At the same time, defensive blends have been suggested to be highly variable at interspecific and intercolonial scales, both in quality and quantity, thus making the frontal gland chemistry an interesting tool for studies on taxonomy and biogeography [66]. These blends are composed of a fascinating diversity of defensive chemicals including alcohols, mono-, sesqui-, di-terpenoid hydrocarbons, ketoaldehydes, fatty acids, macrocyclic lactones, and heterocyclic and aromatic compounds [66].

Current advances in microbiome studies are also bringing new insights into the gut symbiosis revealing how gut microbiomes are shaped by priority effects such as vertical transmission, diet, bacterial transplantation, and antibiotics [111].

2.2 *Actinomycete Symbionts of Termites*

Origins of host association are diversely distributed across symbiotic bacteria emerging independently in at least 11 bacterial phyla including Proteobacteria, Actinobacteria, and Firmicutes [108] and dating back 30–250 million years [31]. In the fungus-growing termite *Odontotermes formosanus* the representative phylotypes were affiliated with four phylogenetic groups, Firmicutes, the Bacteroidetes/Chlorobi group, Proteobacteria, and Actinobacteria of the domain Bacteria [116]. The closest relatives of the actinobacteria inhabiting the gut of *Nasutitermes corniger* regardless of the geographical origin of the termite colony were found to belong to five families of the order Actinomycetales: Propionibacteriaceae, Streptomycetaceae, Cellulomonodaceae, Corynebacteriaceae, and Rubrobacteraceae [76].

Termite gut microbiota investigations using the 16S rRNA gene as a molecular marker were used to characterize the whole bacterial diversity. These studies revealed several 16S rRNA sequences of the Actinobacteria phylum falling into different genera [51, 76]. The analysis of the taxonomic composition showed that the 16S rRNA sequences affiliated with Actinobacteria account for a minor part of the gut bacterial microbiota, although the diversity may have been underestimated in as much as individual taxa present in smaller numbers will not be detected owing to PCR bias in universal primers on the detection of actinobacteria [49, 76]. Moreover, a discrepancy was found between the isolates obtained by cultivation and the dominant phylogenetic groups in the clone libraries [48, 76, 81, 112]. Although culture-independent methods allow an accurate description of dominant phylogenetic groups inhabiting the intestinal tracts of termites, little is known about the community structure and diversity of gut actinobacteria [76].

For a better understanding of the foundational basis of evolution and long-term stability of symbioses (i) the evolutionary dynamics of insect-actinobacteria symbioses (ii) the net benefits for the symbiotic partners and (iii) the selective pressures acting on each partner owing to symbiosis were suggested to be determined by Kaltenpoth [60]. Nutritional symbioses between insects and bacteria are widespread and common, but actinobacteria seem to be rarely involved in such associations [60], instead streptomycetes and other actinobacteria are well adapted to living in symbiosis with invertebrates where they mostly play a protective role. One of the most important examples is antibiotic production [12] and might be used to defend the host's larvae or food source against infections by pathogens [60, 113].

Scheuring and Yu [111] suggested that selective recruitment of mutualists might occur if a host can present a rewarding environment to its symbionts that is simultaneously unattractive to parasites. If the conditions provided by the host are set correctly the host does not need to choose, or never needs to know, the quality of any individual symbiont. As a result, the potential symbionts might evolve to accept the host or to reject it (and remain free-living); according to each symbiont's type [111]. Antibiotic-producing microbiomes (the most abundant class of beneficial microbiome in nature), might be one of the examples of such interactions. Cuticular microbiomes in particular are of interest as they release antibiotics to defend the fungus from parasites. The *Pseudonocardia* link in ants might be an example for long-term vertical transmission resulting in the evolution of new compounds for host protection against parasites. Selective recruitment by ants of multiple and non coevolved actinobacterial genera from the soil, enabling a "multi drug" strategy against parasites has also been suggested [111]. In disease-suppressive soils this partner-choice mechanism might be more effective when at least one actinobacterial symbiont is vertically transmitted or has a high immigration rate [111].

The environment itself where host species evolve might introduce profound differences in the community ecology of a microbiome. Gut microbiome composition might be governed by host species and by host genotype [4, 111].

2.3 Actinomycete-Derived Chemicals in Termites

2.3.1 Antibiotics/Antifungals

Secondary metabolites are molecules of adaptation that evolved for purposes apart from primary metabolism and in contrast to primary metabolites; they are produced by individual species or genera for specific physiological, social, or predatory reasons [96]. These compounds, therefore, are intimately linked with the ecology of the producing organisms [96], such as the adaptive attine ant interactions with *Pseudonocardia*, a member of the order Actinomycetales, for their own protection or sanitation of their nests [91, 114]. An analysis of the chemical potential of this genus identified a single bioactive compound, which proved to be a novel cyclic depsipeptide dentigerumycin and specifically inhibited the growth of pathogenic fungi whereas the cultivar was largely resistant [19, 97]. Moreover, it has been suggested that through production of non specific broad antifungal compounds by *Pseudonocardia* and *Amycolatopsis* species, ants could also damage their fungal garden by subjecting them to such actinobacterial secretions [109, 114]. Both observations indicate specificity or broad-spectrum activities when required by the insect for its own defense or selective cultivation requirements. Fungus-growing ants (Attini, Formicidae) cultivate a basidiomycete fungus in gardens that the ants nurture with dead or fresh plant substrate. The gardens are actually mini-ecosystems of competing, commensal, and mutualistic microbes, including a diverse assembly of filamentous fungi, yeasts, and bacteria [91].

Termites, due to their nesting and foraging activities in soil and on decaying wood, face invasive pressures from pathogenic microbial species. Their social organization requires the sharing of nutrients, symbionts, and pheromones; this then can be conducive to the spread of diseases throughout the colony if such an invasion occurs [41]. Fungal pathogens constitute the most serious disease threat to subterranean termites: viral and bacterial pathogens usually enter the insect through the alimentary tract but many fungal pathogens can directly penetrate the insect cuticle [10]. Among fungal entomopathogens, *Metarhizium anisopliae* is a significant threat to subterranean termites because it is a ubiquitous soil pathogen that can evade the immune system once it has entered its host [3, 10]. As a result, termites, as with other social insects [23–25], have evolved a range of defenses [34, 107] including the antimicrobial peptides called termicins secreted as part of the external antifungal defense strategy [11]. This strategy depends on the active dissemination of antifungal secretions among nestmates [40, 41].

The termicins are constitutively expressed in termite salivary glands and hemocytes, where they may be released into the hemocoel upon infection [75] targeting the cell membrane of fungi and some Gram-positive bacteria [21]. Xu et al. [136] studied *Odontotermes formosanus* (Blattaria: Termitidae) and *Reticulitermes chinensis* (Blattaria: Rhinotermitidae), both termite species living in significantly different habitats, to detect the differences in mRNAs encoding for different numbers of antimicrobial peptides. *O. formosanus* is a fungus-growing termite that constructs a subterranean nest, favoring wet environments and feeding on a variety of materials such as trees, field crops, and the fungi that it cultivates. *R. chinensis,* on the other hand, is an important urban insect pest as well as occurring in the wild and it typically constructs not only subterranean nests but also nests in decayed wood on the ground. It does not cultivate fungi and favors a relatively dry environment; it feeds chiefly on timber-in-service, such as skirting boards, floors, doors, window frames, and furniture. They found differences in the number of termicin genes expressed by these two different termite species and concluded that micro-environmental pressures affected the number of termicin genes expressed indicating functional chemistry taking place in termite defense mechanisms [136].

A range of antifungal compounds have been isolated from actinobacteria associated with termites. Examples include compounds from fungus-growing termites that revealed a high degree of bioactivity inhibiting the invasive fungus *Pseudoxylaria* [130]. New acrylamide and oxazolidin derivatives with antifungal activity from a termite-associated *Streptomyces* species were also isolated by Bi et al. [5].

A qualitative survey of 18 genera from the four largest families, by Siderhurst et al. [119], *Anoplotermes, Amitermes, Cryptotermes, Coptotermes, Gnathamitermes, Heterotermes, Incisitermes, Kalotermes, Margtnitermes, Microcerotermes, Nasutitermes, Neotermes, Paraneotermes, Prorhinotermes, Pterotermes, Reticulitermes, Tenuirostritermes*, and *Zootermopsis*, showed universal norharmane presence. Norharmane is a β-carboline alkaloid with antifungal activity, particularly active against the entomopathogenic fungus *M. anisopliae* and reported to be produced by endosymbionts in termites [14, 118, 119]. Siderhurst et al. [118] indicated that actinomycetes bacteria were the likely candidates as the endosymbionts that biosynthesize norharmane, because they are components of the termite gut and are the only microbes known to produce norharmane, such as the ones reported to be produced by *Nocardia* species [137].

2.3.2 Pheromones and Volatile Compounds

Pheromones are a subclass of semiochemicals, used for communication within a species (intraspecific chemical signals) [135]. Evolution of social behavior by kin selection requires the ability of kin recognition in order to direct altruistic behavior towards relatives and many eusocial insects are known to distinguish nestmates from non-nestmate conspecifics [84]. Discrimination and aggressive responses towards non-nestmates have been frequently observed in a number of termite species and such discrimination systems have been shown to protect colonies that might otherwise be vulnerable to social parasitism by various types of nest invaders. Matsuura [84] indicated that differential intestinal bacteria composition leads to production of colony-specific chemical cues that enable nestmate recognition and the composition of the intestinal bacteria was exclusively colony-specific. Termites that had adsorbed an unfamiliar odor of bacteria sampled from another colony were fiercely attacked by nestmates, indicating the important role intestinal bacteria play in nestmate recognition.

Semiochemically mediated interactions between bacteria and insects, have been reported in relation to insects' response to specific volatiles emitted by specific bacteria hosted by the insect itself (gut, mouthparts, etc.) or present in the natural environment where the insect evolved [6, 78]. Using a cultivation-independent approach Minkley et al. [86] investigated the structure of the bacterial community in the gut of termites from four different colonies of *Hodotermes mossambicus* using 16S rRNA-based terminal restriction fragment length polymorphism (T-RFLP). Their analysis of the bacterial gut microbiota revealed (1) a high consistency of the gut microbiota among nestmates and (2) subtle but distinct differences in community structure between individuals from different colonies. They linked their findings to the bacterial metabolism contributing to the colony odor that can be used as a discriminatory signal. The presence of a colony-specific bacterial community might thus support the hypothesis that the gut microbiota of termites is involved in nestmate recognition [33].

Costa-Leonardo et al. [17] in their review indicated that termite cuticles function as an enormous exocrine gland, producing a mixture of chemical substances, mainly hydrocarbons which are believed to be involved in nestmate recognition. Moreover, they noted that the hydrocarbon cuticular composition differs among termite colonies and these differences might be correlated with intercolonial aggressions. Habbachi et al. [39] investigated the effect of spinosad against chemical communication in the German cockroach, *Blatella germanica* (L). Spinosad is a biopesticide from the soil-dwelling actinomycete *Saccharopolyspora spinosa*, which is used to control a variety of insects and is harmless to mammals and many predatory insects. They reported that a nonlethal dose of spinosad causes adult male and female *B. germanica* to exhibit altered responses to their aggregation pheromone as well as to have a changed cuticular hydrocarbon profile.

Research related to the composition of the frontal gland secretion and its function in the nasutitermitinae subfamily of the termites confirmed the presence of species-specific mixtures of monoterpenes (rarely sesquiterpenes) and diterpenes. These terpenes frequently occurred with other classes of compounds (alcohols, ketones, aromatic compounds, and amides) [122]. These complex secretions from the nasutitermitine frontal gland had different functions including acting as a repellent to incoming enemies, and as an alarm pheromone coordinating defensive activities [122]. Diterpenes are the most characteristic chemical category of the nasutitermitinae including providing novel chemical structures and they contribute to the physical properties of the secretion such as congealment in the air by polymerization [102, 122] as well as being poisonous [122, 126, 128]. Monoterpene hydrocarbons present in the mixture are reported to serve as a solvent to the diterpenes, and due to their hydrophobic properties also helping to dissolve cuticular waxes and enhancing the stickiness and irritancy of the secretion [2, 122]. Antibacterial trinervitadienes from the termite *Nasutitermes triodiae* in Australia have also been reported [138].

Terpenoids comprise the largest, structurally most diverse family of natural products and play important roles in all living organisms. Diterpenoids of bacterial origin are known but rare; however, recent advances in genomics have revealed the biosynthetic potential for terpenoids in bacteria, particularly in the actinomycetes [89, 123]. Researchers have recently isolated diterpenes from a *Streptomyces* species (KO-3988) and described five new diterpenes named oxaloterpins A (1), B (2), C (3), D (4), and E (5). Two new terpenoids, naphterpins B and C from *Streptomyces* sp. (CL-190) were also isolated by [124]. New diterpenes, gifhornenolones A and B from a non-streptomycete actinomycete *Verrucosispora gifhornensis* (YM28-088) were also reported by Shirai et al. [117] as well as from *Mycobacterium tuberculosis* by Prach et al. [101].

Volatile terpenoids are also characteristics of actinomycetes. Geosmin, 2-methylisoborneol, and a range of volatile compounds have been detected from this group of bacteria since the 1960s [35, 59, 105, 110]. Wilkins and Schöller [133] identified volatile compounds from 26 unsequenced streptomycetes. Out of these strains, 21 produced geosmin, nine emitted 2-methylisoborneol, and three released albaflavenone [110, 133]. Utilizing the current advances in molecular techniques Citron et al. [15] investigated the volatiles emitted by sequenced actinomycetes to allow correlation of the genetic information with the production of secondary metabolites. They reported terpenoid volatiles released by 30 actinomycetes known to encode terpene cyclases in their genomes, and concluded that terpenoids are widespread in actinomycetes. They identified 55 putative geosmin synthases, 23 homologues of 2-methylisoborneol synthases, and 98 other sesquiterpene cyclase homologues in actinomycetes.

Another important link derives from an increased understanding of the symbiont communities through analysis of their communication networks. Examples include the reporting of the presence of odorous-gamma butyrolactones in Gram-positive bacteria as signaling molecules [27] that might be involved in shaping the symbiotic actinobacterial communities.

A multitude of antifungal VOCs emitted by bacteria as well as the repression of phytopathogens in soils through VOCs emitted by these microorganisms have been reported. Examples of fungistatic VOCs include 1-octen-3-ol, mono- and sesquiterpenes, nonanal acid, trimethylamine, and dimethyldisulfide which are produced by Actinobacteria [58]. One interesting link was reported by Řezanka et al. [106] and they noted that with the increasing production of avermectins, the synthesis of geosmin was enhanced by more than one order of magnitude. The avermectins produced by *Streptomyces avermitilis* have potent antiparasitic and broad-spectrum activity against nematode and arthropod parasites. Although similar in structure to antibacterial macrolides and antifungal polyenemacrolides, avermectins were reported to lack antibacterial and antifungal activities [57] and might only play a role in parasite defenses employed by termites. Termites might recognize antiparasitic actinomycete-rich environments through geosmin trails and construct their mounds to protect themselves from termite-feeding parasites such as nematodes [95].

2.3.3 Enzymes

Termites formerly classified as the order Isoptera are now included as part of the order Blattaria (or Blattodea; cockroaches) [83]. They are broadly divided into the lower and higher termites [77]. The lower termites are specifically wood-feeding insects, whereas the higher termites have evolved a more diverse diet and gut microbiota. Some feed on wood and well-rotted plant matter, some are exclusively soil-feeders, whereas others grow and feed on cellulolytic fungi [77].

Termites and their symbionts are not only involved in cellulolytic or lignin decomposition activities but also in aromatic hydrocarbons degradation [9, 83, 94]. Lignin is a highly branched, aromatic polymer that is resistant to microbial degradation; very few bacteria are able to degrade lignin and those that are able, are dominated by the actinobacteria. Examples include isolation of various strains of actinobacteria (*Micromonospora* spp. and *Streptomyces* spp.) from the hindgut of various higher termites (*Macrotermes, Odontotermes, Amitermes*, and *Microcerotermes*) [100] and the widespread existence of actinobacteria in both higher and lower termites [131]. Recent studies conducted by utilizing the power of molecular tools on host–symbiont transcriptome confirm host–symbiont collaboration in cellulose/hemicellulose digestion in the termite gut [125].

The biological roles of oxidative enzymes in actinobacteria might be similar to those that are found in fungi. These roles are mainly in degrading phenolic compounds to support a saprophytic life cycle as well as some oxidative enzymes from actinobacteria playing a role in morphogenesis or antibiotic production [77]. Oxidative enzymes are included in the vast group of enzymes known as the oxidoreductases and they catalyze biological oxidation/reduction reactions and play a major role in many chemical and biochemical transformations. The oxidative enzymes, laccase, peroxidase, and tyrosinase, have been detected in actinobacteria and peroxidases have been shown to be one of the key enzymes produced by this group of bacteria during the degradation of lignocellulose compounds [77].

Measurement of β-1, 3-glucanase activity in *Nasutitermes corniger* by Bulmer et al. [10] revealed significant activity in body tissues and secretions including salivary glands and cuticular washes. Other termite species (*Zootermopsis angusticollis*, *Cryptotermes secundus*, and *Rubriceps flavipes*) also showed robust β-1, 3-glucanase activity. Measurement of (1, 4)-glucanase activity was highest in the guts of the wood-feeding termites *Nasutitermes parvonasutus* and *Havilanditermes orthonasus* [53, 79]. The fungal entomopathogen *Metarhizium anisopliae* is a natural termite pathogen and is currently being developed for the biological control of termites and other insect pests. Conidia treated with β-1, 3-glucanases collapse immediately, hampering the successful antagonistic effect on the insect [10]. Innate immunity in termites includes cellular and humoral defenses that are activated by the recognition of pathogen-associated molecular patterns (PAMPs), which are conserved structural features of microbes, including peptidoglycans in bacterial cell walls (e.g., lipopolysaccharide (LPS) in Gram-negative bacteria outer membranes, and β-1, 3-glucans in fungal cell walls). Termite Gram-negative bacteria binding proteins (GNBPs) and β-1, 3-glucan recognition proteins (BGRPs) share sequence homology with bacterial β-1, 3-glucanases. The purified termite GNBP has been shown to exhibit direct antifungal effector activity by breaking down β-1, 3-glucans in fungal cell walls [10].

Extracellular enzymes with glucanase activities are an important component of actinomycete-fungus antagonism. Actinomycetes, in particular streptomycetes, are reported to produce extracellular β-1, 3-, β-1, 4-, and β-1, 6-glucanases [30, 127] and these enzymes can hydrolyze glucans from fungal cell walls resulting in lysis of fungal cells [26]. A streptomycete enzyme system active in lysing *Aspergillus oryzae* and *Fusarium solani* hyphal walls contained chitinase and several β-1, 3 glucanase components were also reported by Skujins et al. [121]. These enzymes were shown to be instrumental in the process of dissolution of hyphal walls. Interestingly, recently endo-β-1, 3-glucanase was reported to be acquired by horizontal gene transfer from bacteria in nematodes indicating multiple independent horizontal gene transfer events that might have helped in shaping the evolution of several different life strategies in nematodes [61]. All these findings might indicate the origins of shared sequence homology in enzymes of both host and the symbiont.

Actinomycetal proteins other than hydrolases were also indicated to be involved in biocontrol of fungi. Examples include the report of an alkaline protease inhibitor (API) as a novel class of antifungal proteins against phytopathogenic fungi by Vernekar et al. [129]. The activity of API was reported to inhibit fungal serine alkaline protease, which is indispensable for the growth of fungi.

3 Diversity of Termite Microbial Symbionts and Actinoflora

3.1 Molecular Analysis of Uncultured Microflora and Metagenomic Approaches

Termite guts are reported to harbor 10^6–10^8 microorganisms comprising 300 species of protists, bacteria, and archaea (which are mostly unique to termites) as a highly structured symbiotic community essential for host survival on recalcitrant materials in their natural environments [47].

Hongoh et al. [50] demonstrated that congeneric termites harbored very similar bacterial gut microbiota, irrespective of the individual, colony, location, and host species. The similarity in bacterial gut microbiota among congeneric termites was also demonstrated by Schmitt-Wagner et al. [112] with the African soil-feeding termites *Cubitermes* spp. (family Termitidae; subfamily Termitinae). They reported moderate to considerably high levels of similarity in microbiota using T-RFLP analysis. Their findings indicated that a high similarity of bacterial gut microbiota within a termite genus may be a general trait for termites leading to a very stable and strong symbiotic relationship.

Hongoh et al. [50] concluded that the majority of gut bacteria from distantly related termites, including the genera *Microcerotermes, Reticulitermes*, and *Cubitermes*, constituted monophyletic clusters that were distinct from other bacterial lineages. This indicates that the majority of gut bacteria may not be allochthonous but rather be autochthonous symbionts that are unique to termites. They also noted that the bacterial gut microbiota was greatly different between the host termite genera *Microcerotermes* and *Reticulitermes*, in contrast to the high similarity within each termite genus. Gut bacteria may have differentiated after acquisition by the ancestors of these termites, displaying differences among dominant bacterial groups such as the genus *Treponema* and the orders Clostridiales and Bacteroidales, but also in other minor phyla such as Actinobacteria, Proteobacteria, "Synergistes," Planctomycetes, and others. A very diverse gut bacterial community might have coevolved with their host termites and have formed a stable symbiotic complex specific to a genus of termites [50].

Fisher et al. [32] examined the diversity of gut bacteria of *Reticulitermes flavipes* using 16S rRNA gene sequencing and amplified rDNA restriction analysis and identified a broad taxonomic range of ribotypes from six phyla within the Domain Bacteria including actinobacteria. Studies conducted by Costa et al. [16] using the phylogenetic analysis of cloned 16S rRNA gene fragments from *Cornitermes cumulans* again identified the presence of actinobacteria. Similarly Shinzato et al. [116] constructed a bacterial 16S gene clone library from the gut microbial community of *Odontotermes mossambicus* and with subsequent RFLP analysis again identified four phylogenetic groups including actinobacteria. Hongoh et al. [48] found for the first time more than 90 % of the phylotypes using 16S rRNA genes and some constituted monophyletic clusters with sequences recovered from the gut of other termite species.

Fall et al. [29] compared the bacterial community structures of the soil-feeding termite (*Cubitermes niokoloensis*) including the mound, termite gut sections, and surrounding soil using PCR-denaturing gradient gel electrophoresis (DGGE) analysis and cloning and sequencing of PCR-amplified 16S rRNA gene fragments. DGGE analysis revealed a drastic difference between the genetic structures of the bacterial communities of the termite gut and the mound. The soil-feeding termite mound was dominated by the Actinobacteria phylum, whereas the Firmicutes and Proteobacteria phyla dominate the gut sections of termites and the surrounding soil, respectively. Phylogenetic analyses revealed a distinct clustering of Actinobacteria phylotypes between the mound and the surrounding soil. The Actinobacteria clones of the termite mound were diverse, distributed among 10 distinct families, and like those in the termite gut environment were lightly dominated by the Nocardioidaceae family.

Phylogenetic analysis of the gut bacterial microflora of the fungus-growing termite *Macrotermes barneyi* was conducted by Zhu et al. [139]. Although many of the clones (95 %) detected were derived from three phyla within the domain Bacteria: Bacteroidetes, Firmicutes and Proteobacteria clones from Deferribacteres, Actinobacteria and Planctomycetes were also found.

The traditional molecular techniques such as DGGE, RFLP, and FISH only revealed information with respect to how they are applied to study the composition, diversity, and dynamics of insect gut symbiotic microbiota [92]. However, advances including "omics" are now bringing new insights into the termite microbial symbiosis [45, 47, 98, 115]. Increasing numbers of genome sequences for many symbionts are now revealing their complete set of genes as well as their functional contribution to their host's metabolism [36, 37]. This metagenome analysis, together with the recent advances in next-generation sequencing, are also providing substantial sequencing information, and in-depth microbial diversity analysis; and modelling of pathways for biological processes are thus now possible [115]. Metagenome sequencing, metatranscriptome and metaproteome methods are currently facilitating studies of system dynamics and gene expression [54, 115]. The integration of different 'omics' level data will soon allow us to understand how the insect gut works as a system to carry out these functions [44, 79, 115].

Another important development following these molecular advances has been the report of numerous cases of bacterial symbionts with extraordinarily small genomes [85, 88]. These organisms were claimed to represent independent lineages from diverse bacterial groups and they carry diminutive gene sets (rivaling some mitochondria and chloroplasts in terms of gene numbers) and lack genes that are considered to be essential in other bacteria [85]. The common features these bacteria share (e.g. fast protein evolution) point to highly degenerate genomes that retain only the most essential functions, often including a considerable fraction of genes that serve the hosts resulting in the review of the currently defined concept of symbiosis and host-associated microbiota [85]. Komatsu et al. [62] were able to induce efficient production of natural products including terpenoids by controlled minimization of the genome of the *S. avermitilis*. These findings might indicate host-specific functional chemistry by symbionts once they are acquired by the insects via minimized genomes.

3.2 Culturing the Representatives of Actinoflora

Lefebvre et al. [76] studied the actinobacterial community structure and putative representative members associated with the gut of the wood-feeding termite, *Nasutitermes corniger* (Motschulsky), using nested PCR-DGGE and 16S rDNA sequence analyses. Regardless of the geographical origin of the termite colony they located members of the families of Propionibacteriaceae, Streptomycetaceae, Cellulomonodaceae, Corynebacteriaceae, and Rubrobacteraceae. They found that 16S rDNA sequences affiliated with the families Streptomycetaceae and Cellulomonodaceae had more than 97 % similarity with the closest isolated strains. Their findings indicated that members of the order Actinomycetales account for the largest proportion of the Actinobacteria phylum inhabiting the gut of the termite *N. corniger* and actinomycetes that have not yet been cultivated are present in their gut (Fig. 1).

Culturing microorganisms, particularly those representing new taxa, is an indispensable requirement for the full description of diversity. However, despite the advances in molecular detection techniques, media and growth conditions routinely used to culture microorganisms still only reveal a fraction of the environmental microflora [70, 99]. Authors agree with Palleroni [99] in that since the 1980s popular methods of molecular analyses of natural microbial communities lacked resolving power in terms of identification at the species level, as well as failing to give sufficient information about the function of newly detected members of the microflora. Recent advances in metagenomics together with next-generation sequencing will now provide enormous sequencing information, allowing in-depth microbial diversity analysis [115]. Moreover, recent advances in metatranscriptomics, and metaproteomics will gradually unveil the true picture of the symbiotic system [46]. Genomics targeting a single species of the unculturable microbial members will also provide an improved understanding of the symbiotic interrelationships among the gut microorganisms as well as revealing the members and functions of the multilayered symbiotic system [46].

If combined with an objective approach, these molecular advances can provide guidance towards design of target-directed culturing methods, thus supplementing their role in surveying the composition of microbial communities and in the characterization of new prokaryotic taxa [42, 99] including termite symbionts. Cultured representatives of termite symbiotic actinoflora will then generate further information on their physiology and growth requirements of these biotechnologically important taxa. Sinma et al. [120] by using a highly selective agar designed to detect rare actinomycetes [43] isolated a novel species of *Saccharopolyspora* from the guts of a *Speculitermes* species. Matsui et al. [83] by using enrichment cultures using carboxymethyl cellulose or filter paper as the sole carbon source identified 23 groups of cellulolytic bacteria including the members of the order Actinomycetales. Kurtböke and French [73] via the use of polyvalent bacteriophages investigated the layers of termite (*Coptotermes lacteus* (Froggatt)) gut actinoflora and reported that members of the family Streptomycetaceae were the dominant species of the gut actinoflora. Abundance and distribution of these taxa were similar to soil layer distribution of actinomycetes, streptomycete species constituting the majority of the species cultured followed by the members of the family Micromonosporaceae (Table 1).

To be able to culture the non-streptomycete fraction of the actinoflora they applied polyvalent streptomycete phages to remove the streptomycete fraction of the actinoflora investigated [73]. Once the plates were clear of streptomycete

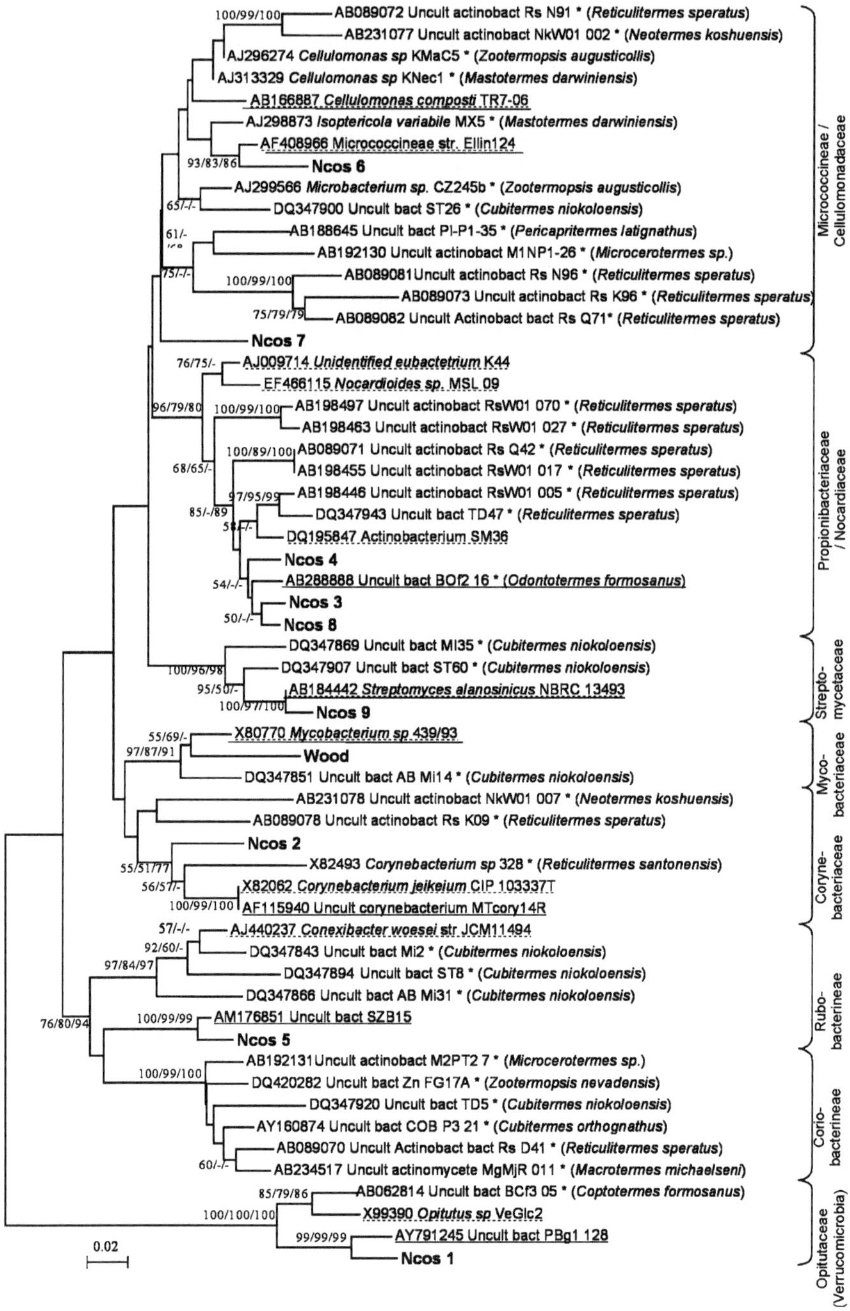

Fig. 1 Phylogenetic tree showing the relationship between the sequences of DGGE bands; *solid line* sequences from uncultured strains; *dashed line* sequences from characterized isolate strains; (*) sequences retrieved in termite gut. The bootstrap percentages are indicated for NJ/MP/ML (©Lefebvre et al. [76])

Table 1 Culturable diversity of actinomycete families from termite guts

Order of dominance[a]	Actinomycete family[b]
1	Streptomycetaceae
2	Micromonosporaceae
3	Nocardioidaceae
4	Nocardiaceae
5	Pseudonocardiaceae
6	Thermomonosporaceae
7	Corynebacteriaceae
8	Mycobacteriaceae
9	Streptosporangiaceae
10	other

[a] Actinobacterial families are presented in decreasing order
[b] Cultured representatives of different actinomycete families from the guts of *C. lacteus* collected in the Sunshine Coast region

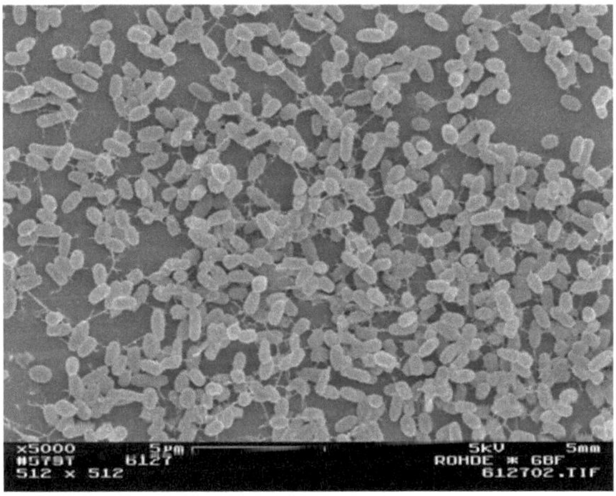

Fig. 2 Electron micrograph of a non-streptomycete actinomycete isolated from the gut of *Coptotermes lacteus* at magnification × 5000 (bar = 5 μm)

species that impede the growth of other actinomycete taxa, representatives of non-streptomycete actinomycetes from other family members of the order Actinomycetales including rare and novel ones were detected (Fig. 2). Polyvalent actinophages combined with the use of other selective inhibitors in the agar (e.g., nalidixic acid used against Gram-negative bacteria) allowed the detection and isolation of other non-streptomycete genera [70] in the order of dominance of termite gut actinoflora.

Tested representatives of these isolates were found to produce hydrolytic enzymes [74], and volatiles (e.g., aldehydes, ketones, terpenoids, geosmin, 2-methylisoborneol Stephen, Hayes, French and Kurtböke, unpublished data) (Fig. 3), as well as antifungal/antimicrobial compounds (Romero-Bonifaz, Grkovic, French, Kurtböke, and Quinn, 2013, unpublished data) produced by symbionts in termites located in the Sunshine Coast Region.

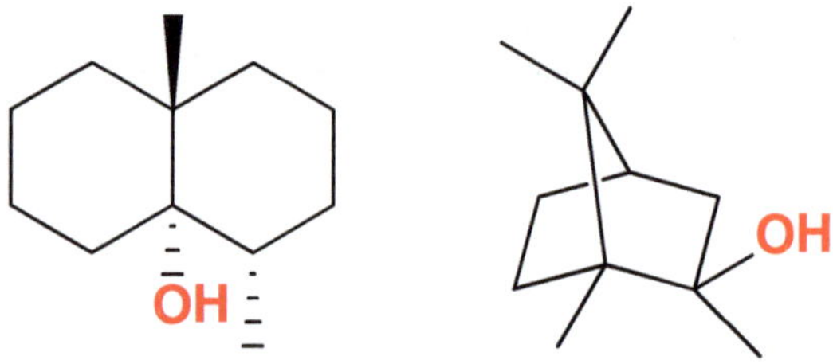

geosmin 2-methylisoborneol

Fig. 3 Structures of sample volatile compounds detected from *C. lacteus* gut-associated *Streptomyces* species

4 Future Prospects

Current worldwide efforts related to insect gut symbiosis indicate that *Streptomyces* species are widespread inhabitants of invertebrate guts and possibly contribute to the degradation of polymeric carbohydrates or to antimicrobial defense [113]. Our observations suggest that termites possibly recognize actinomycete- (in particular streptomycete-) rich environments and construct their mounds in those areas. Function-related recognition and harboring of VOC and anti-insecticidal compounds producing actinomycetes by termites for subsequent use might occur. These actinomycetes might aid termites in their different needs ranging from nestmate recognition to mycofumigation [13] of their nests (Fig. 3).

Termites play a major role in foraging and degradation of plant biomass as well as cultivating bioactive microorganisms for their defense. Current advances in "omics" sciences reveal insights into function-related presence of symbionts [46, 52]. These findings indicate significant potential for the use of these identified functions of termite symbionts and their relevant genes for biotechnology and biodiscovery [82].

Metagenome and metatranscriptome analyses of the gut microbiota are now revealing the presence of diverse functional genes required for fermentation, reductive acetogenesis, and nitrogen fixation [46, 80]. Following the long-term efforts in cultivation of the fastidious microorganisms and through the ecological, physiological, and biochemical studies of the whole insects and cultured gut symbionts these functions have been recognized as essential bacterial activities in the symbiotic system [46]. Functional analysis of the complete genome sequences acquired from intracellular symbionts of the gut indicates their functional roles in the termite gut systems. As further advances become available allowing the use of both meta- and single-species–targeting genomics, transcriptomics, and proteomics our understanding of this highly evolved and complex symbiotic system will increase. Selective isolation will thus play an important role in culturing these functional gene carrier microorganisms for their further use in biotechnology and biodiscovery.

As recently stated by Kurtböke [68–72] the success in recovery of rare actinomycetes, including novel members of the genus *Streptomyces*, will only derive from a sound understanding in ecology, taxonomy, physiology, and metabolism of actinomycetes. If such in-depth understanding is combined with novel information being continuously generated with the aid of advancing molecular information, a powerful knowledge platform revealing the whereabouts as well as taxonomic and chemical identities of previously undetected bioactive actinomycetes will be established.

Acknowledgments The authors thank Professor Manfred Rohde (GBF, Germany) for the electron micrographs of the termite symbiotic actinomycete.

References

1. Abe T, Bignell DE, Higashi M (eds) (2000) Termites: evolution, sociality, symbioses, ecology. Kluwer Academic, Dordrecht/Norwell
2. Baker R, Walmsley S (1982) Soldier defense secretions of the South American termites *Cortaritermes Silvestrii Nasutitermes* SP N.D. and *Nasutitermes Kemneri*. Tetrahedron 38(13):1899–1910
3. Balachander M, Remadevi OK, Sasidharan TO (2013) Dissemination of *Metarhizium anisopliae* infection among the population of *Odontotermes obesus* (Isoptera: Termitidae) by augmenting the fungal conidia with attractants. J Asia-Pacific Entomol 16(3):199–208
4. Benson AK, Kelly SA, Legge R, Ma F, Low SJ, Kim J, Zhang M, Oh PL, Nehrenberg D, Hua K, Kachman SD, Moriyama EN, Walter J, Peterson DA, Pomp D (2010) Individuality in gut microbiota composition is a complex polygenic trait shaped by multiple environmental and host genetic factors. Proc Nat Acad Sci USA 107:18933–18938
5. Bi SF, Li F, Song YC, Tan RX, Ge HM (2011) New acrylamide and oxazolidin derivatives from a termite-associated *Streptomyces* sp. Nat Prod Commun 6(3):353–355
6. Bode HB (2009) Insects: true pioneers in anti-infective therapy and what we can learn from them. Angew Chem Int Ed 48(35):6394–6396
7. Breznak JA (2000) Ecology of prokaryotic microbes in the guts of wood- and litter-feeding termites. In: Abe T, Bignell DE, Higashi M (eds) Termites: evolution, sociality, symbioses, ecology. Kluwer Academic, Dordrecht, pp 209–231
8. Brune A, Ohkuma M (2010) Role of the termite gut microbiota in symbiotic digestion. In: Bignell DE, Roisin Y, Lo N (eds) Biology of termites: a modern synthesis. Springer, Dordrecht, pp 439–475
9. Bugg TDH, Ahmed M, Hardiman EM, Singh R (2010) The emerging role for bacteria in lignin degradation and bio-product formation. Curr Opin Biotechnol 22(3):394–400
10. Bulmer MS, Bachelet I, Raman R, Rosengaus RB, Sasisekharan R (2009) Targeting an antimicrobial effector function in insect immunity as a pest control strategy. PNAS 106(31):12652–12657
11. Bulmer MS, Lay F, Hamilton C (2010) Adaptive evolution in subterranean termite antifungal peptides. Insect Mol Biol 19(5):669–674
12. Carr G, Poulsen M, Klassen JL, Hou Y, Wyche TP, Bugni TS, Currie CR, Clardy J (2012) Microtermolides A and B from termite-associated *Streptomyces* sp. and structural revision of Vinylamycin. Org Lett 14(11):2822–2825
13. Chen J, Henderson G, Grimm CC, Lloyd SW, Laine RA (1998) Termites fumigate their nests with naphthalene. Nature 392:558–559
14. Chouvenc T, Su N-Y, Elliott ML (2008) Interaction between the subterranean termite *Reticulitermes flavipes* (Isoptera: Rhinotermitidae) and the entomopathogenic fungus *Metarhizium anisopliae* in foraging arenas. J Econ Entomol 101:885–893
15. Citron CA, Gleitzmann J, Laurenzano G, Pukall R, Dickschat JS (2012) Terpenoids are widespread in actinomycetes: a correlation of secondary metabolism and genome data. ChemBioChem 13(2):202–214

16. Costa PS, Oliveira PL, Chartone-Souza E, Nascimento AMA (2012) Phylogenetic diversity of prokaryotes associated with the mandibulate nasute termite *Cornitermes cumulans* and its mound. Biol Fertil Soils 49:567–574
17. Costa-Leonardo AM, Casarin FE, Lima JT (2009) Chemical communication in Isoptera. Neotrop Entomol 38(1):1–6
18. Costa-Leonardo AM, Laranjo LT, Janei V, Haifig I (2013) The fat body of termites: functions and stored materials. J Insect Physiol 59(6):577–587
19. Crawford JM, Clardy J (2011) Bacterial symbionts and natural products. Chem Commun 47(27):7559–7566 (Camb)
20. Crespi BJ (1994) Three conditions for the evolution of eusociality: are they sufficient? Insectes Soc 41:395–400
21. Da Silva P, Jouvensal L, Lamberty M, Bulet P, Caille A, Vovelle F (2003) Solution structure of termicin, an antimicrobial peptide from the termite *Pseudacanthotermes spiniger*. Protein Sci 12(3):438–446
22. de la Cruz MN, Junior HM, Oliveira DF, Costa-Lotufo LV, Ferreira AG, Alviano DS, Rezende CM (2013) Chemical composition and biological activities of soldiers of the Brazilian termite species, *Nasutitermes macrocephalus* (Isoptera: Natutitermitinae). Nat Prod Commun 8(1):69–74
23. Dillon RJ, Charnley AK (1988) Inhibition of *Metarhizium anisopliae* by the gut bacterial flora of the desert locust: characterisation of antifungal toxins. Can J Microbiol 34:1075–1082
24. Dillon RJ, Charnley AK (2002) Mutualism between the desert locust *Schistocerca gregaria* and its gut microbiota. Res Microbiol 153:503–509
25. Dillon RJ, Vennard CT, Buckling A, Charnley AK (2005) Diversity of locust gut bacteria protects against pathogen invasion. Ecol Lett 8:1291–1298
26. Doumbou CL, Hamby Salove MK, Crawford DL, Beaulieu C (2001) Actinomycetes, promising tools to control plant diseases and to promote plant growth. Phytoprotection 82(3):85–102
27. Dunn AK, Handelsman J (2002) Toward an understanding of microbial communities through analysis of communication networks. Antonie Van Leeuwenhoek 81:565–574
28. Engel P, Moran NA (2013) The gut microbiota of insects—diversity in structure and function. FEMS Microbiol Rev 37:699–735
29. Fall S, Hamelin J, Ndiaye F, Assigbetse K, Aragno M, Chotte JL, Brauman A (2007) Differences between bacterial communities in the gut of a soil-feeding termite (*Cubitermes niokoloensis*) and its mounds. Appl Environ Microbiol 73(16):5199–5208
30. Fayad K, Simao-Beaunoir AM, Gauthier A, Leclerc C, Mamady H, Beaulieu C, Brzezinski R (2001) Purification and properties of a *β*-1, 6-glucanase from *Streptomyces* sp. EF-14, an actinomycete antagonistic to Phytophthora spp. Appl Microbiol Biotechnol 57(1–2): 117–123
31. Feldhaar E (2011) Bacterial symbionts as mediators of ecologically important traits of insect hosts. Ecol Entomol 36:533–543
32. Fisher M, Miller D, Brewster C, Husseneder C, Dickerman A (2007) Diversity of gut bacteria of *Reticulitermes flavipes*as examined by 16S rRNA gene sequencing and amplified rDNA restriction analysis. Curr Microbiol 55:254–259
33. French JRJ, Ahmed (Shiday) BM (2010) The challenge of biomimetic design for carbon-neutral buildings using termite engineering. Insect Sci 17(2):154–162
34. Fuller CA (2007) Fungistatic activity of freshly killed termite, *Nasutitermes acajutlae*, soldiers in the Caribbean. J Insect Sci 7:1–8
35. Gerber NN, Lechevalier HA (1965) Geosmin, an earthy-smelling substance isolated from actinomycetes. Appl Microbiol 13(6):935–938
36. Gil R, Latorre A, Moya A (2010) Evolution of prokaryote-animal symbioses from a genomics perspective. In: Hackstein JHP (ed) (Endo)symbiotic methanogenic archaea, vol 19., Microbiology MonographsSpringer, Berlin, pp 207–233

37. Gosalbes JM, Latorre A, Lamelas A, Moya A (2010) Genomics of intracellular symbionts in insects. Int J Med Microbiol 300:271–278
38. Grimaldi D (2001) Insect evolutionary history from Handlirsch to Hennig, and beyond. J Paleontol 75:1152–1160
39. Habbachi W, Bensafi H, Adjami Y, Ouakid ML, Farine JP, Everaerts C (2009) Spinosad affects chemical communication in the German cockroach, *Blatella germanica* (L). J Chem Ecol 35(12):1423–1426
40. Hamilton C, Bulmer MS (2012) Molecular antifungal defenses in subterranean termites: RNA interference reveals in vivo roles of termicins and GNBPs against a naturally encountered pathogen. Dev Comp Immunol 36:372–377
41. Hamilton C, Lay F, Bulmer MS (2011) Subterranean termite prophylactic secretions and external antifungal defenses. J Insect Physiol 57:1259–1266
42. Handelsman J (2004) Metagenomics: application of genomics to uncultured microorganisms. Microbiol Mol Biol R 68(4):669–685
43. Hayakawa M, Nonomura H (1987) Efficacy of artificial humic acid as a selective nutrient in HV agar used for the isolation of soil actinomycetes. J Ferment Technol 65(6):609–616
44. He S, Ivanova N, Kirton E, Allgaier M, Bergin C, Scheffrahn RH, Kyrpides NC, Warnecke F, Tringe SG, Hugenholtz P (2013) Comparative metagenomic and metatranscriptomic analysis of hindgut paunch microbiota in wood- and dung-feeding higher termites. PLoS ONE 8(4):e61126. doi:10.1371/journal.pone
45. Hongoh Y (2010) Diversity and genomes of uncultured microbial symbionts in the termite gut. Biosci Biotechnol Biochem 74:1145–1151
46. Hongoh Y (2011) Toward the functional analysis of uncultivable, symbiotic microorganisms in the termite gut. Cell Mol Life Sci 68(8):1311–1325
47. Hongoh Y, Ohkuma M (2008) Diversity and genome analysis of symbiotic microbiota in termite guts. J Environ Biotechnol 8(1):29–34
48. Hongoh Y, Ohkuma M, Kudo T (2003) Molecular analysis of bacterial microbiota in the gut of the termite *Reticulitermes speratus* (Isoptera; Rhinotermitidae). FEMS Microbiol Ecol 44:231–242
49. Hongoh Y, Yuzawa H, Ohkuma M, Kudo T (2003) Evaluation of primers and PCR conditions for the analysis of 16S rRNA genes from a natural environment. FEMS Microbiol Lett 221(2):299–304
50. Hongoh Y, Deevong PN, Inoue T, Moriya S, Trakulnaleamsai T, Ohkuma M, Vongkaluang C, Noparatnaraporn N, Kudo T (2005) Intra- and inter- specific comparisons of bacterial diversity and community structure support co-evolution of gut microbiota and termite host. Appl Environ Microbiol 71(11):6590–6599
51. Hongoh Y, Ekpornprasit L, Inoue T, Moryia S, Trakulnaleamsai S, Ohkuma M, Noparatnaraporn N, Kudo T (2006) Intracolony variation of bacterial gut microbiota among castes and ages in the fungus-growing termite *Macrotermes gilvus*. Mol Ecol 15(2):505–516
52. Hongoh Y, Sharma VK, Prakash T, Noda S, Taylor TD, Kudo T, Sakaki Y, Toyoda A, Hattori M, Ohkuma M (2008) Complete genome of the uncultured termite group 1 bacteria in a single host protist cell. Proc Nat Acad Sci USA 105:5555–5560
53. Hughes AL (2012) Evolution of the βGRP/GNBP/β-1, 3-glucanase family of insects. Immunogenetics 64(7):549–558
54. Hussain A, Li Y-F, Cheng Y, Liu Y, Chen C-C, Wen S-Y (2013) Immune-related transcriptome of *Coptotermes formosanus* Shiraki workers: the defense mechanism. PLoS ONE 8(7):e69543
55. Husseneder C (2010) Symbioses in subterranean termites: a review of insights from molecular studies. Environ Entomol 39(2):378–387
56. Husseneder C, Ho H-Y, Blackwell M (2010) Comparison of the bacterial symbiont composition of the formosan subterranean termite from its native and introduced range. Open Microbiol J 4:53–66
57. Ikeda H, Omura S (1997) Avermectin biosynthesis. Chem Rev 97(7):2591–2610

58. Insam H, Seewald MS (2010) Volatile organic compounds (VOCs) in soils. Biol Fert Soils 46(3):199–213
59. Jachymova J, Votrub J, Viden I, Rezenka T (2002) Identification of streptomyces odour spectrum. Folia Microbiol (Praha) 47(1):37–41
60. Kaltenpoth M (2009) Actinobacteria as mutualists: general healthcare for insects? Trends Microbiol 17(12):529–535
61. Kikuchi T, Shibuya H, Jones JT (2005) Molecular and biochemical characterization of an endo-β-1,3-glucanase from the pinewood nematode *Bursaphelenchus xylophilus* acquired by horizontal gene transfer from bacteria. Biochem J 389:117–125
62. Komatsu M, Uchiyama T, Omura S, Canec DE, Ikeda H (2010) Genome-minimized Streptomyces host for the heterologous expression of secondary metabolism. PNAS 107(6):2646–2651
63. König H, Varma A (eds) (2006) Intestinal microorganisms of termites and other invertebrates. Springer, New York
64. König H, Fröhlich J, Hertel H (2006) Diversity and lignocellulolytic activities of cultured microorganisms. In: König H, Varma A (eds) Intestinal microorganisms of termites and other invertebrates. Springer, New York, pp 271–301
65. König H, Fröhlich J, Berchtold M, Wenze M (2002) Diversity and microhabitats of the hindgut flora of termites. Recent Res Dev Microbiol 6:125–156
66. Krasulová J, Hanus R, Kutalová K, Šobotník J, Sillam-Dussès D, Tichý M, Valterová I (2012) Chemistry and anatomy of the frontal gland in soldiers of the sand termite *Psammotermes hybostoma*. J Chem Ecol 38:557–565
67. Kumari R, Sachdev M, Sharma S, Prasad R, Giang PH, Garg AP, Varma A (2006) Microbiology of termite hill (mound) and soil. In: König H, Varma A (eds) Intestinal microorganisms of termites and other invertebrates. Springer, New York
68. Kurtböke DI (ed) (2003) Selective isolation of rare Actinomycetes. Queensland Complete Printing Services, Nambour, Queensland. ISBN 0-646-42910-8
69. Kurtböke DI (2009) Use of phage-battery to isolate industrially important rare actinomycetes. In: Adams HT (ed) Contemporary trends in bacteriophage research. Nova Science, New York, pp 119–149
70. Kurtböke DI (2011) Exploitation of phage battery in the search for bioactive actinomycetes. Appl Microbiol Biotechnol 89(4):931–937
71. Kurtböke DI (2012) A phage-guided route to discovery of bioactive rare actinomycetes. In: Kurtböke DI (ed) Bacteriophages. InTech, Rijeka, pp 237–256. ISBN 978-953-51-0272-4
72. Kurtböke DI (2012) Biodiscovery from rare actinomycetes: an eco-taxonomical perspective. Appl Microbiol Biotechnol 93(5):1843–1852
73. Kurtböke DI, French JRJ (2007) Use of phage battery to investigate the actinofloral layers of termite gut microflora. J Appl Microbiol 103:722–734
74. Kurtböke DI, French JRJ (2008) Actinobacterial resources from termite guts for regional bioindustries. Microbiol Aust 29(1):42–44
75. Lamberty M, Zachary D, Lanot, Bordereau C, Robert A, Hoffmann JA, Bulet P (2001) Insect immunity—constitutive expression of a cysteine-rich antifungal and a linear antibacterial peptide in a termite insect. J Biol Chem 276:4085–4092
76. Lefebvre T, Miambi E, Pando A, Diouf M, Rouland-Lefèvre C (2009) Gut-specific actinobacterial community structure and diversity associated with the wood-feeding termite species, *Nasutitermes corniger* (Motschulsky) described by nested PCR-DGGE analysis. Insect Soc 56:269–276
77. Le Roes-Hill M, Goodwin C, Burton S (2009) Phenoxazinone synthase: what's in a name? Trends Biotechnol 27(4):248–258
78. Leroy PD, Sabri A, Verheggen FJ, Francis F, Thonart P, Haubruge E (2011) The semiochemically mediated interactions between bacteria and insects. Chemoecology. doi:10.1007/s00049-011-0074-6
79. Li ZQ, Liu BR, Zeng WH, Xiao WL, Li QJ, Zhong JH (2013) Character of cellulase activity in the guts of flagellate-free termites with different feeding habits. J Insect Sci 13:37

80. Liu N, Zhang L, Zhou H, Zhang M, Yan X, Wang Q, Long Y, Xie L, Wang S, Huang Y, Zhou Z (2013) Metagenomic insights into metabolic capacities of the gut microbiota in a fungus-cultivating termite (*Odontotermes yunnanensis*). PLoS ONE 8(7):e69184
81. Long Y-H, Xie L, Liu N, Yan X, Li M-H, Fan M-Z, Wang Q (2010) Comparison of gut-associated and nest-associated microbial communities of a fungus-growing termite (*Odontotermes yunnanensis*). Insect Sci 17:265–276
82. Margulis L, Fester R (1991) Symbioses as a source of evolutionary innovation: speciation and morphogenesis. MIT Press, Cambridge, MA
83. Matsui T, Tokuda G, Shinzato N (2009) Termites as functional gene resources. Recent Pat Biotechnol 3:10–18
84. Matsuura K (2001) Nestmate recognition mediated by intestinal bacteria in a termite, *Reticulitermes speratus*. Oikos 92:20–26
85. McCutcheon JP, Moran NA (2012) Extreme genome reduction in symbiotic bacteria. Nat Rev Microbiol 10:13–26
86. Minkley N, Fujita A, Brune A, Kirchner WH (2006) Nest specificity of the bacterial community in termite guts (*Hodotermes mossambicus*). Insect Soc 53:339–344
87. Moran NA (2006) Symbioses. Curr Biol 16(20):R866–R871
88. Moran NA, McCutcheon JP, Nakabachi A (2008) Genomics and evolution of heritable bacterial symbionts. Annu Rev Genet 42:165–190
89. Motohashi K, Ueno R, Sue M, Furihata K, Matsumoto T, Dairi T, Omura S, Seto H (2008) Studies on terpenoids produced by actinomycetes: oxaloterpins A, B, C, D, and E, diterpenes from *Streptomyces* sp. KO-3988. J Nat Prod 70(11):1712–1717
90. Moya A, Pereto J, Gil R, Latorre A (2008) Learning how to live together: genomic insights into prokaryote-animal symbioses. Nat Rev Genet 9:218–229
91. Mueller UG, Ishak H, Lee JC, Sen R, Gutell RR (2010) Placement of attine ant-associated *Pseudonocardia* in a global *Pseudonocardia* phylogeny (Pseudonocardiaceae, Actinomycetales): a test of two symbiont-association models. Antonie Van Leeuwenhoek 98:195–212
92. Nakajima H, Hongoh Y, Usami R, Kudo T, Ohkuma M (2005) Spatial distribution of bacterial phylotypes in the gut of the termite *Reticulitermes speratus* and the bacterial community colonizing the gut epithelium. FEMS Microbiol Ecol 54(2):247–255
93. Nalepa CA, Bignell DE, Bandi C (2001) Detritivory, coprophagy, and the evolution of digestive mutualisms in *Dictyoptera*. Insect Soc 48:194–201
94. Ni J, Tokuda G (2013) Lignocellulose-degrading enzymes from termites and their symbiotic microbiota. Biotechnol Adv 31(6):838–885
95. Nguyen KB, Smart GC Jr (1994) *Neosteinernema longicurvicauda* n. gen., n. sp. (Rhabditida: Steinernematidae), a parasite of the termite *Reticuldermes flavipes* (Koller). J Nematol 26(2):162
96. O'Brien J, Wright GD (2011) An ecological perspective of microbial secondary metabolism. Curr Opin Biotech 22:552–558
97. Oh DC, Poulsen M, Currie CR, Clardy J (2009) Dentigerumycin: a bacterial mediator of an ant-fungus symbioses. Nat Chem Biol 5(6):391–393
98. Ohkuma M, Brune A (2011) Diversity, structure, and evolution of the termite gut microbial community. In: Bignell DE, Roisin Y, Lo N (eds) Biology of termites: a modern synthesis. Springer Science + Business Media B.V, Dordrecht, pp 413–438
99. Palleroni NJ (1997) Prokaryotic diversity and the importance of culturing. A Van Leeuw 72:3–19
100. Pasti MB, Belli ML (1985) Cellulolytic activity of actinomycetes isolated from termites (Termitidae) gut. FEMS Microbiol Lett 26(1):107–112
101. Prach L, Kirby J, Keasling JD, Alber T (2010) Diterpene production in *Mycobacterium tuberculosis*. FEBS J 277:3588–3595
102. Prestwich GD (1979) Chemical defense by termite soldiers. J Chem Ecol 5:459–480
103. Prestwich GD (1984) Defense mechanisms of termites. Annu Rev Entomol 29:201–232

104. Quennedey A (1984) Morphology and ultrastructure of termite defense glands. In: Hermann HR (ed) Defensive mechanisms in social insects. Praeger, New York, pp 151–200
105. Řezanka T, Libalova D, Votruba J, Viden I (1994) Identification of odorous compounds from *Streptomyces avermitilis*. Biotechnol Lett 16(1):75–78
106. Řezanka T, Sobotka M, Prell A, Sigler K (2007) Relationship between volatile odorous substances and production of Avermectins by *Streptomyces avermitilis*. Folia Microbiol 52(1):26–30
107. Rosengaus RB, Guldin MR, Traniello JFA (1998) Inhibitory effect of termite fecal pellets on fungal spore germination. J Chem Ecol 24:1697–1706
108. Sachs JL, Skophammer RG, Regus JU (2011) Evolutionary transitions in bacterial symbioses. PNAS 108(2):10800–10807
109. Scherlach K, Graupner K, Hertweck C (2013) Molecular bacteria-fungi interactions: effects on environment, food, and medicine. Annu Rev Microbiol 67:375–397
110. Schöller CEG, Gürtler H, Pedersen R, Molin S, Wilkins K (2002) Volatile metabolites from actinomycetes. J Agric Food Chem 50:2615–2621
111. Scheuring I, Yu DW (2012) How to assemble a beneficial microbiome in three easy steps. Ecol Lett 15:1300–1307
112. Schmitt-Wagner D, Friedrich MW, Wagner B, Brune A (2003) Axial dynamics, stability, and interspecies similarity of bacterial community structure in the highly compartmentalized gut of soil-feeding termites (*Cubitermes* spp.). Appl Environ Microbiol 69:6018–6024
113. Seipke RF, Kaltenpoth M, Hutchings MI (2011) *Streptomyces* as symbionts: an emerging and widespread theme? FEMS Microbiol Rev 36(4):862–876
114. Sen R, Ishak HD, Estrada D, Dowd SE, Hong E, Mueller UG (2009) Generalized antifungal activity and 454-screening of *Pseudonocardia* and *Amycolatopsis* bacteria in nests of fungus-growing ants. Proc Nat Acad Sci USA 106:17805–17810
115. Shi W, Syrenne R, Sun J-Z, Yuan JS (2010) Molecular approaches to study the insect gut symbiotic microbiota at the 'omics' age. Insect Sci 17:199–219
116. Shinzato N, Muramatsu M, Matsui T, Watanabe Y (2007) Phylogenetic analysis of the gut bacterial microflora of the fungus-growing termite *Odontotermes formosanus*. Biosci Biotechnol Biochem 71(4):906–915
117. Shirai M, Okuda M, Motohashi K, Imoto M, Furihata K, Matsuo Y, Katsuta A, Shizuri Y, Seto H (2010) Terpenoids produced by actinomycetes: isolation, structural elucidation and biosynthesis of new diterpenes, gifhornenolones A and B from *Verrucosispora gifhornensis* YM28-088. J Antibiot 63(5):245–250
118. Siderhurst MS, James DM, Blunt TD, Bjostad LB (2005) Endosymbiont biosynthesis of norharmane in *Reticulitermes* termites (Isoptera: Rhinotermitidae). Sociobiology 45(3):687–705
119. Siderhurst MS, James DM, Blunt TD, Bjostad LB (2005) Antimicrobial activity of the termite (Isoptera) alkaloid norharmane against the entomopathogenic fungus *Metarhizium anisopliae*. Sociobiology 46(3):563–577
120. Sinma K, Ishida Y, Tamura T, Kitpreechavanich V, Tokuyama S (2011) *Saccharopolyspora pathumthaniensis* sp. nov., a novel actinomycetes isolated from termite guts (*Speculitermes* sp.). J Gen Appl Microbiol 57(2):93–100
121. Skujins JJ, Potgieter HJ, Alexander M (1965) Dissolution of fungal cell walls by a streptomycete chitinase and β-(1→ 3) glucanase. Arch Biochem Biophys 111(2):358–364
122. Šobotník J, Jirošová A, Hanus R (2010) Chemical warfare in termites. J Insect Physiol 56(9):1012–1021
123. Smanski MJ, Peterson RM, Huang SX, Shen B (2012) Bacterial diterpene synthases: new opportunities for mechanistic enzymology and engineered biosynthesis. Curr Opin Chem Biol 16(1):132–141
124. Takagi H, Motohashi K, Miyamoto T, Shin-ya K, Furihata K, Seto H (2005) Studies on terpenoids produced by actinomycetes isolation and structural elucidation of antioxidative agents, Naphterpins B and C. J Antibiot 58(4):275–278

125. Tartar A, Wheeler MM, Zhou X, Coy MR, Boucias DG, Scharf ME (2009) Parallel metatranscriptome analyses of host and symbiont gene expression in the gut of the termite *Reticulitermes flavipes*. Biotechnol Biofuels 2:25. doi:10.1186/1754-6834-2-25
126. Thorne BL (1997) Evolution of eusociality in termites. Annu Rev Ecol Syst 28:27–54
127. Tiunova NA, Kobzeva N, Zaĭkina IV, Eliakova LA, Nazarova NI (1983) 1, 3-Beta-glucanases of actinomycetes. Mikrobiologiia 52(4):586
128. Valterová I, Křeček J, Vrkoč J (1989) Intraspecific variation in the defense secretions of *Nasutitermes ephratae* soldiers and the biological activity of some of their components. Biochem Syst Ecol 17(4):327–332
129. Vernekar JV, Ghatge MS, Deshpande VV (1999) Alkaline protease inhibitor: a novel class of antifungal proteins against phytopathogenic fungi. BiochemBiophys Res Commun 262(3):702–707
130. Visser AA, Nobre T, Currie CR, Aanen DK, Poulsen M (2012) Exploring the potential for symbionts in fungus-growing termites. Microb Ecol 63:975–985
131. Watanabe Y, Shinzato N, Fukatsu T (2003) Isolation of actinomycetes from termites' guts. Biosci Biotechnol Biochem 67(8):1797–1801
132. Wier A, Dolan M, Grimaldi D, Guerrero R, Wagensberg J, Margulis L (2002) Spirochete and protist symbionts of a termite (*Mastotermes electrodominicus*) in Miocene amber. Proc Nat Acad Sci USA 99:1410–1413
133. Wilkins K, Schöller C (2009) Volatile organic metabolites from selected *Streptomyces* strains. Actinomycetologica 23:27–33
134. Woyke T, Teeling H, Ivanova NN, Huntemann M, Richter M, Gloeckner FO, Boffelli D, Anderson IJ, Barry KW, Shapiro H, Szeto E, Kyrpides NC, Mussmann M, Amann R, Bergin C, Ruehland C, Rubin EM, Dubilier N (2006) Symbioses insights through metagenomic analysis of a microbial consortium. Nature 443:950–955
135. Wyatt TD (2003) Pheromones and animal behavior: communication by smell and taste. Cambridge University Press, Cambridge
136. Xu P, Shi M, Lai R, Chen XX (2012) Differences in numbers of termicins expressed in two termite species affected by fungal contamination of their environments. Genet Mol Res 11(3):2247–2257
137. Yomosa K, Hirota A, Sakai H, Isogai A (1987) Isolation of harman and norharman from *Nocardia* sp. and their inhibitory activity against plant seedlings. Agric Biol Chem Tokyo 51(3):921–922
138. Zhao C, Rickards RW, Trowell SC (2004) Antibiotics from Australian terrestrial invertebrates Part 1: antibacterial trinervitadienes from the termite *Nasutitermes triodiae*. Tetrahedron 60:10753–10759
139. Zhu Y, Li J, Liu H, Yang H, Xin S, Zhao F, Zhang X, Tian Y, Lu X (2012) Phylogenetic analysis of the gut bacterial microflora of the fungus-growing termite *Macrotermes barneyi*. Afr J Microbiol Res 6(9):2071–2078

Adv Biochem Eng Biotechnol (2015) 147: 137–183
DOI: 10.1007/10_2014_276

Published Online: 15 October 2014

Bioresources for Control of Environmental Pollution

Barindra Sana

Abstract Environmental pollution is one of the biggest threats to human beings. For practical reasons it is not possible to stop most of the activities responsible for environmental pollution; rather we need to eliminate the pollutants. In addition to other existing means, biological processes can be utilized to get rid of toxic pollutants. Degradation, removal, or deactivation of pollutants by biological means is known as bioremediation. Nature itself has several weapons to deal with natural wastage and some of them are equally active for eliminating nonnatural pollutants. Several plants, microorganisms, and some lower eukaryotes utilize environmental pollutants as nutrients and some of them are very efficient for decontaminating specific types of pollutants. If exploited properly, these natural resources have enough potential to deal with most elements of environmental pollution. In addition, several artificial microbial consortia and genetically modified organisms with high bioremediation potential were developed by application of advanced scientific tools. On the other hand, natural equilibria of ecosystems are being affected by human intervention. Rapid population growth, urbanization, and industrialization are destroying ecological balances and the natural remediation ability of the Earth is being compromised. Several potential bioremediation tools are also being destroyed by biodiversity destruction of unexplored ecosystems. Pollution management by bioremediation is highly dependent on abundance, exploration, and exploitation of bioresources, and biodiversity is the key to success. Better pollution management needs the combined actions of biodiversity conservation, systematic exploration of natural resources, and their exploitation with sophisticated modern technologies.

Keywords Bioremediation · Biodiversity · Exploration · Natural resources · Environmental tools

---

B. Sana (✉)
Division of Bioengineering, School of Chemical and Biomedical Engineering,
Nanyang Technological University, Singapore 637457, Singapore
e-mail: barindrasana@yahoo.com

Contents

1 Introduction 138
2 Methods for Exploration and Development of New Bioresources 140
2.1 Nonmolecular Bioprospecting 141
2.2 Molecular Bioprospecting Techniques 142
2.3 Genetic Engineering Techniques 145
3 Bioresources in Controlling Environmental Pollutions: Recent Advances 147
3.1 Microbial Resources 148
3.2 Plants and Phytoremediation 159
3.3 Nonmicroscopic Lower Eukaryotes 170
4 In-Field Situation of Bioremediation: Limitations and Challenges 173
5 Conclusion and Future Perspectives 175
References 176

1 Introduction

The ever-growing global population and industrialization are causing severe pressure on the atmosphere of our planet. An increase in the number of people calls for an increase in resources, which come directly or indirectly from nature at the cost of a balanced ecosystem. Manufacturing industries, mining, refineries, and power plants are well-known sources of toxic and hazardous chemicals that pollute air, water, and land. Vehicular pollutants, domestic and municipal wastage, agricultural and fish farm wastage, and pesticides are also a considerable source of pollutants. The excessive release of solid waste, wastewaters, industrial sludge, and slurries are deteriorating water and soil quality, and increased gaseous emissions cause air pollution and increase the atmospheric level of greenhouse gases. Most of the pollutants can be absorbed or degraded by natural activities of plants and microorganisms but the Earth is losing its intrinsic remediation power due to the destruction of natural ecosystems. As a result, the atmosphere is being contaminated with cumulative pollutants. Increased demand of land and continuous extraction of natural resources also make significant contributions to the destruction of ecosystems. Natural CO_2 absorption is sharply decreasing due to extreme deforestation and destruction of marine ecosystems by destructive fishing techniques and oil/chemical spilling. However, the problems can still be addressed by using natural resources; the processes are generally known as bioremediation or phytoremediation, where the biological agents are used to clean up polluted environments. Bioremediation is an ecofriendly technology that does not use any acid, alkali, or toxic chemicals. In addition, it works at lower temperature and pressure that consume less energy than conventional chemical processes.

Bioremediation techniques use microbial reactions (metabolisms) for removing contaminants from polluted soil, sediment, and water. Microorganisms can catalyze several types of reactions including hydrolysis, cleavage, oxidation-reduction, substitution, dechlorination, dehydrogenation, and dehydrohalogenation [134].

Biological decontamination techniques typically rely on enhancement of biodegradation, biotransformation, or biosorption of the contaminants by promoting the growth of specific bacteria, fungi, microalgae, or a mixed microbial consortium that can use the pollutants as energy or carbon sources and convert them into nonhazardous or less hazardous compounds [6]. Bioremediation by stimulating the growth and metabolism of an indigenous microbial community is known as biostimulation. Microbial activities can be stimulated by addition of growth-promoting substances (nutrients or some special chemicals), by enhancement of oxygen availability, by controlling physical parameters, or by using a combination of these techniques [6, 58, 121]. The indigenous microbial inhabitants of the contamination sites are often well adapted to survive at the physicochemical condition of the contaminated fields. In addition, their metabolism often depends on utilization of the contaminants as nutrient or electron acceptor. In fact, indigenous microorganisms are working unnoticed, day and night, to keep the environment free of contaminants. However, introduction of selected nonnative microorganisms is a useful strategy when the indigenous microorganisms cannot remove certain pollutants. Biological decontamination by addition of one or more external microbial species is known as bioaugmentation. Natural or synthetic microbial consortia and even specially designed genetically engineered microorganisms can be used for bioaugmentation when no known natural species shows the desired activity.

Bioremediation can be performed in situ by artificially increasing the desired microbial activity at the contaminated sites or by ex situ methods where the contaminated material is removed from the site and decontaminated by some special treatment, for example, using bioreactors. In situ bioremediation techniques include bioventing, biostimulation, bioaugmentation, biosparging, and some composting methods. In situ techniques are preferred over ex situ techniques due to fewer equipment requirements and economic feasibility. However, a variety of physical, chemical, and biological factors of contaminated sites determine the metabolic activity of microbial populations and hence their survival and bioremediation ability [134]. Ex situ bioremediation is a relatively expensive method that needs treatment plants and several pieces of equipment. Application of this technique is not restricted by physicochemical conditions of the polluted site and it is extremely useful if the pollutants need (chemical) pretreatment prior to bioconversion. Although this technique is not practicable for very large-scale pollution, the technique is advantageous for recovering valuable substances from the pollutants.

Plants are often used for remediation of heavy-metal–polluted soil and water. Certain plant species can absorb heavy metals and accumulate them in their biomass, which are then harvested and processed for safe disposal of heavy metals or even for their extraction and reuse. They are also able to degrade a few organic and inorganic compounds including petroleum hydrocarbons and pesticides. Plants play a most important role in controlling air pollution by absorbing toxic gases including several greenhouse gases. The use of green plants for decontaminating polluted soil, water, or air is known as phytoremediation. Plants are also used for aeration of polluted soil or water to facilitate microbial bioremediation. In many cases, successful pollutant removal depends on the combined action of plants and microorganisms. Genetic engineering techniques are also applied for development of transgenic plants with enhanced phytoremediation capability. Phytoremediation techniques are useful in removing pesticides, solvents, explosives, and crude oil contaminants, and also for long-term restoration of large contaminated areas such as mining grounds.

2 Methods for Exploration and Development of New Bioresources

Natural resources have unimaginable potential for establishing and maintaining perfect ecological equilibrium in favor of a sustainable ecosystem. The systematic search for discovery and development of new sources of chemical compounds, genes, microorganisms, macroorganisms, or any other valuable products from natural resources is termed bioprospecting. It is an umbrella term that includes a range of processes and techniques starting from very primitive prehistoric practices to modern molecular and genomic techniques. Screening of many bioremediating agents was made possible by bioprospecting versatile ecosystems. In addition to bioprospecting natural resources, genetic engineering techniques are also applied for development of suitable bioremediation agents. This section focuses on currently used most common bioprospecting techniques and their relevance in pollution control. The techniques are diagrammatically presented in Fig. 1.

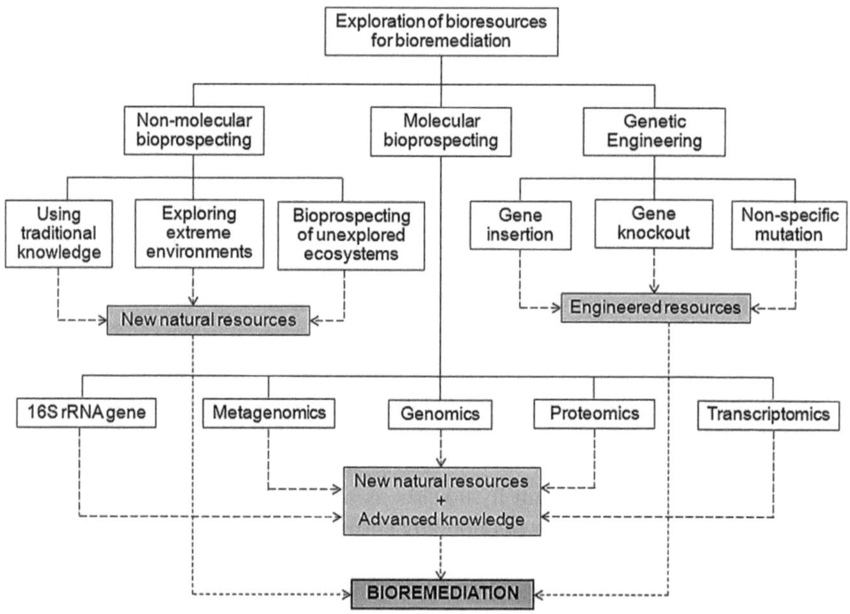

Fig. 1 Diagrammatic presentation bioresource exploration for bioremediation

2.1 Nonmolecular Bioprospecting

In traditional bioprospecting practices, natural resources were randomly screened for desired chemical or biological activities without any scientific understanding and the crude products were often used without further purification. However, traditional knowledge played a significant role in isolating several useful compounds from known species of plants, animals, and microorganisms [29, 40, 113]. Most common nonmolecular bioprospecting programs use assay-based screening techniques to isolate the targeted chemical compound, microorganism, plant, or animal from a rationally selected source, although identification of the microorganism or the active compound is usually performed by molecular techniques. The exploration sites (ecosystems) are usually selected based on a set of physical, chemical, and biological parameters whereas active biomolecules are screened in plant, animal, or microbial species most likely to have the desired activity; for example, hot springs can be explored for thermophilic microorganisms and actinomycetes are frequently screened for novel antimicrobial compounds [19]. The natural inhabitants of an ecosystem need to adapt themselves to its physical and chemical conditions, which make extreme environments a potential source of novel compounds [103]. For the same reason, polluted environments are a rational source of pollutant-degrading organisms; intrinsic plant or microbial species of a contaminated site may be most active in its bioremediation. Exploring previously unexplored unique ecosystems also increases the chance of discovering novel organisms and their metabolites; several novel compounds are being isolated from newly discovered marine animal, plants, macroalgae, and microorganisms [40, 71, 74]. Despite significant advancement of molecular and genetic techniques, nonmolecular bioprospecting is still the most successful method for discovering novel bioresources.

The success of activity-based nonmolecular microbial bioprospecting is generally limited by the fact that the microorganism needs to grow and produce the targeted metabolite under laboratory conditions. Nonculturable microorganisms constitute a significant fraction of most microbial communities and they have equal potential for producing any targeted product. In addition, many valuable microbial metabolites are produced by unknown stress-induced metabolic pathways or by interspecies interaction of the natural microbial community [85, 92, 114]. These products would remain undiscovered without application of modern molecular tools, which help to study interspecies interaction, nonculturable microbial communities, and their metabolic pathways [46, 54, 143]. Animal and plant products isolated by nonmolecular bioprospecting techniques often face challenges for large-scale production, economic feasibility, and long-term viability. Their continuous uninterrupted isolation from the wild is nearly impossible and large-scale artificial cultivation of many species is either impossible or impracticable. However, these problems can be more conveniently solved by application of genetic engineering techniques and bioprospecting can still be carried out with nonmolecular approaches.

2.2 *Molecular Bioprospecting Techniques*

Environmental bioprospecting got a new tempo with the recent development of molecular biology, bioinformatics, and high-throughput screening techniques [31, 34, 42, 142, 150]. Molecular techniques are used for understanding the composition of total microbial communities including nonculturable microorganisms. Genomics and proteomics approaches are useful for understanding what genes or metabolic pathways are involved in a particular activity such as biodegradation.

2.2.1 16S rRNA Gene Analysis Approach

The sequence-based identification of microbial species caused significant advancement in the field of microbial bioprospecting [67, 113, 122]. The 16S rRNA gene sequence is used for identification of microbial isolates. This approach is useful for understanding what microorganisms are present in an ecosystem or exactly which species exert the desired activity. 16S rRNA genes can function as an "evolutionary clock" because their nucleotide sequences are conserved within a microbial species. The16S rRNA gene sequence of an isolate can be compared with a database, which establishes phylogenetic location of the newly isolated microorganism in the microbial kingdom. Identification and classification of microbial isolates are performed using isolated DNA from a pure culture whereas microbial diversities in environmental samples are estimated through metagenomic approaches. As a result, the 16S rRNA gene analysis approach has the potential to identify new microbial species (including nonculturable species) that play an active role in environmental bioremediation, which are otherwise impossible to identify by nonmolecular techniques. This method has added significantly to the field of microbial bioremediation because it can be used to understand changes of microbial composition by comparing the species present at different stages of bioremediation. In addition, understanding phylogenetic locations of pollutant-degrading microbial species may be helpful for designing artificial microbial consortia for bioremediation of specific pollutants. The 16S rRNA gene analysis approach is very effective in identifying microorganisms but phylogenetic knowledge of bioremediating microbes does not necessarily predict physiological or functional aspects of a new isolate. Metabolic functions of certain microbial strains are sometimes predicted by comparing those of nearly related species although they may not be true in all cases [76]. 16S rRNA genes are directly related only to microbial evolution and straightforward prediction of functional aspects may cause gross misinterpretation of the phylogenetic information. Use of this technique should be limited only to phylogenetic characterization and identification of microorganisms.

2.2.2 Functional Genomics and Proteomics Approaches

Functional genomics and proteomics are great tools for understanding which genes and metabolic pathways are involved in bioremediation. These tools are extremely useful for understanding the dynamic aspects of gene expression and protein function at the level of transcription, translation, protein–DNA, and protein–protein interactions. Functional genomics and proteomics studies give a complete picture of how biological function arises from the information encoded in an organism's genome. Similar studies are also possible with the genome, transcriptome, or proteome of an entire community, which are usually covered under the subfields of functional metagenomics, metatranscriptomics, or metaproteomics. All these studies use advanced sequencing and high-throughput technologies to establish the function of genomes at the levels of DNA, RNA transcripts, and protein products, which explain the relationship between an organism's genomic structure and its phenotypic response to environmental changes.

The major advantages of the functional genomics approach come from the fact that it expands the scope of biological studies from investigating a single gene expression to investigating expression of all the genes of a biological system. Quantifying the total gene expression in a microbial cell is useful in understanding upregulation and downregulation of genes in response to a toxic pollutant, which in turn provide information regarding the mechanisms involved in the defense, detoxification, or adaptation to the polluted surroundings. Cellular changes, including induction and expressions of regulatory proteins/enzymes in response to external stimuli such as aromatic hydrocarbons and heavy metals, were studied in detail using the techniques of functional genomics and proteomics [115, 150–152]. These studies revealed that pollutant degradation is a complex phenomenon involving several auxiliary proteins (such as heat shock proteins and membrane proteins) in addition to the catabolic enzymes. Identification of these proteins and their corresponding genes would be useful in improving the bioremediation capability of microorganisms by genetic engineering [83]. In addition, the specific pollutant-degrading capability of a single microorganism or a microbial community can be assessed by analyzing the total proteome or transcriptome of the microorganism(s) grown in the presence of the pollutant; upregulation of the relevant catabolic enzyme(s) will indicate potential bioremediation application of the microbial species or community. However, microbial degradation of pollutants is extremely complex at the molecular level and application of functional genomics in the field of environmental biotechnology is still in its infancy.

2.2.3 Metagenomic Approaches

Metagenomics is a very useful technique for understanding the entire microbial community of a polluted site or a site undergoing bioremediation treatment. It is the technique for studying metagenomes, the collective genetic material isolated directly from an environmental sample without culturing the microorganisms. This technique provides direct access to the genome of an entire microbial community whereas traditional cultivation-based genomics can analyze only culturable microorganisms and thus miss a significant fraction of the (nonculturable) microbial population. Metagenomics is extremely relevant in bioremediation because vast majorities of microbial populations of contaminated sites are adapted to extreme environments and cannot be cultured in any defined laboratory condition. A better understanding of how microbial communities cooperatively cope with toxic pollutants helps improve bioaugmentation or biostimulation strategies.

Metagenomics commonly refers to construction and analysis (screening) of metagenomic libraries. Although a range of techniques is used for metagenomic library construction, a typical protocol can be divided into the following steps: (1) extraction of environmental DNA and generation of DNA fragments of appropriate size, (2) ligation of the fragments (insert) into appropriate cloning vectors (e.g., cosmid, fosmid, or bacterial artificial chromosome) depending on their size, (3) transformation of the recombinant construct into a suitable host cell, and (4) sequencing of the clones. The result is a metagenomic library, consisting of thousands of cells carrying the DNA fragments from the metagenome. The next step is screening the clones containing specific sequences that are responsible for particular activities or exploring functional and genetic diversities. Metagenome analysis can also reveal unknown DNA sequences that describe novel functions of environmental microorganisms, impossible to discover by culture-based techniques. However, metagenome isolation from a chemically complex, highly undefined polluted environment (such as industrial sludge, wastewater, or acid mine drainage) itself is a challenging job due to interference of inhibitory contaminants. Extraction of total metagenomic DNA for successful representation of all microbial genomes involves vigorous extraction methods that cause DNA shearing to low-size fragments [37]. Moreover, directly extracted metagenomic DNA from some heavily polluted environments that contains very low cell densities may not be sufficient for subsequent library construction; it needs PCR-independent whole genome amplification techniques [25, 127]. Such a technique for whole-genome amplification using a minute quantity of metagenomic DNA may introduce some amplification bias but give access to information that would otherwise remain inaccessible [4].

Nonetheless, screening or analysis of a metagenomic library is also a challenging and laborious job. Two different screening techniques are used in the bioprospecting of metagenomic libraries: function-driven screening for an expressed trait, and sequence-driven screening for a specific DNA sequence. Function-driven screenings are often based on heterologous expression of the desired trait and selection is dependent on successful expression of the functional gene at a detectable level. The success rate of this approach decreases when the signals are detected by low-throughput screening techniques [127]. Fortunately, recent applications of several automated techniques such as automated colony picking, pipetting robotics, use of microtiter plates, and informatics-assisted data management have improved the success rate of function-driven metagenomic library screening [75]. The function-driven screening "hit" rate can also be increased by simultaneous use of several expression hosts, by enrichment of the community genome with desired traits prior to the DNA isolation for metagenomic library construction, substrate-induced gene expression, and by application of novel high-throughput screening strategies [104]. In contrast, the screening hit of sequence-driven analysis is much higher due to the availability of sophisticated high-throughput molecular tools. A sequence of interest can be detected by designing suitable PCR primers (based on conserved DNA sequences) or by hybridization with target-specific probes (such as group-specific 16S rRNA-targeted oligonucleotide probes). Both PCR-based and hybridization techniques are dependent on information from databases and, therefore, these techniques can only be applied for the identification of new members of known gene families [38]. Also linking a metabolic function to specific microbial species needs phylogenetic and functional genes to be inserted in the same construct; this is possible only by constructing large DNA insert libraries.

Shotgun sequencing and screening of cloned libraries is another useful and highly sophisticated approach to understanding the entire metagenome of any environment. Using this method it is possible to understand which microbes are present in the ecosystem and what metabolic processes are possible in the microbial community [43]. However, predicting correct metabolic pathways using this massive amount of data is very challenging especially for complex environmental samples. Accuracy of the sequence-driven approach is also very much dependent on the reliability of the information available in various sequence databases. In practice, a combination of function-driven and sequence-driven approaches may be the best solution for understanding a complex and dynamic microbial ecosystem of a site undergoing bioremediation treatment.

2.3 Genetic Engineering Techniques

Natural plants and microbial species are very effective in biological remediation of various polluted sites but they often fail to deal with some complex pollutants, especially when the pollutants are present in high concentration [52]. With the advancement of science, newer chemicals and complex polymers are being synthesized every day that cannot be degraded by any natural organism. Genetic engineering can play some role here. Plants and microorganisms can be engineered to enhance the bioremediating efficacy of already active bioremediating organisms [137]. For example, heavy-metal phytoremediating plants can be engineered to increase the number of metal transporters, to enhance intracellular ligand production that keeps accumulated metalloids in a safe form, or to biochemically transform the absorbed metals to less toxic or volatile derivatives. Novel plant and microbial strains with the desired bioremediation properties can be developed by engineering catabolic enzyme affinity and specificity, by modifying catabolic pathways, and by improvement of genetic stability. Several engineered organisms were reported to degrade the pollutants that are difficult to break down using native species [41, 65].

Genetic engineering is the biotechnological process for modifying the genome of a living organism, which is typically done by insertion or manipulation of one or more gene in the genome of the organism. Insertion is a widely used genetic engineering technique in bioremediation and usually practiced by transferring one or more genes from one organism to another [144]. Insertion of foreign gene(s) is essential when the most suitable organism (as per other criteria) does not have the desired activity and the source organism (that contains the gene) could not be used in that particular application due to some practical limitations. The most important consideration for inserting a foreign gene is to determine which gene to add. The answer is directly related to what novel function is desired in the target organism or what problem is to be solved using the engineered organism. The next point is where to find the gene or how to search for it. Screening relevant genetic libraries is often helpful to find the correct gene. Potentially any gene from any organism or even a synthetic gene that encodes the desired protein can be introduced in the organism but it must be properly linked to the expression machinery of the host. This raises the question of where to insert the gene and how. The problem is relatively simple if the gene can be introduced, maintained, and expressed using an extrachromosomal vector (such as plasmid). However, the stability of the extra-chromosomal DNA and availability of expression machinery are among the other important concerns [59]. The integration site should be carefully selected when the foreign gene is to be inserted within the organism's chromosome; it should not interrupt any native functional gene. A range of techniques is available for making a DNA construct (insert + vector), for inserting the construct into the target cell, and for selecting the properly transformed cells, but the researcher needs to decide which one is most suitable for the experimental system.

The function of an indigenous gene can be modulated by inserting, deleting, or replacing one or a few nucleotides. The process is called genetic mutation. Correctly designed (site-directed) mutation may cause a significant change of physical properties, structure, or function of the engineered gene product (protein). Site-directed mutations are achieved by using the finely tuned molecular biology tools that can selectively change only the targeted nucleotide(s) in the entire genome. Mutation can also be introduced in nonselective random sites of the genome by physical or chemical stress [26, 51, 101]. This technique is very useful for developing or improving the performance of bioremediating microorganisms. Usually a microbial consortia or pure culture is exposed to a small dose of chemical or radiation for a certain period of time that causes mutation in the microorganism and some mutants may show enhanced pollutant degradation [32, 49, 82]. A set of dose and exposure periods can be tried to achieve the desired activity of the microorganisms and the mutation can be studied in detail by screening (identifying) the mutant genes using molecular biology techniques. Gene knock-out is another process for deletion or inactivation of indigenous genes. This process may be accomplished through gene targeting by insertion of a specific DNA construct that inactivates the target gene by homologous recombination or by introducing engineered nucleases (such as zinc-finger nuclease) that can target a specific DNA sequence in the gene to be activated and disable its expression. The inserted construct or nuclease should not affect the function of any other gene. Selection of the correct genes and understanding their function is very important for genetic engineering by deletion or inactivation.

Genetically engineered organisms have huge potential but there are some serious concerns regarding their negative impacts on the environment and human health, especially if released in nature. The ecological balance can be threatened if the engineered organisms preferentially grow over the indigenous species because all indigenous organisms have some role to play in maintaining the ecosystem. Gene flow is the biggest concern in releasing genetically modified microorganisms in the environment. Engineered genes can be horizontally transferred to wild microorganisms if genetically modified microorganisms are introduced in the environment for in situ bioremediation. Antibiotic resistance is a commonly used marker to identify accurate insertion of a gene but the consequence would be dangerous if the antibiotic resistance gene is spread in wild bacteria and pathogens. Predicting ecological consequences or health hazards from genetically modified organisms is an extremely difficult job for scientists and yet nothing can be confirmed in advance, which causes additional concern over the potential risks of the genetically modified organism. The univocal debate is treating all genetically engineered organisms in the same line irrespective of their spreading potential, horizontal gene transfer capability, or regeneration capability in nature. However, many concerns are genuine and all potential risks must be completely eliminated before releasing any engineered organisms into the environment. One good approach is the use of "suicide genes," the genes that cause bacteria death after complete degradation of the toxic chemicals [96, 98]. This technique would be useful in addressing most of the problems associated with the use of genetically engineered microorganisms for bioremediation.

3 Bioresources in Controlling Environmental Pollutions: Recent Advances

Bioremediation has been practiced for a long time but recent developments are different in terms of a systematic approach and use of sophisticated modern technologies. Research is intensified mainly in the area of discovery and development of suitable bioremediating organisms and their bioremediation-related characterization. Plants, microorganisms, and some lower eukaryotes have shown potential bioremediation activity. The following section focuses on recent research (2008–2013) on bioremediation using different species of microorganisms, plants, and nonmicroscopic lower eukaryotes and their potential application in pollutant removal from contaminated soil, water, and air.

3.1 Microbial Resources

Microorganisms are the key players in decomposition of different kind of waste and thus extremely useful in restoration of polluted soil and water. Several microbial processes have also been established for treatment of industrial solid waste, sludge, and slurries. Microorganisms produce diverse enzymes to degrade complex natural and industrial waste and use them as nutrients. They can easily come in contact with the most number of contaminant molecules due to their small size and fast reproduction in suitable environments. Microbial degradation of organic matter also supplies the plant's nutrients and thus plays an important role in ecological nutrient recycling.

To date, bioremediation is practiced mostly by enhancing growth of the native or nonnative natural microbial communities, although researchers are investigating some synthetic microbial consortia for favoring degradation of certain pollutants. Many efforts are directed in developing engineered microbial species or communities that are especially suitable for degrading specific contaminant(s). Research is also focused on the understanding of reactions behind bioremediation and requirements of nutritional supplements [62, 78, 134]. Microbial process development for ex situ bioremediation has also attracted significant attention in recent years. The following sections focus on recent developments of microbial bioremediation for controlling soil, water, and air pollution. The findings are summarized in Table 1.

3.1.1 Soil and Solid Waste Treatment

Bioremediation is successfully applied in in situ and ex situ treatment of contaminated soil. In situ degradation of contaminants in the subsurface depends on the type of contaminants, the type of microorganisms, and the physicochemical conditions. Soil contains several types of microorganisms including bacteria, actinobacteria, cyanobacteria, fungi, microalgae, and protozoa. Usually bacteria are far more numerous than any other soil microbes and are the most focused group for bioremediation, mainly due to their rapid growth and fast metabolic rate. However, other microbial communities can also catalyze bioremediation reactions independently and in many cases they serve as essential components of the bioremediating consortia.

Numerous microbial strains were reported to degrade petroleum hydrocarbons from contaminated soils. Several bioremediation techniques were used individually as well as in combination. Laboratory-scale bioremediation of petroleum-contaminated soil was studied by enhancing growth of the indigenous microorganisms of the contaminated site with and without addition of a petroleum-degrading *Pseudomonas aeruginosa* strain [61]. Up to 94 % *n*-octane was removed from the soil sample after 191 days of treatment. The rate and extent of the bioremediations were not substantially changed by bioaugmentation with *P. aeruginosa*. This observation suggests that the indigenous microbial community may be the best choice for decontaminating a polluted site, but only after working out the proper biostimulation protocol.

Table 1 Summary of recent research (2008–2013) on microbial bioremediation of soil, water, and air

Active microorganism(s)	Target pollutants	Bioremediation methods	Reference
For bioremediation of contaminated soil and solid waste			
Pseudomonas aeruginosa	Petroleum hydrocarbons	Biostimulation	[61]
Gordonia alkanivorans CC-JG39, *Rhodococcus erythropolis* CC-BC11, *Acinetobacter junii* CC-FH2, *Exiguobacterium aurantiacum* CC-LSH-4, and *Serratia marcescens* KH1	Aromatic hydrocarbons	Bioaugmentation and biostimulation	[73]
Arthrobacter, *Pseudomonas*, *Rhodococcus*, *Bacillus*, *Exiguobacterium*, and *Delftia*	C_{12}–C_{20} hydrocarbons, isoprenoids	Bioaugmentation	[8]
UV-induced mutant of *Acinetobacter* sp. YC-X2, *Kocuria* sp. YC-X4, and *Kineococcus* sp. YC-X7	Viscous oil hydrocarbon	Bioaugmentation	[33]
Ascomycota, Actinomycetes, Proteobacteria, Firmicutes, and Chloroflexi	Oil refinery sludge	Biostimulation	[108]
Flavobacterium and *Aspergillus*	Petroleum hydrocarbon	Biostimulation	[112]
Bacillus sp., *Chromobacterium* sp., *Enterobacter* sp., *Achremonium* sp., and *Aspergillus* sp. and *Verticillium* sp.	Naphthalene, phenanthrene, anthracene, pyrene, dibenzo[a] anthracene, benzo[a] pyrene	Bioaugmentation	[118]
Rhodococcus ruber Em1	PAH	Bioaugmentation	[128]
Sphingomonas sp., *Sphingomonas wittichii* RW1, *Pseudomonas veronii* PH-03, *Paenibacillus* sp. VSE5L, *Phanerochaete chrysosporium* DSM 6909, *Phanerochaete chrysosporium* DSM 1556, *Irpex* sp. KW3, *Trametes* sp. CH2, and *Fusarium* sp. VSO7	Polychlorinated dibenzo-*p*-dioxins and dibenzofurans	Ex situ bioremediation	[93]
Trichoderma longibrachiatum and *Byssochlamys spectabilis*	PAH	Ex situ bioremediation	[109]
Achromobacter, *Alcaligenes*, *Pseudomonas*, and *Pusillimonas*	PAH, Arsenic	Ex situ bioremediation	[136]

(continued)

Table 1 (continued)

Active microorganism(s)	Target pollutants	Bioremediation methods	Reference
Alcaligenes sp., *Pseudomonas* sp., *Pandorea* sp., and *Paenibacillus* sp.	PAH, Cadmium	Ex situ bioremediation	[135]
Anthracophyllum discolor	2,4-dichlorophenol, 2,4,6-trichlorophenol, pentachlorophenol	Bioaugmentation	[5, 30, 44]
β-Proteobacteria, γ-Proteobacteria, Ascomycota, and Basidiomycota	Pentachlorophenol	Biostimulation	[30]
Kocuria rhizophila, *Microbacterium resistens*, *Staphylococcus equorum*, and *Staphylococcus cohnii*	Lindane	Bioaugmentation	[2]
Streptomyces sp. M7	Lindane	Bioaugmentation	[21]
Ochrobacterum, *Burkholderia*, *Pseudomonas*, and *Arthrobacter*	Endosulfan	Bioaugmentation, Ex situ bioremediation	[68]
Bacillus subtilis, Pseudomonas aeruginosa, and *Saccharomyces cerevisiae*	Chromium (Cr)	Ex situ bioremediation	[20]
Graphium putredinis, *Fusarium solani*, *Fusarium sp.*, and *Penicillium chrysogenum*	Cd, Cr, Ni, Pb, and Zn	Ex situ bioremediation	[138]
For bioremediation of contaminated water, sludge, and slurry			
Firmicutes	Tetrachloro-ethylene	Natural attenuation	[24]
β-proteobacteria		Biostimulation	
γ-Proteobacteria, Actinobacteria, Firmicutes, *Penicillium*, *Candida*, *Geotrichum*, *Pichia*, and *Cladosporium Aschochyta*	Polyphenols	Ex situ bioremediation	[92]
Thiomicrospira sp., *Achromobacter* sp., *Cyclobacterium linum*, and *Nitromonas halophila*	Total organic carbon (TOC)	Ex situ bioremediation	[110]
Pseudomonas, Bacillus, *Pannonibacter*, and *Ochrobacterum*	TOC	Ex situ bioremediation	[69]
Proteobacteria, *Chroococcus*, *Lyngbya*, and *Nitzschia*	Orthophosphate, ammonium, nitrite, and nitrate	Bioaugmentation	[145]
Scenedesmus sp. AMDD	Dissolved nitrogen and phosphorous	Ex situ bioremediation	[84]

(continued)

Table 1 (continued)

Active microorganism(s)	Target pollutants	Bioremediation methods	Reference
Scenedesmus acutus PVUW12	Dissolved nitrogen	Ex situ bioremediation	[45]
For remediation of air pollution			
Nostoc commune, *Leptolyngbya thermalis*, and *Gloeotila* sp.	Hydrocarbon	Ex situ bioremediation	[7]
Engineered *Chlorella* species	CO_2, NO, SO_2	Bioaugmentation	[35]
Chlorella vulgaris	CO_2	Bioaugmentation	[47]

As a member of the microbial consortia, several *Pseudomonas* species also contributed in the bioremediation of petroleum hydrocarbons, polyaromatic hydrocarbons (PAH), and organochlorine pesticides [8, 68, 136]. However, most studies established bioaugmentation as an efficient bioremediation technique and in many cases it is extremely useful in combination with biostimulation. Biodegradation of petroleum hydrocarbon of an oil storage site was studied with more than one bioaugmentation and biostimulation technique, individually and in combination [73]. Bioaugmentation with a defined microbial consortium containing two diesel-degrading strains (*Gordonia alkanivorans* CC-JG39 and *Rhodococcus erythropolis* CC-BC11) and three oil-degrading strains (*Acinetobacter junii* CC-FH2, *Exiguobacterium aurantiacum* CC-LSH-4, and *Serratia marcescens* KH1) removed ~65 % aromatic hydrocarbon after 140 days of treatment and the polar components were preferably decomposed by the addition of a kitchen waste consortia supplemented with low-level nutrient. Treatment with the nutrient-supplemented kitchen waste compost itself degraded more than 80 % of the hydrocarbon contaminants.

Bioremediation of heavy metal and diesel oil cocontaminated soils was also possible by bioaugmentation using an optimized microbial formula developed with selected indigenous microbial strains [8]. After 42 days of treatment in laboratory conditions it removed 75 % of the total diesel oil hydrocarbons present in the heavy-metal cocontaminated polluted soil, including 100 % removal of C_{12}–C_{20} hydrocarbons and 60 % removal of isoprenoids. Interestingly, partial biodegradation of diesel oil hydrocarbons enhanced growth of a group of minor indigenous strains that actively participated in the bioremediation process. This observation demonstrates the significance of interspecies interactions (of indigenous microorganisms) in microbial bioremediation techniques. Possibly the smaller hydrocarbons (metabolites) produced by partial degradation of long-chain hydrocarbons served as a new carbon source and favored emergence of these native strains that were previously undetected due to their low abundance. The most active microorganisms in this experiment were identified as different species of the genus *Arthrobacter*, *Pseudomonas*, *Rhodococcus*, *Bacillus*, *Exiguobacterium,* and *Delftia*. In a similar study, 85 % C_{18} *n*-alkanes, 50 % C_{24} *n*-alkanes, and 60 % of an unresolved complex mixture of petroleum hydrocarbons were removed by a tailormade microbial formula developed with 12 isolated allochthonous strains at optimized physical condition [124].

Bioremediation of a viscous oil-contaminated soil was facilitated by ultraviolet-induced mutation of a microbial consortium constructed with seven oil-degrading microbial strains (including *Acinetobacter* sp. YC-X2, *Kocuria* sp. YC-X4, and *Kineococcus* sp. YC-X7) isolated from crude-oil-contaminated soil [33]. The mutation showed improved performance of the microbial consortia and the best-performing mutant itself was able to remove 52.42 % hydrocarbon from the viscous oil. The enhanced bioremediation activity could be explained by increased activity of catabolic enzymes by the ultraviolet-induced mutation; enhancement of catabolic enzyme activity by ultraviolet-induced mutation was reported previously [119].

Long-term in situ bioremediation of an aged recalcitrant hydrocarbon-contaminated soil showed that the addition of sewage sludge is a useful strategy for restoration of hydrocarbon-polluted soils [108]. The semi-arid area was exposed to oil refinery sludge for more than 10 years before the experimental sites were treated over a period of 8 months after mixing with fresh sewage or sewage compost. The microbial counts of the treated sites were significantly higher compared to those of an untreated site. Fungus of the phylum Ascomycota and bacteria of phyla Actinomycetes, Proteobacteria, Firmicutes, and Chloroflexi were the predominant group of microorganisms in the bioremediation sites. The highest hydrocarbon degradation was noticed in the fresh sewage-treated sites that also showed the highest microbial population.

In situ and ex situ bioremediation of petroleum products were studied with the aerobic microorganisms isolated from petroleum-contaminated soil [141]. A selected microbial community showed extremely fast biodegradation of benzene (914 μM/l/day), toluene (771 μM/l/day), ethylbenzene (644 μM/l/day), and xylene (673 μM/l/day). Up to 84 % benzene, 86 % toluene, 80 % ethylbenzene, and 82 % of xylene were degraded in batch cultures under laboratory conditions although the values were slightly lower in the in situ experiments. In another study, isolated aerobic zymogenous microorganisms of an oil-contaminated site showed crude oil biodegradation ability in laboratory experiments using a mixture of paraffinic types of oils as the substrate [123]. GC-MS analyses were performed at a 15-day interval to quantify *n*-alkanes, isoprenoids, phenanthrene, and their derivatives in the treated oil mixture, which revealed that the zymogenous microorganisms most efficiently degrade the *n*-alkanes and isoprenoids followed by phenanthrene and methylphenanthrene but perform very poorly in polycyclic alkane biodegradation.

In another study, primary hydrocarbon-degrading microorganisms were isolated from a hydrocarbon-polluted soil by using diesel oil as the sole carbon source [112]. The major populations of the hydrocarbon-degrading community were identified as members of the genera *Flavobacterium* and *Aspergillus*. It was possible to reduce the total petroleum hydrocarbon (TPH) content of the contaminated soil from 61,000 to 1,800 mg/kg after 15 days of treatment with this microbial community by mimicking a laboratory-scale heap leaching process (an industrial mining process) using column and piles.

Polycyclic aromatic hydrocarbons are one of the most widespread organic pollutants found mainly in soil and sediment. Fossil fuels and their derivatives are the main source of PAH contamination although it may come from various natural products (steroids, hydrocarbons, etc.) and some pesticides. Several researchers demonstrated successful microbial bioremediation of PAH-contaminated soil. Mao et al. reported nearly 36 % PAH removal from a contaminated soil (containing $\sim$10 mg PAH per kg of dry soil) after 56 days of treatment with a PAH-degrading microbial consortia isolated from a PAH-contaminated site [79]. Molecular phylogeny identified the most abundant populations of the consortium as close relatives of the *Mesorhizobium*, *Alcaligenes,* and *Bacillus* species. Degradation of several PAH (including naphthalene, phenanthrene, anthracene, pyrene, dibenzo[a]anthracene, and benzo[a]pyrene) was studied in forest soil microcosms with and without bioaugmentation using bacteria and fungi isolated from a diesel-contaminated site [118]. Five bacteria strains such as three *Bacillus* sp., one *Chromobacterium* sp. 4015, and an *Enterobacter* sp. and three filamentous fungi strains *Achremonium* sp., *Aspergillus* sp., and *Verticillium* sp. were used for bioaugmentation. The indigenous microbes rapidly responded to the PAH addition in the forest soil and utilized the low-molecular-weight components (such as naphthalene, phenanthrene, and anthracene) as the energy source; biodegradation of these compounds was not significantly changed by bioaugmentation.

In contrast, the high-molecular-weight PAHs (such as pyrene, benz[a]anthracene, and benz[a]pyrene) were very slowly degraded by this native microbial consortia but significant improvement of their biodegradation was achieved by bioaugmentation with one *Aspergillus* species isolated from the diesel-contaminated site. A pilot-scale ex situ investigation established potential long-term and large-scale bioremediation of heavily PAH-contaminated soil by biostimulation of the native microbial consortia as well by bioaugmentation with the PAH-degrading and bioemulsifier-producing *Rhodococcus ruber* Em1 strain [128]. After 175 days 26.82 % of total PAHs and 35.36 % of 4–6 ring PAHs were removed by biostimulating a soil sample containing 375-mg PAH per kg of dry soil, and 33.9 % and 11.0 % degradation of the respective chemicals was achieved by bioaugmentation. However, a combination of biostimulation and bioaugmentation removed 43.9 % total PAHs and 55.0 % of 4–6 ring PAHs after 175 days of treatment of the same test soil, which suggests that in some bioremediation processes, a combination of more than one strategy may be more useful than either one.

Microbial bioremediation was also useful in the treatment of varieties of PAH-contaminated solid waste. A mixture of 4 bacterial and 5 fungal dioxin-degrading strains showed successful biodegradation of polychlorinated dibenzo-*p*-dioxins and dibenzofurans from contaminated municipal solid-waste incinerator fly ash [93]. After 21 days solid-state fermentation with the above-mentioned microbial mixture at laboratory conditions, 68.7 % elimination of these contaminants was achieved. The bacterial strains included *Sphingomonas* sp., *Sphingomonas wittichii* RW1, *Pseudomonas veronii* PH-03, and *Paenibacillus* sp. VSE5L whereas the fungal strains were comprised of *Phanerochaete chrysosporium* DSM 6909, *P. chrysosporium* DSM 1556, *Irpex* sp. KW3, *Trametes* sp. CH2, and *Fusarium* sp. VSO7.

The microbial mixture also removed 66.8 % of the 2,3,7,8-substituted congeners from the fly ash. Analysis showed that all the bacterial and fungal strains were well maintained in the reaction condition. In a study, PAH-degrading microorganisms were isolated from PAH-contaminated lab waste containers [109]. Two fungal strains that showed the best PAH-degrading activity were identified as *Trichoderma longibrachiatum* and *Byssochlamys spectabilis*. After 6–9 days treatment in a liquid culture, *T. longibrachiatum* showed 97 % benz[a]anthracene degradation in a 100-μM solution. High-level degradation of this substrate was also achieved by using the immobilized *T. longibrachiatum* strain in an expanded bed bioreactor operated in continuous mode. Bioremediation of the PAH pollutants, which is cocontaminated with heavy metals may be difficult due to their combined inhibitory effect on microbial growth. However, a few microbial consortia were successfully used for bioremediation of heavy-metal—and PAH-cocontaminated soils [135, 136]. Simultaneous bioremediation of PAH and arsenite cocontamination was possible by biodegradation of PAH and oxidation of arsenite using a heterotrophic bacterial consortium isolated from the soil of an aged coking plant site [136]. After 48 h of aerobic incubation, about 71.4 % phenanthrene and 96.2 % arsenite were removed from a liquid culture containing 200 and 60 mg/l of phenanthrene and arsenite, respectively. The major populations of the bioremediating consortia in the cocontaminated soil were identified as *Achromobacter*, *Alcaligenes*, *Pseudomonas,* and *Pusillimonas* species. Interestingly, composition of the microbial consortia at the end of bioremediation was highly dependent on the nature of the contaminants. *Achromobacter* and *Pseudomonas* species dominated the consortia when contaminated only with phenanthrene but when only arsenite was present, *Alcaligenes* and *Pseudomonas* were the dominant species. In the presence of both contaminants, the bacteria from the genus γ-*Proteobacteria* and β-*Proteobacteria* were abundant in the bioremediating sample.

In the other study, a bacterial consortium was developed for high-molecular-weight PAH (HMW PAH) degradation in the cadmium (Cd)- and PAH-cocontaminated samples [135]. The consortium was composed of four bacterial species, such as *Alcaligenes* sp., *Pseudomonas* sp., *Pandorea* sp., and *Paenibacillus* sp. It was able to utilize HMW PAHs such as benzo[a]pyrene and pyrene as the sole carbon source and completely degrade low-molecular-weight PAHs, phenanthrene, and anthracene after 60 days of treatment in the presence or absence of cadmium (Cd). The presence of cadmium even stimulated degradation of several PAH. In the absence of Cd it was able to remove 100 and 89 % of benzo[a]pyrene and pyrene, respectively, whereas in the presence of Cd 100 and 94 % of the respective PAH were degraded.

Indiscriminate use of agricultural pesticides causes continuous contamination of farming lands, which can further pollute surface water and even underground water when washed out by rain. Several studies focused on microbial biodegradation of the chemicals present in pesticides or insecticides. Bioaugmentation with the white-rot fungi *Anthracophyllum discolor* successfully removed an organochlorine pesticide, pentachlorophenol from contaminated soil [30]. Several indigenous microbial species including some bacteria from β-Proteobacteria and γ-Proteobacteria phylum and some fungi from Ascomycota and Basidiomycota phyla actively took part in pentachlorophenol degradation when biostimulated by utilizing wheat straw residue as nutrient. Biodegradation of pentachlorophenol was further enhanced by combined effect of bioaugmentation by *A. discolor* and biostimulation by wheat straw residue. *A. discolor* was also reported to degrade various PAHs and other chlorophenols including 2,4-dichlorophenol, 2,4,6-trichlorophenol, and pentachlorophenol [5, 44].

Microbial bioremediation of another frequently used organochlorine pesticide, lindane was also reported by a few researchers. Abhilash et al. isolated four lindane-degrading microbial strains from the rhizosphere of selected plants of a lindane-contaminated site and identified the isolates as *Kocuria rhizophila*, *Microbacterium resistens*, *Staphylococcus equorum*, and *Staphylococcus cohnii* [2]. After acclimatizing in high lindane-containing media (5–100 mg/l) the isolates were able to remove lindane from soil samples containing up to 100 mg/kg of the contaminant. The highest lindane removal activity was reported in *S. cohnii* (subspecies urealyticus); this strain removed 100 and 67 % of the contaminants after 45 days of treatment of two soil samples containing 5 and 100 mg/kg lindane, respectively. Bioremediation of lindane-contaminated soil using the *Streptomyces* sp. M7 helped germination and growth of maize plantation in the contaminated soil [21]. The microorganism was resistant to 0.4 mg lindane/kg soil and at optimum conditions it removed 68 % of the pollutant from the soil sample contaminated with 0.1 mg lindane/kg soil.

In situ bioremediation of an endosulfan-contaminated site was performed by isolating the indigenous microorganisms and inoculating them in the contaminated site after increasing their number under laboratory conditions [68]. Several endosulfan-degrading bacteria were isolated from the contaminated soil and by molecular phylogeny some of the isolates were identified as *Ochrobacterum* sp., *Burkholderia* sp., *Pseudomonas* sp., and *Arthrobacter* sp. After 12 weeks of treatment, a consortium consisting of a 1:1 mixture of two *Pseudomonas* species was able to degrade 80 % of α-endosulfan and 65 % of β-endosulfan from the contaminated site containing 50-mg endosulfan per kg of soil. However, all isolated strains showed better bioremediation performance in optimum laboratory conditions.

Heavy metals are dangerous for human health and their presence in soil or water is a serious concern as they can easily enter the human body through contaminated vegetables or drinking water. Serious heavy-metal toxicity may be caused by arsenic, cadmium, chromium, lead, and mercury poisoning, which may cause permanent damage to kidney, liver, nervous system, skin, bone, or teeth. Recent studies showed that heavy metals can be removed from contaminated sites by bioremediation, phytoremediation, or by combination of both. Benazir et al. studied the efficacy of a few microbial consortia in removing chromium from an industrial effluent [20]. A consortium containing *Bacillus subtilis*, *P. aeruginosa*, and *Saccharomyces cerevisiae* successfully reduced the chromium content to <1 % of its original concentration. In another study, 51 heavy-metal-resistant microbes were isolated from three different composts developed using different raw materials such as horticulture waste, sewage sludge, and municipal solid waste [138]. Most of the microorganisms were able to remove several heavy metals including Cd, Cr, Ni, Pb, and Zn. Some of the isolates were able to remove more than 90 % of Pb and other metals were removed within the range of 20–60 %. Molecular phylogeny identified the best performing isolates as *Graphium putredinis*, *Fusarium solani*, *Fusarium sp.*, and *Penicillium chrysogenum*. Most metals were removed predominantly by intracellular accumulation within the microorganisms, with the exception of Ni that was equally removed by extracellular absorption and intracellular accumulation.

3.1.2 Sludge, Slurry, and Water Treatment

Several in situ and ex situ bioremediation techniques are used for treatment of industrial wastewater, sludge, and slurry. Natural attenuation and biostimulation approaches were studied for treatment of groundwater contaminated with tetrachloroethylene, a carcinogenic chlorinated hydrocarbon that is often used for dry cleaning [24]. Analysis of the naturally attenuated consortia showed the presence of some typical dehalogenating bacteria including *Firmicutes*. Phylogenetic analysis identified the presence of multiple species from different phyla in the natural attenuation consortium but the biostimulated consortium was dominated by species closely related to the class β-proteobacteria. The indigenous microorganisms showed high potential for biodegradation of polyphenolic contaminants in "two-phase olive mill waste" [91]. The investigation in a lab-scale bioreactor suggested that the bioremediation activity was highly dependent on aeration and the nutrient content of the culture. The maximum biodegradation rate was observed in aerated bioreactors with nutrient-supplemented media. The rate of biodegradation was also time dependent: the contaminant was degraded very quickly at the beginning of the process and slowed down after 7 days fermentation. About 36 % polyphenols were degraded in the first 7 days whereas only 54 % degradation was achieved after 55 days fermentation. The predominant microorganisms in the culture were identified as members of 14 different bacterial genera of the phyla γ-Proteobacteria, Actinobacteria, and Firmicutes that are associated with the plant biomass and, several genera of fungi including *Penicillium*, *Candida*, *Geotrichum*, *Pichia*, *Cladosporium*, and *Aschochyta*.

A continuous-flow two-stage bioreactor was used for decontaminating fractionated drainage water from an oil terminal that contained emulsified oil and water-soluble hydrocarbons [110]. Some of the predominant bioremediating microorganisms were identified as *Thiomicrospira* sp., *Achromobacter* sp., *Cyclobacterium linum*, and *Nitromonas halophila*. The first bioreactor (1 m^3) was fed with fractionated drainage water at a flow rate of 50 l/h after inoculating with a 5-l mixed culture containing the strains isolated from the oil refinery and its efflux was then fed into the second bioreactor (2 m^3). The process removed 100 % ammonia, 93 % sulfate, and 90 % of total organic carbon (TOC) from the drainage water. Total organic carbon of the drainage water decreased faster than the expected rate and the microbial population study showed a smaller number of bacteria and more phages in the second bioreactor than the first one. These two observations were explained by formation of a phage-driven microbial loop where bacteriophages induce bacterial cell lysis that is followed by degradation of released bacterial compounds.

Lignin-, tannic acid-, xylan-, and cellulose-degrading strains were screened from the intrinsic microbial community of an agro-based pulp mill effluent and six microbial consortia were developed using 14 isolates [69]. Each consortia contained four strains, one for each substrate-degrading activity. These consortia could reduce 35–45 % chemical oxygen demand (COD) of 40 % black liquor solution (pulp and paper effluent diluted in backwater at 2:3 ratio). The performance of the most active consortia was improved by addition of nitrogen and phosphorous in the effluent and 65–66 % COD-reduction was possible after optimizing physical parameters (temperature, pH, and agitation) in an ex vivo testing. 16S rRNA gene sequencing identified the strains of this consortium as members of *Pseudomonas, Bacillus, Pannonibacter*, and *Ochrobacterum* genus.

Effluent of a municipal wastewater treatment plant was decontaminated by a novel technology developed using constructed microbial mats on low-density polyester [145]. Various bacterial, cyanobacterial, and microalgal communities were grown on the polyester support and resulted in continuous, self-sufficient microbial mats utilizing the nutrients present in the wastewater. The mats were dominated by cyanobacteria such as *Chroococcus* sp. and *Lyngbya* sp., diatoms of the genus *Nitzschia*, and, bacteria of the subclass *Proteobacteria*. This technique successfully removed 94 % orthophosphate, 79 % ammonium, 78 % nitrite, and 83 % nitrate with 48-h treatment of the wastewater effluent. Some researchers studied simultaneous wastewater bioremediation and the energy production perspective of microalgae [45, 84]. A microalga, *Scenedesmus* sp. AMDD showed efficient decontamination of secondary effluent obtained from a municipal wastewater treatment plant [84]. When studied in a batch photobioreactor, it removed 90 % of dissolved nitrogen and phosphorous after about 6.5 days treatment and biomass yields ranged from 0.23 to 0.65 kg/m^3 wastewater. An approximate twofold increase of biomass productivity was achieved when the wastewater was treated in a 2-l continuous chemostat and that also removed >99 % dissolved nitrogen and phosphorous from the same sample.

Another photosynthetic microalga, *Scenedesmus acutus* PVUW12, was able to remove total nitrogen content only by 3 days of treatment of the wastewater sample collected from an urban purifier plant containing 18.8 mg/l nitrate [45]. The experiment was carried out in a vertical column photobioreactor and the algae produced a substantial amount of triglycerides (28.8 % of dry biomass) when left for another 20 days. Algae were also reported to be active in bioremediating organic material of highly polluted piggery wastewater ponds [55]. The predominant microalgae were identified as *Chlamydomonas* sp., *Ankistrodesmus* sp., *Protoderma* sp., *Selenastrum* sp., *Chlorella* sp., *Oocystis* sp., *Achnanthes* sp., *Nitzschia* sp., and *Microspora* sp. Diversity of the microalgae population of the ponds was reported to be dynamic depending on seasonal change of temperatures and solar radiation. At ideal conditions, an average 76 % COD and 88 % TKN (total Kjeldahl nitrogen) removals were achieved by 10 days of treatment of tenfold and 20-fold diluted swine manure in two high rate algal ponds. However, due to high buffer capacity of the piggery wastewater, the consortia showed very low phosphorous removal efficiency.

The potential of the microbial fuel cell (MFC) technology was studied for nitrate removal from polluted groundwater [100, 139, 148]. A two-chamber MFC was designed for simultaneous carbon and nitrogen removal using an acetate-adopted microbial consortium obtained from another MFC of a water management center [139]. By feeding the nitrate-rich water to the cathode chamber, the MFC was able to remove 0.41 kg nitrate per cubic meter of cathode chamber in each day while producing about 35 W of power per cubic meter of cathode chamber. The MFC was more energy efficient than the conventional nitrate removal process due to energy production and minimized aerobic consumption of organic carbon that reduced the aeration costs. The nitrate removal potential of another MFC-generated microbial consortium was studied in a single-chamber MFC [100]. This system reduced the nitrate concentration of a groundwater sample from 28.32 ± 6.15 to 12.14 ± 3.59 mg/l, without any nitrite accumulation. Indigenous microorganisms were used for simultaneous carbon and nitrogen removal from groundwater using another single-chamber MFC with a rotating biocathode [148]. The system was able to remove 85.7 ± 7.4 % total organic carbon and 91.5 ± 7.2 % total nitrogen from a groundwater sample, with a maximum power output of 585 mW/m^3. Although the process was started with the indigenous microorganisms, denitrifying bacteria emerged as the dominant group after 40 days of treatment.

Mining has an extremely destructive effect on the biosphere, which is often associated with deforestation, soil erosion, and water and soil pollution with various contaminants. Long-term bioremediation and phytoremediation may be useful in removing mine-associated pollutants. Sulfate and heavy-metal content of an acid mine drainage water were successfully reduced in an ex situ bioremediation technique using sulfate-reducing bacteria [53]. The effluent was treated by immobilizing the sulfate-reducing bacteria in an anaerobic bioreactor and enhancing their growth by feeding with a mixture of grass cutting and rumen fluid biomass. Efficient heavy-metal removal from another acid mine drainage sample was achieved by biofilm formation of an indigenous algae–fungi–bacteria consortium within a photorotating biological contactor [95]. A biofilm was developed by 60 days batch mode operation of a biocontactor with the microbial consortium; it successfully removed 20–50 % of various heavy metals from the highly contaminated acid mine drainage after 10 weeks of continuous treatment. Removal of various metals was on the order of Cu > Ni > Mn > Zn > Sb > Se > Co > Al. The biofilm was dominated by the algae *Ulothrix* species.

3.1.3 Polluted Air Treatment

Microbial bioremediation of air pollution is the least studied field as compared with soil or water pollution. Airborne phototrophic microorganisms and hydrocarbon-utilizing heterotrophic bacteria were isolated from the dust samples collected at the

15-m height of Kuwait city air [7]. Three phototrophs were identified by molecular phylogeny as *Nostoc commune*, *Leptolyngbya thermalis*, and a chlorophyte of the genus *Gloeotila*. Each of them was associated with unique consortia of oil-vapor degrading bacteria that may be useful in in situ bioremediation of atmospheric hydrocarbon pollutants. The phototrophs may potentially serve as a source of nutrients and growth-enhancing metabolites for the heterotrophic consortia. Novel approaches were taken for in situ flue gas bioremediation and simultaneous microalgal biomass production [35, 47]. On-site bioremediation of carbon dioxide, nitrogen oxide, and sulphur dioxide of a coke oven flue gas was achieved by directly passing the flue gas through a photobioreactor containing a culture of heat- and CO_2-tolerant engineered *Chlorella* species [35]. The system successfully increased the algal biomass accompanied by efficient capture of CO_2, NO, and SO_2 from the flue gas. About 60 % CO_2, 70 % NO, and 50 % SO_2 content of the coke oven flue gas was removed by this process.

The other study used *Chlorella vulgaris* to absorb CO_2 present in a municipal waste incineration flue gas and simultaneously decrease biomass production cost [47]. Growth and CO_2 fixation rate of the algal culture was higher using the flue gas (containing 10–13 % CO_2 and 8–10 % O_2) than using a mixture containing equivalent proportions of pure CO_2 and air. The biomass produced using untreated flue gas had mercury content slightly higher than the limit of the European Union foodstuff legislation. However, the mercury content of the algal biomass falls below the above-mentioned limit when the flue gas is treated in a simple activated carbon column prior to passing through the cultivation unit.

3.2 Plants and Phytoremediation

The pollutant-removal technique using green plants and plant-associated microorganisms is known as phytoremediation. This technique can clean up various kinds of pollutants including heavy metals, pesticides, and petroleum hydrocarbons. Phytoremediation is extremely useful in large-scale in situ decontamination of soil, surface water, and groundwater. It is also effective for removal of particulate matters and organic or inorganic toxic gases from polluted air. Most phytoremediation processes depend on absorption/adsorption of pollutants and their accumulation in plant biomass that reduces mobility of the chemicals. Also plants can biotransform several contaminants directly by their metabolism or with the help of associated microorganisms. Literally any plant can contribute in preventing environmental pollution but phytoremediation ability varies from species to species and it is highly dependent on the nature of the pollutants. Selection of suitable plant species is the key to successful phytoremediation of a polluted environment. Plant–microorganism symbiosis is also helpful for bioremediation of some pollutants because microbial metabolisms enhance the mobility of certain polluting chemicals that are otherwise not bioavailable to plants. Several plant species are already characterized for their capabilities for growing in contaminated soil or wastewater and eliminating a particular type of contaminant. The following section focuses on recent developments and the current research trend in the area of plant-assisted environmental remediation. The findings are summarized in Tables 2 and 3.

3.2.1 Polluted Soil Treatment

Mining-related activities may cause serious heavy-metal contamination of surrounding soil, surface water, and groundwater. The plant species naturally growing in the mining areas need to be adapted to the heavy-metal-contaminated environments and potentially they are able to accumulate a high concentration of heavy metals. Several studies were conducted to understand phytoremediation ability of these native plants. Zn, Pb, and Cd accumulation in various tissues of predominant native species growing in an abandoned mining site were analyzed to understand their heavy-metal uptake capacity from soils, sediments, and mine tailings [17]. The highest metal concentration was observed in the aerial parts of *Inula viscosa, Euphorbia dendroides*, and *Poa annua* species with the average values of Zn: 1.68, 1.02, 1.40; Pb: 0.42, 0.24, 0.08; Cd: 0.028, 0.0077, 0.019 g/kg dry biomass, respectively. *Thlaspi caerulescens* species was revealed as a Zn–Cd hyperaccumulator in a similar study conducted with 31 native plant species of an abandoned Pb–Zn mining site [18]. It was able to accumulate more than 18 g/kg dry biomass of Zn. Shoots accumulated the highest concentrations of all metals as compared to other tissue; Zn was present in highest concentration followed by Pb and Cd. A similar approach was taken for determining Cu, Zn, Fe, and Mg accumulation in spontaneously growing native plants in an iron and copper mining area [94].

Higher metal accumulations were reported in the species grown in higher metal-containing soil. Results showed that metal accumulation varies between species and between different tissues of the same plant. Out of all species studied, *Chenopodium botrys* accumulated the highest concentration of Cu (0.183 and 0.150 mg/g) and Mn (0.177 and 1.288 mg/g) in root and shoot tissues, whereas *Verbascum speciosum* accumulated as much as 9.226.3 and 15.343 mg/g Fe in the root and shoot tissues, respectively. *Scariola orientalis* accumulated the highest concentration of Zn (1.208 mg/g) in the root tissues; but when compared in the aerial part *Stipa barbata* showed the highest Zn accumulation (329.3 mg/kg). Strontium (Sr) was reported to be preferentially accumulated in the shoots of three plant species such as *Euphorbia macroclada, Verbascum cheiranthifolium,* and *Astragalus gummifer* [117]. *E. macroclada* showed the highest Sr accumulation and most efficient translocation of the contaminant from root to shoot. A field study in an iron mine area revealed several plant species as accumulating higher concentrations of As, Cd, Cr, Fe, Mn, Mo, Ni, Pb, Si, and Zn when grown in the mine area than in normal soil [80]. When measured in the aerial parts, the highest metal accumulations were found in *Epilobium fragilis* (As), *Carthamus oxyacantha* (Cd, Fe, Mn, and Pb), *Verbascum speciosum* (Cu), *Centaurea iberica* (Mo), *Salvia spinosa* (Ni and Cr), *Glaucium grandiflorum* (Se), and *Malva neglecta* (Zn) species. In contrast, when the metal concentrations were measured at the roots, the highest concentrations were reported

Table 2 Summary of recent research (2008–2013) on plant-mediated bioremediation (Phytoremediation) of soil, water, and air

Active microorganism(s)	Target pollutants	Preferred accumulation site	Reference
For bioremediation of contaminated soil and solid waste			
Inula viscosa, *Euphorbia dendroides*, and *Poa annua*	Zn, Pb, and Cd	Aerial parts	[17]
Thlaspi caerulescens	Zn, Cd	Shoot	[18]
Chenopodium botrys	Cu, Mn	Root and shoot	[94]
Verbascum speciosum	Fe	Root and shoot	
Scariola orientalis	Zn	Root	
Stipa barbata	Zn	Aerial part	
Euphorbia macroclada, *Verbascum cheiranthifolium*, and *Astragalus gummifer*	Sr	Shoot	[117]
Epilobium fragilis	As	Aerial part	[80]
Carthamus oxyacantha	Cd, Fe, Mn, Pb		
Verbascum speciosum	Cu		
Centaurea iberica	Mo		
Salvia spinosa	Ni, Cr		
Glaucium grandiflorum	Se		
Malva neglecta	Zn		
Euphorbia cheiradenia	As	Root	
Stipa barbata	Cd, Pb, Cr		
Euphorbia macroclada	Cu		
Centaurea iberica	Fe		
Reseda lutea	Mo		
Salvia spinosa	Ni, Zn		
Xanthium strumarium	Se		
Pistacia lentiscus, *Scrophularia bicolor*	Pb, Zn	Root	[15]
Tagetes patula	Benzo[a]pyrene	Root	[131]
Tagetes patula	Cd	Aerial part	
Chromolaena odorata	Zn, Cd, Ni	Root	[11]
Chromolaena odorata	Total petroleum hydrocarbon (TPH)	—	
Plantago major	Imidacloprid	Roots, leaves	[105]
Nicotiana tabacum cv. Xanthi	Methyl parathion	—	[140]
Sorghum bicolor, *Linum usitatissumum*	TPH	—	[116]
Zea mays, *Festuca arundinacea*	TPH	—	[146]

(continued)

Table 2 (continued)

Active microorganism(s)	Target pollutants	Preferred accumulation site	Reference
For bioremediation of polluted water, sludge, and slurry			
Wolffia globosa	As	Whole plant	[149]
Eichhornia crassipes, Lemna minor	As	Whole plant	[9]
Eichhornia crassipes, Lemna minor, and *Spirodela polyrrhiza*	As, Mg	Root	[86]
Ceratophyllum demersum, Lemna gibba	Pb, Cr	Whole plant	[1]
Ceratophyllum demersum, Echinochloa pyramidalis, Eichhornia crassipes, Myriophyllum spicatum, Phragmites australis, and *Typha domingensis*	Cu, Zn	Root	[50]
	Pb	Leaf	
Eleocharis acicularis	In	Root	[56]
	Ag, Pb, Cu, Cd, Zn	Shoot	
Callitriche cophocarpa	Cr	Shoot	[13]
Oryza sativa, Brachiaria mutica, Eichhornia crassipes	Cr	Root	[87]
Portulaca tuberosa, Portulaca oleracea	Cu, Ni, Hg, Pb	Root	[48]
For remediation of air pollution			
Zamioculcas zamiifolia	Benzene, toluene, ethylbenzene, and xylene (BTEX)	Leaf	[126]
Scindapsus aureus, Asparagus setaceus, Sansevieria trifasciata, Chlorophytum comosum, Aglaonema commutatum, Scindapsus pictus, Gasteria gracilis, and *Philodendron sodiroi*	Formaldehyde	—	[153]
Alstonia scholaris, Anthocephalus indicus, Cassia auriculata, Cassia siamea, Lagerstroemia speciosa, Mimusops elengi, Peltophorum inerme, and *Tabebuia aurea*	Dust	Leaf surface	[39]

in *Euphorbia cheiradenia* (As), *Stipa barbata* (Cd, Pb, and Cr), *Euphorbia macroclada* (Cu), *Centaurea iberica* (Fe), *Reseda lutea* (Mo), *Salvia spinosa* (Ni and Zn), and *Xanthium strumarium* (Se) species.

Remarkably, many plant species that accumulate very high concentrations of metals only in roots showed relatively bad phytoremediation performance. This observation suggests that mobilization of the absorbed metals from root to shoot is an important phytoremediation strategy of some plant species. However,

Table 3 Summary of recent research (2008–2013) on soil bioremediation by combined action of plants and microorganisms

Plant	Microorganism	Target pollutants	References
Zea mays	*Bacillus mycoides* and *Micrococcus roseus*	Cd	[81]
Zea mays	*Streptomyces* species MC1	Cr	[99]
Alnus firma	*Bacillus thuringiensis* GDB-1	Pb, Zn, As	[14]
Polygonum avicular	*Alternari, Aspergillus terreus, Bipolaris, Fusarium acuminatum, Fusarium reticulatum,* and *Rhizoctonia*	Total petroleum hydrocarbon (TPH)	[88]
Amaranthus retroflenus	*Alternaria, Fusarium acuminatum, Fusarium equiset, Fusarium reticulatum*, and *Rhizoctonia*		
Poa sp.	*Alternaria, Aspergillus terreus, Fusarium acuminatum, Fusarium reticulatum, Mucur*, and *Penicillinium*		
Noea mucronata	*Alternaria, Oloclodiumb*		
Alhaji cameleron	*Alternaria, Aspergillus terreus, Paecilomyces*		
Alhaji cameleron	*Alternaria, Aspergillus terreus, Paecilomyces*		
Crozophora heirosololymitrana	*Alternaria, Aspergillus terreus, Biopolaris, Fusarium reticulatum*		
Convolvulus arvensis	*Alternaria, Mucur*		
Acacia angustissima, Acacia auriculiformis, Acacia holosericea, Acacia mangium, Mimosa artemisiana, Mimosa caesalpiniifolia, Samanea saman	Nitrogen-fixing bacteria and arbuscular mycorrhizal fungi	TPH	[22]
Cortaderia selloana	Commercial bioaugmentation product (MicroSolv-400; Trademark, Environmental Leverage)	Petroleum hydrocarbons	[36]
Rizophora mangle	Plant-associated microorganisms	TPH	[89]
Avicennia schaueriana	Plant-associated microorganisms	TPH	[90]
Mirabilis Jalapa	Rhizospheric microorganisms	Petroleum (saturated hydrocarbons)	[97]

uncontrolled spreading of metal ions in plants' aerial parts may exert heavy-metal toxicity. Pb and Zn phytoremediation of an abandoned mining site was made possible by using two Mediterranean plant species *Pistacia lentiscus* and *Scrophularia bicolor* [15]. The toxicity was partially overcome by using different combinations of compost, chemical fertilizer, and zeolites. All amendments showed an increase in survival of *P. lentiscus*. However, survival of *S. bicolor* was improved only in the amendments done with zeolite or a combination of zeolite and fertilizer. *P. lentiscus* accumulated the metal ions mostly in root and showed less metal accumulation ability than *S. bicolor*; yet it is more suitable for phytostabilization and environmental restoration due to higher resistance to toxicity and higher biomass production. The phytoremediation potential of castor oil plants (*Ricinus communis*) for decontamination of boron and heavy-metal-cocontaminated land was studied in conjugation with organic matter amendments using peat and filter cake [3]. A high concentration (0.626 g/kg dry biomass) of boron was accumulated in the castor oil shoot when the plant was grown in the presence of filter cake but no substantial heavy-metal accumulation (Cd, Cr, Cu, Ni, Pb, or Zn) was noticed at any condition.

The combined action of specific plant and microbial species was reported to enhance the bioremediation profile of heavy-metal-contaminated soil as compared to the individual action of the same plant or the microorganism. Artificial symbiosis was also developed by inoculating certain microbial species in the soil for cultivating the targeted plant species. This approach was practiced for improving bioremediation efficiency of a Cd-contaminated soil by cultivation of maize plants (*Zea mays* L.) inoculated with two growth-promoting rhizobacteria such as *Bacillus mycoides* and *Micrococcus roseus* [81]. Growth and nutrient uptake of the plant were substantially increased by the bacterial treatments. Bioremediation of Cr-contaminated soil was also successfully enhanced by cooperative action of maize plant and *Streptomyces* species MC1 [99]. The presence of *Streptomyces* sp. MC1 caused 57 % increase of the plant growth and 46 % increase of chromium accumulation in the plant biomass as well as a 96 % decrease of Cr content of the soil sample. A similar approach was taken for bioremediation of a heavy-metal–contaminated mine tailing by inducing plant–microorganism symbiosis [14]. The *Bacillus thuringiensis* GDB-1 strain isolated from roots of *Pinus sylvestris* showed an efficient heavy-metal bioremediation property. Plant growth and nodule formation of the hyperaccumulator *Alnus firma* was increased significantly when this microbial strain was introduced in the soil for growing *A. firma* seedlings. Phytoremediation ability of mine-tailing–contaminated soil was also improved by this symbiosis of *A. firma* and *B. thuringiensis* GDB-1.

An ornamental plant, *Tagetes patula*, was studied for phytoremediation of benzo [a]pyrene and heavy-metal–cocontaminated soil [131]. Low concentration ($\leq$10 mg/kg) of benzo[a]pyrene facilitated the plant growth that caused 10.0–49.7 % increase in the plant biomass as compared to the control. Accumulations of this chemical in plant tissues were almost directly proportional to its concentration in the soil. When grown in the benzo[a]pyrene cocontaminated soil, this plant accumulated a high concentration of Cd but showed very low Cu and Pb absorption efficacy. However, plant growth and benzo[a]pyrene uptake were inhibited by the presence of these heavy metals. Another plant species, *Chromolaena odorata*, showed phytoremediation ability of heavy metal and crude oil co-contaminated soil [11]. After 180 days of treatment, *C. odorata* removed 62 % Cd, 47 % Ni, and 63 % Zn from the experimental soil samples containing 2 mg/kg of each metal ion. The plant grew normally in the soil containing up to 50 g/kg crude oil or 2 g/kg Zn but little adverse effect was noticed in experiments containing 2 g/kg Cd, although plant growth was adversely affected in the soil treated with 1 g/kg or more Ni. The plant removed about 82 % crude oil in the absence of any heavy metal; this value reduced slightly in the presence of heavy-metal cocontaminants. A bioremediation process was developed for decontaminating heavy-metal–contaminated river sediments where plants were used for conditioning dredged sludge [147]. Reed canary grass (*Phalaris arundinacea*) was the most suitable plant species for the conditioning process optimized in laboratory and pilot-scale experiments. Practical feasibility of the proposed conditioning was demonstrated in large scale by conditioning 1,400 m^3 of dredged sludge using the reed canary grass.

In addition to heavy-metal removal, plants are also used for their pesticide and insecticide phytoremediation property. Both living plant and dry biomass of *Plantago major* were successfully used for insecticide (imidacloprid) removal from soil and water [105]. The viable plant removed up to 95 % imidacloprid from water when it was treated for up to 10 days, and accumulation of the compound in roots, leaves, and fruits reached the maximum levels of 15.74, 37.21, and 5.74 μg/g, respectively. The biosorption capacity of dry roots, fruits, and leaves were measured as 7.94, 6.31, and 2.51 μg/g of dry biomass. Experiments showed that a gram-negative microorganism associated with this plant played an important role in biodegradation of imidacloprid inasmuch as the microbe uses this molecule as a carbon and nitrogen source. In the presence of this microorganism 93.34 % reduction of imidacloprid concentration was achieved in 48 h as compared with 31.9 % imidacloprid degradation in the control without the microorganism. A transgenic tobacco plant was developed for biodegradation of organophosphorus compounds by introducing a bacterial organophosphorus hydrolase gene from *Pseudomonas pseudoalcaligenes* that was successfully expressed in the tobacco plants [140]. The transgenic plant was able to degrade the organophosphorous pesticide (methyl parathion) as it can secrete the organophosphorus hydrolase enzyme. After 14 days of growth, the transgenic plant degraded more than 99 % methyl parathion when grown in the presence of 0.02 % (v/v) of the compound. Presence of the pesticide increases shoot and root biomass of the transgenic plant in comparison to that of wild-type tobacco plants.

Plants are also used for long-term phytoremediation of petroleum- and hydrocarbon-contaminated soil. Several field studies were conducted in petroleum-contaminated oil refineries. In most studies, plant-associated microbial consortia were reported to play an important role. One such study isolated seven petroleum-resistant plant species for potential bioremediation of petroleum-polluted soils by combined action of plants and their root-associated fungal strains [88]. Using morphological characterization, the plants were identified as *Alhaji cameleron*,

Amaranthus retroflexus, *Convolvulus arvensis*, *Chrozophora hierosolymitana*, *Noea mucronata*, *Poa* sp., and *Polygonum aviculare*. Each plant species contains distinct microbial consortia in their roots and also shows different bioremediation potential. Out of a total 11 root-associated fungal strains identified in these plants' roots, only *Altenaria* sp. was present in the root of all 7 plant species whereas others were distributed specifically in certain species. Other root-associated fungi were identified as members of the genus *Alternaria*, *Bipolaris*, *Fusarium*, *Penicillinium*, *Rhizoctonia*, *Aspergillus*, *Macrophomina*, and *Mucur*. All these fungal species were able to grow in culture media containing 1 % (v/v) petroleum and a few species were resistant up to 10 % (v/v) petroleum- containing media. Bioremediation tests confirmed contribution of both the plants and fungal consortia in bioremediation of petroleum-polluted soils. Several leguminous plants of oil-contaminated areas, their symbiotic nitrogen-fixing bacteria and arbuscular mycorrhizal fungi (AMF) showed similar behaviors for phytoremediation of petroleum-contaminated soil [22]. The study with four *Acacia* sp., two *Mimosa* sp., and a *Samanea* sp. showed that there is no direct relation between petroleum resistance and phytoremediation capability of these plants but association between a plant and its corresponding symbiotic microorganisms played a critical role in the phytoremediation process.

Samanea saman and its symbiotic microorganisms showed the highest level of petroleum removal in the soils contaminated with up to 70 g petroleum per kg soil. Pampas grass (*Cortaderia selloana*) was able to remove petroleum hydrocarbons from the contaminated soil of an oil refinery with and without soil amendments; the process was associated with activity of intrinsic microbial consortia [36]. The phytoremediation performance improved when the soil was amended by treatments with a surfactant and a commercial bioaugmentation formulation. Soil amendments especially helped in hydrocarbon removal from the deeper layers of soil. Petroleum-contaminated mangrove sediment was biologically remediated using two mangrove plant species *Rizophora mangle* and *Avicennia schaueriana* and the plant-associated microorganisms [89, 90]. The microorganisms growing in the *R. mangle*'s rhizosphere contributed to the high phytoremediation efficiency of the plant. Maximum petroleum hydrocarbons were removed from the contaminated sediments after 3 months of the plantation, when the largest bacterial count was observed in the plants' rhizosphere. Nearly 87 % petroleum hydrocarbon was removed from the contaminated soil individually by both plants, and growth of *R. mangle* was even better in the contaminated soil as compared to those grown in reference sediments, which suggests a good adaptation of this plant species for utilizing petroleum contaminants.

In a greenhouse experiment, a commonly used ornamental plant *Mirabilis Jalapa* showed extremely high tolerance to petroleum hydrocarbons [97]. The plant is more effective in degrading saturated hydrocarbon fraction than the other components of petroleum. It can grow normally in soil contaminated with up to 10 g petroleum/kg soil and remove up to 63.2 % of total petroleum hydrocarbons after 127 days of treatment. The soil microbial consortia were also reported to be adaptive to similar petroleum concentrations and their role in petroleum degradation was evidenced by removal of up to 37.92 % contaminated hydrocarbons by natural attenuation in the absence of *M. Jalapa* plantation. Sorghum (*Sorghum bicolor*) and common flax (*Linum usitatissumum*) were also used for decontamination of petroleum-contaminated soil [116]. When grown in high-petroleum-containing soil (40 g/kg) of an oil refinery, sorghum and common flax removed 9.5 and 18.5 g TPH from each kg of contaminated soil, respectively. Growth of both plant species was inhibited by petroleum contamination of the soil. However, plants' growth and bioremediation activity were improved by an organic fertilizer amendment. An organic fertilizer amendment was also effective for growth enhancement of maize (*Zea mays*) and tall fescue (*Festuca arundinacea*) plants grown in the presence of 35 g petroleum/kg soil [146]. Both plants were able to decrease TPH concentration of the highly contaminated aged soil. Better performance was observed in the tall fescue plantation, which removed 96.3 % of the initial hydrocarbons after 120 days of treatment.

3.2.2 Polluted Water Treatment

Arsenic and other heavy-metal pollution in drinking water is causing a range of health complications, especially in third-world countries. Their presence in surface water or groundwater is also a matter of serious concern as they can easily enter the food chain through vegetables grown with the contaminated water. A number of aquatic plant species showed potential application for the remediation of toxic heavy metals including As, Cd, Cr, Cu, Hg, Pb, and Zn. Several aquatic plants were investigated to understand their potential application in heavy-metal phytoremediation and some species were reported to accumulate a high level of arsenic from contaminated water [102]. A water hyacinth species (*Eichhornia crassipes*) and several duckweed species (family: Araceae) were reported to be very efficient in polluted water bioremediation. An arsenic-tolerant rootless duckweed species, *Wolffia globosa,* was able to accumulate both arsenate and arsenite compounds up to a level of 1,000 mg As/kg dry biomass [149]. Only arsenate uptake was reported to be suppressed by the presence of phosphate compounds. Interestingly, this species accumulated the element predominantly in the form of arsenite irrespective of the presence of arsenate or arsenite in the contaminated water.

Comparable bioaccumulation capabilities were reported in a water hyacinth (*E. crassipes*) and a lesser duckweed (*Lemna minor*) species [9]. However, the As removal efficacy of the water hyacinth was higher due to a faster growth rate. Hg and As removal from a coal mine effluent was studied using three aquatic macrophytes including a water hyacinth species, *E. crassipes,* and two duckweed species, *L. minor* and *Spirodela polyrrhiza* [86]. *E. crassipes* showed the best performance followed by *L. minor* and *S. polyrrhiza*. After 21 days of treatment, *E. crassipes* was able to remove 71 % Hg and 80 % As from the coal mine effluent containing about 0.007 mg/l Hg and 0.05 mg/l As. Studies revealed accumulation of a high proportion of metal in the root due to low metal transportation efficiency of these species. However, metal accumulation deteriorated the N, P, K, chlorophyll, and protein content in these macrophytes. After 15 days of treatment in laboratory conditions, two aquatic macrophytes, *Ceratophyllum demersum* and *Lemna gibba*, successfully removed 95 and 96 % lead, and 84 and 92 % chromium, respectively, from a metal solution containing both metal ions [1]. None of the plants showed any symptom of metal toxicity or reduction in growth in the presence of these metal contaminants. Cd, Cu, Pb, and Zn removal from water and river sediments were investigated using six indigenous macrophytes of the Nile River ecosystem such as *C. demersum*, *Echinochloa pyramidalis*, *E. crassipes*, *Myriophyllum spicatum*, *Phragmites australis*, and *Typha domingensis* [50]. Heavy-metal concentration in water, sediment, and plant biomass showed the same trend: Zn was present in the highest concentration followed by Cu, Pb, and Cd. In all six species, Cu and Zn were preferentially accumulated in roots whereas the highest Pb concentration was observed in plants' leaves. Tissue-specific Cd accumulation varied across the plant species.

The total metal accumulation capability of the investigated species was reported as: *C. demersum* > *E. crassipes* > *M. spicatum* > *E. pyramidalis* > *T. domingensis* > *P. australis*. Another aquatic macrophyte, *Eleocharis acicularis*, was reported to accumulate multiple metal ions [56]. Out of all the test metals, only indium (In) was accumulated in the plants' root whereas Ag, Pb, Cu, Cd, and Zn were preferentially accumulated in the shoots. Field cultivation experiments suggested potential application of this macrophyte in heavy-metal phytoremediation of mining sites [111]. Heavy-metal accumulation in the plant's shoots increases logarithmically with their concentration in the soil and the highest concentrations in the shoot were reported as 20.2 g Cu, 14.2 g Zn, 1.74 g As, 0.894 g Pb, and 0.239 g Cd per kg of dry biomass.

In another innovative approach, 10 heavy-metal–tolerant macrophytes were isolated by constructing small-scale wetlands for bioremediation of wastewater of an electroplating plant [130]. The isolates were identified as *P. australis*, *Typha orientalis*, *Lythrum salicaria*, *Arundo donax*, *Typha minima*, *Juncus effusus*, *Pontederia cordata*, *Cyperus alternifolius*, *Acorus calamus*, and *Iris pseudacorus*. *P. australis*, *A. calamus*, *T. minima*, and *L. salicaria* were revealed as the most suitable and promising plant species for heavy-metal phytoremediation because of their high heavy-metal accumulation capabilities. The aquatic macrophyte, *Callitriche cophocarpa*, demonstrated extremely high affinity to Cr^{3+} and Cr^{6+} ions with average accumulations of 28.3 and 7.3 g/kg dry biomass, respectively [13]. After 5 days of treatment, it removed all of the chromium of a test solution containing 0.5 mM Cr^{3+} ion. The major fraction of Cr^{3+} was strongly bound as metallo-organic compounds and 57 % of Cr^{6+} was accumulated as an easily mobilizable compound.

An in situ study was conducted to assess potentials of three plant species such as rice (*Oryza sativa* L.), paragrass (*Brachiaria mutica*), and water hyacinth (*E. crassipes*) for Cr^{6+} removal from mine wastewater [87]. Total Cr accumulation increases with the plants' age and Cr^{6+} content of the test soil. The water hyacinth removed the highest amount (24–54 %) of the contaminant from mine water followed by the paragrass species (18–33 %). The water hyacinth showed good Cr^{6+} transportation efficiency from root to shoot, however, the total accumulation rate was faster in paragrass (8.29 mg/kg dry biomass/day). Dwivedi et al. analyzed heavy-metal accumulation in two flowering plant species, *Portulaca tuberosa* and *Portulaca oleracea*, collected from several fields irrigated with industrial effluent and tube wellwater [48]. Higher heavy-metal accumulation was reported in the plants grown in effluent irrigated areas. When compared across different tissues, maximum accumulation was reported in the roots and the least accumulation was in the flowers. Although *P. oleracea* showed better performance, both species accumulated a significantly high concentration of test metals including Cu, Ni, Hg, and Pb.

3.2.3 Polluted Air Treatment

Plants are well known for absorbing greenhouse gases. They are also useful in bioremediation of toxic organic and inorganic gases. Although plant-assisted decontamination of soil and water is dependent on absorption by roots, leaves are the main receptor of gas and particulate solids from air. The benzene, toluene, ethylbenzene, and xylene (BTEX) removal potential of Zanzibar Gem plant (*Zamioculcas zamiifolia*) was studied in contaminated indoor air [125]. The plant was able to absorb all four gases and smaller molecules were absorbed faster than large molecules. After 72 h of exposure, the respective absorption rates of benzene, toluene, ethylbenzene, and xylene were reported as 0.96 ± 0.01, 0.93 ± 0.02, 0.92 ± 0.02, and 0.86 ± 0.07 mmol/m^2 of *Z. zamiifolia* leaf. Study revealed that 80 % benzene, 76 % toluene, 75 % ethylbenzene, and 73 % xylene were removed by stomata whereas 20 % benzene, 23 % toluene, 25 % ethylbenzene, and 26 % xylene were removed by cuticles. The BTEX did not affect the photosynthesis of the plants and showed no apparent toxicity.

In another study, 30 species of Araceae, Agavaceae, and Liliaceae families were screened for their formaldehyde-removal ability from air [153]. When exposed to initial formaldehyde concentration of 15 mg/m^3 for 7 days, 10 plant species (*Scindapsus aureus*; *Asparagus setaceus;* *Sansevieria trifasciata* cv. Hahnii; *Chlorophytum comosum*; *Aglaonema commutatum* cv. White Rajah, cv. Red Narrow, cv. Treubii; *Scindapsus pictus* cv. Argyraeus; *Gasteria gracilis*, and *Philodendron sodiroi* cv. Wendimbe) were able to absorb more than 99 % formaldehyde of the 1.0 m × 1.0 m × 0.8 m experimental chamber. The plants were least affected by formaldehyde pollution and they absorbed most of the formaldehyde during the first 3 days of the experiment. Plants also help in controlling air pollution by accumulating dust particles on their leaf surfaces. The dust-trapping ability of 15 plant species growing around a steel factory was assessed from the dust load on leaf surfaces and leaf surface morphology [39]. High dust-capturing capacity was reported in eight plant species such as *Alstonia scholaris*, *Anthocephalus indicus*, *Cassia auriculata*, *Cassia siamea*, *Lagerstroemia speciosa*, *Mimusops elengi*, *Peltophorum inerme*, and *Tabebuia aurea*. The dust-trapping capacity of different plant species was highly dependent on their leaf surface morphology.

3.3 *Nonmicroscopic Lower Eukaryotes*

Some lower eukaryotes such as earthworms and polychaetes can contribute in decontamination of biological solid wastage of domestic, municipal, and industrial origin. They are also reported to be active in removing toxic heavy metals, oil, petroleum hydrocarbon, and other chemical contaminants from polluted soil. Earthworms are known to aerate soils and improve bioavailability of contaminants by bioturbation, which in turn eases the bioremediation and phytoremediation of the contaminants. This section focuses on some recent research activities regarding the use of some nonmicroscopic lower eukaryotes in decontamination of environmental pollutants.

3.3.1 Treatment of Soil Pollution

The potential of the earthworm species *Eudrilus Eugenia* was investigated for vermi-assisted bioremediation of petroleum-hydrocarbon-contaminated soils [10]. Introduction of this earthworm showed enhanced bioremediation of petroleum-contaminated soil of a mechanical workshop as compared to the bioremediation of the same soil samples without worms. It also lowered carbon and nitrogen content of the soil samples. However, earthworm survival and composition of microbial species were dependent on the concentration and nature of the petroleum hydrocarbon contaminants. In all cases, the earthworm count was reported to be much lower in petroleum-contaminated soil than in the control. Several earthworm species were reported to contribute to PAH bioremediation. Another earthworm species, *Pontoscolex corethrurus*, was reported to enhance benzo[a]pyrene degradation when it was used to treat contaminated soils already amended with legume, *Mucuna pruriens*, or the grass, *Brachiaria humidicola* [57]. After 112 days of treatment, the earthworm alone was able to remove 26.6 % of the contaminant from the sterile soil sample containing 100 mg/kg benzo[a]pyrene; the amendments with *B. humidicola* and *M. pruriens* increased this value to 35.7 and 34.2 %, respectively. This earthworm also removed 36.1 % of the contaminant from the unsterilized soil whereas the autochthonous microorganism itself could remove only 9.1 % of the contaminants over same treatment period.

Enhancement of microbial pyrene degradation was also reported by the presence of another earthworm species, *Eisenia foetida* [129]. A microbial degradation study in the presence and absence of the earthworm showed that introduction of the earthworm enhanced pyrene removal significantly from both freshly contaminated and aged soils. After 14 days of incubation with the *E. foetida*, 45.5–91.0 % pyrene was removed, which was about two to three times higher than the sample treated for the same time without addition of any earthworm.

Earthworms also helped in bioremediation of another PAH, fluoranthene, from contaminated soil [60]. Fluoranthene removal by individual and combined actions of an earthworm population (consisting mostly of *Eisenia fetida*) and ryegrass

(*Lolium multiflorum*) cultivation was investigated over a period of 10 weeks. More than 60 % of the contaminant was removed by the first 2 weeks and above 80 % contaminant was removed at the end of the experiment by combined action of the earthworm and the ryegrass. However, the ryegrass and earthworm individually removed nearly 75 % fluoranthene whereas the indigenous microorganisms were able to remove 70 % of the contaminant over the same timeframe. Only 0.01–1.20 % of the removed fluoranthene was accumulated in the ryegrass and earthworm biomasses, which gave evidence of the role of indigenous microorganisms in fluoranthene degradation. Experimental observations suggest that the earthworm and the ryegrass cultivation enhanced soil fluoranthene removal mainly by enhancing polyphenol oxidase activity of the microorganisms.

Two ecological earthworm species (*E. foetida* and *Amynthas robustus*) were reported to enhance biodegradation of DDT present as a soil contaminant [72]. The experiment was conducted over a period of 360 days and faster DDT degradation was observed in the soil with higher earthworm density. Analysis of the degradation products suggests that the anaerobic reductive dechlorination was the main degradation pathway for the first half of the study, and the aerobic dechlorination process was gradually increased during the second half of incubation.

Sizmur et al. [120] studied the role of an earthworm species (*Lumbricus terrestris*) in heavy-metal mobilization from contaminated soils. Analysis of chemical form and concentration of the metals showed higher concentrations of water-extractable heavy-metal compounds in the earthworm-treated soil than the earthworm-free control. The bioremediation ability of the earthworm was dependent on soil composition and concentration of the heavy-metal contaminants. In another study, heavy-metal- and hydrocarbon-cocontaminated soil was bioremediated by combined action of two plants (*Paulownia tomentosa* and *Cytisus scoparius*), an earthworm species (*E. fetida*), and organic matter (horse manure) amendment [77]. The process was scaled up to carry out the bioremediation of soil polluted with municipal waste. Bioremediation performances of the plants were improved by application of the organic matter and the earthworm. Plants and the native soil microorganisms were more effective in reducing the heavy-metal level in soil and the earthworm contributed mainly in the decontamination of organic pollutants (hydrocarbons).

Bioremediations of organically enriched fish farm sediments were enhanced by artificial mass culture of the polychaete *Capitella* sp. I [63, 70]. A long-term study suggested fast decomposition of the organic matter during the rapid population growth of the polychaete. The technique is promising for minimizing negative environmental impacts of fish farms. The results indicated that the organic matter was finally decomposed by the local microbial community. Also the microbial population increased during the fast-increasing period of the polychaete species, and simultaneous enhancement of the bioremediation rate confirmed combined action of the polychaete and the indigenous microbial population in the organic matter biodegradation. In contrast to the cooperative bioremediation by *Capitella* sp. and native microbial consortia, another polychaete, *Sabella spallanzanii*, was reported to prevent bacterial pathogen enrichment in the aquaculture waste by removal of bacterioplankton from water [126]. Experimental results showed that *S. spallanzanii* was able to filter, accumulate, and remove most of the bacterial species from the waste including potential human pathogens. However, the bacteria-filtering capability of the polychaete was dependent on environmental parameters and significant variation in performance was reported with seasonal changes.

3.3.2 Treatment of Water Pollution

Earthworms were also useful for vermi-stabilization of industrial wastewaters [132, 133]. Vermi-stabilization of wastewater sludge from the milk-processing industry was studied by using the earthworm *E. fetida* in combination with cow dung [133]. Significant reduction in pH and organic carbon was achieved after 90 days of bioremediation. Concentrations of exchangeable cations such as K^+ and Ca^{2+}, and extractable trace metals such as Fe, Mn, and Zn also increased with this treatment. Growth and cocoon formation of the earthworm were stimulated by the addition of cow dung. A better growth and reproduction pattern was achieved in the vermibeds containing 40–60 % sludge, whereas a high sludge concentration increased the mortality of the earthworm. Physical parameters and chemical composition of another vermibed were changed in a similar way when the same earthworm species and cow dung mixture was used for bioremediation of aerobically treated distillery sludge from a sugar industry [132]. The fastest growth rate and maximum individual live weight of the earthworms were observed in the vermibeds containing 20 % distillery sludge whereas better reproduction success was noticed in the beds treated with 40 % distillery sludge; earthworm mortality increased in the vermibeds with 60 % or more sludge content (Table 4).

Table 4 Summary of recent research (2008–2013) on soil and water bioremediation with nonmicroscopic lower eukaryotes

Active organism(s)	Target pollutants	Reference
For bioremediation of polluted soil and solid waste		
Eudrilus Eugenia	Petroleum hydrocarbon	[10]
Pontoscolex corethrurus	Benzo[a]pyrene	[57]
Eisenia foetida	Pyrene	[129]
Eisenia fetida	Fluoranthene	[60]
Eisenia foetida, Amynthas robustus	DDT	[72]
Lumbricus terrestris	Cu, Pb, Zn	[120]
Eisenia fetida	Organic hydrocarbons	[77]
Capitella sp. I	Organic pollutants	[63, 70]
Sabella spallanzanii	Bacteria	[126]
For bioremediation of contaminated water		
Eisenia fetida	Organic pollutants	[132, 133]

4 In-Field Situation of Bioremediation: Limitations and Challenges

Although there are a few limiting factors, bioremediation has the ability to restore most contaminated environments. In fact, microorganisms and plants have been cleaning the environment by degrading environmental wastage for billions of years. Particularly, in situ bioremediation can be regarded as skillful and extended use of natural microbial activities. Recent scientific advances and development of several bioinformatics tools and high-throughput techniques are extremely useful in understanding the bioremediation processes at the molecular level. The use of metagenomic tools and culture-independent molecular techniques is helping the understanding of the structure and dynamics of the microbial community. As a result, bioremediation is no longer considered as a function of a single microorganism; rather, newer bioremediation technologies are being developed based on microbial consortia. Researchers already developed genetically modified organisms and synthetic microbial communities with the special ability of surviving in unfavorable environmental conditions and degrading complex contaminants. However, many hurdles are yet to be overcome before their in-field use in various polluted sites.

Bioremediation is a well-studied field today but not well practiced yet. Although the bioremediation market is increasing sharply, current application of this technology is confined to a small fraction of the very large waste management market. This review shows that microorganism and plant species of different taxonomic groups were characterized for their potential bioremediation ability, but most of them are yet to be commercialized. Bioremediation is used mainly in long-term decontamination of petroleum-/PAH-/metal-contaminated soils, cleaning up of chemical spills from soil or water, large-scale water treatment for decreasing BOD and COD values, heavy metal removal from groundwater by phytoremediation, and in activated sludge plants [16, 23, 27, 66, 106, 107]. Several bioremediation companies are now providing a range of products and services for microbial decontamination of polluted soil and water, which include a supply of biostimulating formulations, designing of bioremediation programs, field management and supervision, remediation performance review, and even development of customized site-specific microbial consortia. The biostimulating formulations usually contain macro- or micronutrients, oxygen-releasing compounds, hydrogen-releasing compounds, other oxidizing or reducing agents, electron acceptors, electron donors, or surfactants. However, these companies rarely disclose the composition of microbial consortia or the exact formulation of bioremediating nutrients.

Oil Spill Eater International, Corp. provides bioremediation services for the cleaning up of various environments including brackish water, ocean water, fresh water, and intertidal zones by application of bioremediation techniques (http://osei.us/). Another bioremediation company, Remediation and Natural Attenuation Services, supplies various bioremediation formulations for enhancing the growth of indigenous microbial species (http://www.rnasinc.com/). Some of the products work by providing fast and slow release electron donors, by enhancing buffering capacity of soils, and by increasing dissolved oxygen concentration of water. EOS Remediation is a bioremediation company that provides various biostimulation and bioaugmentation products (http://www.eosremediation.com/). They revealed use of *Dehalococcoides* sp. and other dechlorinating bacteria for biodegradation of chlorinated hydrocarbons. This microorganism was also revealed as a component of Bio-Dechlor Inocculum® Plus, a bioaugmentation product supplied by another bioremediation company, Regenesis (http://www.regenesis.com/). This product claims to stimulate dechlorination of several compounds such as tetrachloroethene, trichloroethene, dichloroethene, and vinyl chloride.

Large-scale bioremediation technology was successfully deployed in the 1989 *Exxon Valdez* oil spill in Prince William Sound, Alaska, and the BP *Deepwater Horizon* oil spill in 2010 in the Gulf of Mexico [12]. The *Exxon Valdez* oil spill was degraded by biostimulating the indigenous microbial community by addition of nitrogenous fertilizers, whereas bioavailability of the petroleum hydrocarbons were enhanced by the addition of a dispersant such as Corexit 9500 to cope with the bioremediation of the *Deepwater Horizon* oil spill. Significantly higher numbers of hydrocarbon-degrading microorganisms were reported at the *Exxon Valdez* contaminated sites within the path of the oil slick than at reference sites [28]. A total of 24 oil-degrading bacterial strains were isolated from a *Deepwater Horizon* oil spill that contaminated beach sands at Pensacola Beach, Florida [64]. Isolated bacterial strains were identified primarily as hydrocarbon-degrading genera such as *Alcanivora*x, *Marinobacter*, *Pseudomonas*, and *Acinetobacter* of the *γ-Proteobacteria* family and *Rhodobacteraceae* sp. of the *Alphaproteobacteria* family.

In-field use of bioremediation is limited by several factors including slowness of the system, challenges of scale-up, unfavorable physical parameters of actual sites, uneven distribution of pollutants, government policies and regulatory hurdles, unawareness of environmental protection, and economic liabilities. Biological processes are extremely selective and in most cases bioremediation cannot eliminate all contaminants of a site, which needs additional chemical processes. This made bioremediation only an efficient pretreatment procedure of the final decontaminating process and often users prefer to use single-step faster chemical processes rather than wasting time with a comparatively slow pretreatment step, which in turn saves a small fraction of the total expenditure. Bioremediation often takes a much longer time than other treatment options and is really impracticable in industries that produce tons of wastage in a single day and also for municipal wastage. As a result, land filling and incineration methods are still the most popular methods for municipal waste management, and they are also being practiced for treatment of some industrial wastage, especially in underdeveloped and developing countries. Although bioaugmentation and biostimulation techniques enhance the bioremediation rate, it is far from the requirements of most industries.

Scale-up is yet another big challenge. Maintenance of optimum conditions in large-scale ex situ processes is often more difficult than laboratory-scale or pilot-plant testing. Site factors are also an important consideration for in situ bioremediations. Bioremediation formulations that are highly effective in in-house testing may lose efficiency due to uncertain physical and chemical conditions of the application site. The physical conditions of the treatment sites are ever-changing due to environmental parameters, microbial metabolism, and the increasing amount of biodegradation products, which are tough to mimic in a small-scale experimental set-up. In this context, process optimization for ex situ bioremediation techniques is also a relatively neglected field in comparison to basic science studies of bioremediation. In-house bioremediation processes need to be performed in controlled environments, which requires sophisticated techniques and trained manpower. Also there are several practical problems for ex situ bioremediation of large amounts of waste materials, which includes transport burden, risks of spreading, and handling hazard. Bioremediation technology has the potential of replacing nonecofriendly waste management strategies in the future but it needs to go a long way before overcoming the present obstacles. Intensified interdisciplinary research may show the way for converting cumulative fundamental knowledge to successful bioremediation practices.

5 Conclusion and Future Perspectives

Environmental pollution is the biggest threat to survival of lives on the planet Earth and with the progress of human civilization we are worsening the situation. Pollution has several direct effects on human and animal health. For a long time, science and technologies have been used in such ways that pollute the environment but we can use scientific knowledge and technological advancements to get rid of environmental pollution. Bioremediation is one such technology for decontaminating polluted environments by utilizing the ability of living organisms to degrade toxic chemicals. A large number of living organisms show potential application in bioremediation of soil, water, and air pollution. These organisms come from a diverse phylogenetic origin and often interactions between more than one micro- or macro-organism are essential for biodegradation of a pollutant. Bioremediation is used for degrading certain toxic pollutants but cannot yet replace existing industrial waste management techniques for various practical reasons. Scientists are making continuous efforts to understand existing bioremediating organisms and improve their bioremediation capability. At the same time, research is also being focused on discovery and development of more efficient bioremediating agents. Most sophisticated techniques are being used to understand the dynamic chemical and microbial compositions of bioremediation sites, which would be helpful for understanding bioremediation methods at the molecular level.

Instead of efficient pollutant degradation ability of several plants and microbial species, bioremediation is not a panacea for treatment of all pollution-related problems, at least for now. A gap exists between advances in laboratory research and in-field industrial applications. Capabilities of natural or engineered microbial consortia need to be integrated with appropriate process designs to provide efficient bioremediations at the commercial scale. Heterogeneity, complexity, and the dynamic nature of the contaminated environment are other major hurdles to successful application of bioremediation techniques. In fact, bioremediation strategies should be developed based on physicochemical and biological composition of each contaminated site. Assessment of pollutant degradation kinetics is essential for setting up industrial bioremediation processes; it is yet another inadequately studied parameter. Although there is significant progress in overcoming some of the challenges, efforts need to be intensified in understanding problems behind industrial bioremediation and in searching for their practical solutions.

References

1. Abdallah MAM (2012) Phytoremediation of heavy metals from aqueous solutions by two aquatic macrophytes, *Ceratophyllum demersum* and *Lemna gibba* L. Environ Technol 33:1609–1614
2. Abhilash PC, Srivastava S, Singh N (2011) Comparative bioremediation potential of four rhizospheric microbial species against lindane. Chemosphere 82:56–63
3. Abreu CAD, Coscione AR, Pires AM, Paz-Ferreiro J (2012) Phytoremediation of a soil contaminated by heavy metals and boron using castor oil plants and organic matter amendments. J Geochem Explor 123:3–7
4. Abulencia CB, Wyborski DL, Garcia JA et al (2006) Environmental whole-genome amplification to access microbial populations in contaminated sediments. Appl Environ Microbiol 72:3291–3301
5. Acevedo F, Pizzul L, Castillo MDP, Cuevas R, Diez MC (2011) Degradation of polycyclic aromatic hydrocarbons by the Chilean white-rot fungus *Anthracophyllum discolor*. J Hazard Mater 185:212–219
6. Agarwal SK (1998) Environmental biotechnology. APH Publishing Corporation, New Delhi
7. Al-Bader D, Eliyas M, Rayan R, Radwan S (2012) Air–dust-borne associations of phototrophic and hydrocarbon-utilizing microorganisms: promising consortia in volatile hydrocarbon bioremediation. Environ Sci Pollut Res 19:3997–4005
8. Alisi C, Musellar R, Tasso F et al (2009) Bioremediation of diesel oil in a co-contaminated soil by bioaugmentation with a microbial formula tailored with native strains selected for heavy metals resistance. Sci Total Environ 407:3024–3032
9. Alvarado S, Guedez S, Lue-Meru MP et al (2008) Arsenic removal from waters by bioremediation with the aquatic plants Water Hyacinth (*Eichhornia crassipes*) and Lesser Duckweed (*Lemna minor*). Biores Technol 99:8436–8440
10. Ameh AO, Mohammed-Dabo IA, Ibrahim S, Ameh JB (2013) Earthworm-assisted bioremediation of petroleum hydrocarbon contaminated soil from mechanic workshop. Afr J Environ Sci Technol 7:531–539
11. Atagana HI (2011) Bioremediation of co-contamination of crude oil and heavy metals in soil by phytoremediation using *Chromolaena odorata* (L) king & HE Robinson. Water Air Soil Pollut 215:261–271
12. Atlas RM (2011) Oil biodegradation and bioremediation: a tale of the two worst spills in US history. Environ Sci Technol 45:6709–6715
13. Augustynowicz J, Kyziol-Komosinska J, Smolen S, Waloszek A (2013) Study on chromium-binding capacity of *Callitriche cophocarpa*in an aquatic environment. Arch Environ Contam Toxicol 64:410–418
14. Babu AG, Kim JD, Oh BT (2013) Enhancement of heavy metal phytoremediation by *Alnus firma* with endophytic *Bacillus thuringiensis* GDB-1. J Hazard Mater 250–251:477–483

15. Bacchetta G, Cao A, Cappai G, Casti CM, Fercia ML, Lonis R, Fola F (2012) A field experiment on the use of *Pistacia lentiscus* L. and *Scrophularia canina* L. subsp. *Bicolor* (Sibth. et Sm.) Greuter for the phytoremediation of abandoned mining areas. Plant Biosys 146:1054–1063
16. Baird WE, Panda J, Lang WJ (2002) Success story. Pollut Eng 34:18–19
17. Barbafieri M, Dadia C, Tassi E, Bretzel F, Fanfani L (2011) Uptake of heavy metals by native species growing in a mining area in Sardina, Italy: discovering native flora for phytoremediation. Int J Phytoremed 13:985–997
18. Barrutia O, Artexe U, Hernandez A et al (2011) Native plant communities in an abandoned Pb-Zn mining area of northern Spain: implication for phytoremediation and germ plasm preservation. Int J Phytoremed 13:256–270
19. Beattie AJ, Hay M, Magnusson B, Nys R, Smeathers J, Vincent JFV (2011) Ecology and bioprospecting. Austral Ecol 36:341–356
20. Benazir JF, Suganthi R, Rajvel D, Pooja MP, Mathithumilan B (2010) Bioremediation of chromium in tannery effluent by microbial consortia. Afr J Biotechnol 9:3140–3143
21. Benimeli CS, Fuentes MS, Abate CM, Amoroso MJ (2008) Bioremediation of lindane-contaminated soil by Streptomyces sp. M7and its effects on *Zea mays* growth. Int Biodeterior Biodegrad 61:233–239
22. Bento RA, Saggin-Junior OJ, Pitard RM et al (2012) Selection of leguminous trees associated with symbiont microorganisms for phytoremediation of petroleum-contaminated soil. Water Air Soil Pollut 223:5659–5671
23. Bestawy EE, Helmy S, Hussein H, Fahmy M, Amer R (2013) Bioremediation of heavy metal-contaminated effluent using optimized activated sludge bacteria. Appl Water Sci 3:181–192
24. Bhowmik A, Asahino A, Shiraki T, Nakamura K, Takamizawa K (2009) In situ study of tetrachloroethylene bioremediation with different microbial community shifting. Environ Technol 30:1607–1614
25. Binga EK, Lasken RS, Neufeld JD (2008) Mini-review: something from (almost) nothing: the impact of multiple displacement amplification on microbial ecology. ISME J 2:233–241
26. Bjedov I, Tenaillon O, Gerard B et al (2003) Stress induced mutagenesis in bacteria. Science 300:1404–1409
27. Boopathy R (2000) Factors limiting bioremediation technologies. Bioresour Technol 74:63–67
28. Braddock JF, Lindsdorm JE, Brown RJ (1995) Distribution of hydrocarbon-degrading microorganisms in sediments from Prince William Sound, Alaska, following the *Exxon Valdez* oil spill. Mar Pollut Bull 30:125–132
29. Buenz EJ, Schnepple DJ, Bauer BA, Riddle JM, Motley TJ (2004) Techniques: bioprospecting historical herbal texts by hunting for new leads in old tomes. Trends Pharmacol Sci 25:494–498
30. Cea M, Jorquera M, Ribilar O, Langer H, Tortella G, Diez MC (2010) Bioremediation of soil contaminated with pentachlorophenol by Anthracophyllum discolor and its effect on soil microbial community. J Hazard Mater 181:315–323
31. Chakraborty R, Wu CH, Hazen TC (2012) Systems biology approach to bioremediation. Curr Opin Biotechnol 23:483–490
32. Chaudhary VK, Borah D (2011) Isolation and molecular characterization of hydrocarbon degrading bacteria from tannery effluent. Int J Plant Anim Env Sci 1:110–114
33. Chen J, Yang Q, Huang T, Zhang Y, Ding R (2011) Enhanced bioremediation of soil contaminated with viscous oil through microbial consortium construction and ultraviolet mutation. World J Microbiol Biotechnol 27:1381–1389
34. Chikere CB (2013) Application of molecular microbiology techniques in bioremediation of hydrocarbons and other pollutants. Br Biotechnol J 3:90–115
35. Chiu SY, Kao CY, Huang TT et al (2011) Microalgal biomass production and on-site bioremediation of carbon dioxide, nitrogen oxide and sulfur dioxide from flue gas using Chlorella sp. Cultures. Bioresour Technol 102:9135–9142

36. Couto MNPFS, Pinto D, Basto MCP, Vasconcelos TSD (2012) Role of natural attenuation, phytoremediation and hybrid technologies in the remediation of a refinery soil with old/recent petroleum hydrocarbons contamination. Environ Technol 33:2097–2104
37. Cowan D, Meyer Q, Stafford W, Muyanga S, Cameron R, Wittwer P (2005) Metagenomic genediscovery: past, present and future. Trends Biotechnol 23:321–329
38. Daniel R (2005) The metagenomics of soil. Nat Rev Microbiol 3:470–478
39. Das S, Prasad P (2012) Particulate matter capturing ability of some plant species: implication for phytoremediation of particulate pollution around Rourkela steel plant, Rourkela, India. Nat Environ Pollut Technol 11:657–665
40. Demunshi Y, Chugh A (2010) Role of traditional knowledge in marine bioprospecting. Biodivers Conserv 19:3015–3033
41. Deng Q, Li QB, Lu YH, He N, Jiang J (2005) Genetic engineering of *E. coli* SE5000 and its potential for Ni^{2+} bioremediation. Proc Biochem 40:425–430
42. Desai C, Pathak H, Madamwar D (2010) Advances in molecular and "-omics" technologies to gauge microbial communities and bioremediation at xenobiotic/anthropogen contaminated sites. Biores Technol 101:1558–1569
43. Deutschbauer AM, Chivian D, Arkin AP (2006) Genomics for environmental microbiology. Curr Opin Biotechnol 17:229–235
44. Diez MC, Gallardo F, Tortella G, Rubilar O, Navia R, Bornhardt C (2012) Chlorophenol degradation in soil columns inoculated with *Anthracophyllum discolor* immobilized on wheat grains. J Envirol Manage 95:S83–S87
45. Doria E, Longoni P, Scibila L, Iazzi N, Cella R, Nielsen E (2012) Isolation and characterization of a *Scenedesmus acutus* strain to be used for bioremediation of urban wastewater. J Appl Phycol 24:375–383
46. Douglas AE (2013) Microbial brokers of insect-plant interactions revisited. J Chem Ecol 39:952–961
47. Douskova I, Doucha J, Livansky K (2009) Simultaneous flue gas bioremediation and reduction of microalgal biomass production costs. Appl Microbiol Biotechnol 82:179–185
48. Dwivedi S, Mishra A, Kumar A et al (2012) Bioremediation potential of genus Portulaca L. collected from industrial areas in Vadodara, Gujarat, India. Clean Techn Environ Policy 14:223–228
49. Eisenstadt E, Warren JA, Porter J, Atkins D, Millert JH (1982) Carcinogenic epoxides of benzo[alpyrene and cyclopenta[b]pyrene induce base substitutions via specific transversions. Proc Nat Acad Sci USA 79:1945–1949
50. Fawze MA, Badr NE, El-Khatib A, Abo-El-Kassem A (2012) Heavy metal biomonitoring and phytoremediation potentialities of aquatic macrophytes in river Nile. Environ Monit Assess 184:1753–1771
51. Foster PL (2007) Stress induced mutagenesis in bacteria. Crit Rev Biochem Mol Biol 42:373–397
52. Garbisu C, Alkorta I (1999) Utilization of genetically engineered microorganisms (GEMs) for bioremediation. J Chem Technol Biotechnol 74:599–606
53. Greben HA, Baloyi J, Sigama J, Venter SN (2009) Bioremediation of sulphate rich mine effluents using grass cuttings and rumen fluid microorganisms. J Geochem Explor 100:163–168
54. Grosch R, Dealtry S, Schreiter S, Berg G, Mendonca-Hagler L, Smalla K (2012) Biocontrol of *Rhizoctonia solani*: complex interaction of biocontrol strains, pathogen and indigenous microbial community in the rhizosphere of lettuce shown by molecular methods. Plant Soil 361:343–357
55. Godos ID, Blanco S, Garcia-Encina PA, Becares E, Munoz R (2009) Long-term operation of high rate algal ponds for the bioremediation of piggery waste waters at high loading rates. Bioresour Technol 100:4332–4339
56. Ha NTH, Sakakibara M, Sano S (2011) Accumulation of Indium and other heavy metals by *Eleocharis acicularis*: an option for phytoremediation and phytomining. Bioresour Technol 102:2228–2234

57. Hernandez-Castellanos B, Ortiz-Ceballos A, Martinez-Hernandez S et al (2013) Removal of benzo (a) pyrene from soil using an endogeic earthworm *Pontoscolex corethrurus* (Müller, 1857). Appl Soil Ecol 70:62–69
58. Hess A, Zarda B, Hahn D, Hanner A, Stax D (1997) In situ analysis of denitrifying toluene and m-xylene-degrading bacteria in a diesel fuel contaminated laboratory aquifer column. Appl Environ Microbiol 63:2136–2141
59. Imanaka T, Aiba S (1981) A perspective on the application of genetic engineering: stability of recombinant plasmid. Ann N Y Acad Sci 369:1–14
60. Jing Y-P, Liu M-Q, Yin Q-P, Li H-X, Hu F (2013) Effects of earthworms and ryegrass on the removal of fluoranthene from soil. Pedosphere 23:523–531
61. Karamalidis AK, Evangelou EC, Karabika E, Koukkou AI, Drainas C, Voudrias EA (2010) Laboratory scale bioremediation of petroleum-contaminated soil by indigenous microorganisms and added *Pseudomonas aeruginosa* strain Spet. Bioresour Technol 101:6545–6552
62. Karigar CS, Rao SS (2011) Role of microbial enzymes in the bioremediation of pollutants: a review. Enzym Res 1 (Article ID 805187)
63. Kinoshita K, Tamaki S, Yoshioka M et al (2008) Bioremediation of organically enriched sediment deposited below fish farms with artificially mass-cultured colonies of a deposit-feeding polychaete *Capitella* sp. I. Fish Sci 74:77–87
64. Kostka JE, Prakash O, Overholt WA et al (2011) Hydrocarbon-degrading bacteria and the bacterial community response in Gulf of Mexico beach sands impacted by the *Deepwater Horizon* oil spill. Appl Env Microbiol 77:7962–7974
65. Kotrba P, Najmanova J, Macek T, Ruml T, Mackova M (2009) Genetically modified plants in phytoremediation of heavy metal and metalloid soil and sediment pollution. Biotechnol Adv 27:799–810
66. Kulshreshtha S (2012) Current trends in bioremediation and biodegradation. J Bioremed Biodegrad 3:7
67. Kumar M, Khanna S (2009) Diversity of 16S rRNA and dioxygenase genes detected in coal-tar-contaminated site undergoing active bioremediation. J Appl Microbiol 108:1252–1262
68. Kumar M, Lakshmi CV, Khanna S (2008) Microbial biodiversity and in situ bioremediation of endosulfan contaminated soil. Indian J Microbiol 48:128–133
69. Kumar V, Dhall P, Kumar R, Singh YK, Kumar A (2012) Bioremediation of agro-based pulp mill effluent by microbial consortium comprising autochthonous bacteria. Sci World J 2012:127014
70. Kunihiro T, Miyazaki T, Uramoto Y (2008) The succession of microbial community in the organic richfish-farm sediment during bioremediation by introducing artificially mass-cultured colonies of a small polychaete, Capitella sp. I. Mar Pollut Bull 57:68–77
71. Leal MC, Madeira C, Brandao CA, Puga J, Calado R (2012) Bioprospecting of marine invertebrates for new natural products—a chemical and zoogeographical perspective. Molecules 17:9842–9854
72. Lin Z, Li X-M, Li Y-T, Huang D-Y, Dong J, Li FB (2012) Enhancement effect of two ecological earthworm species (*Eisenia foetida* and*Amynthas robustus*) on removal and degradation processes of soil DDT. J Environ Monit 14:1551–1558
73. Liu PWG, Chang TC, Whang LM, Kao CH, Pan PT, Cheng SS (2011) Bioremediation of petroleum hydrocarbon contaminated soil: Effects of strategies and microbial community shift. Int Biodeterior Biodegradation 65:1119–1127
74. Liu X, Ashforth E, Ren B et al (2010) Bioprospecting microbial natural product libraries from the marine environment for drug discovery. J Antibiot 63:415–422
75. Lorenz P, Eck J (2005) Metagenomics and industrial applications. Nat Rev Microbiol 3:510–516
76. Ma HW, Zeng AP (2004) Phylogenetic comparison of metabolic capacities of organisms at genome level. Mol Phylogenet Evol 31:204–213
77. Macci C, Doni S, Peruzzi E, Ceccanti B, Masciandaro G (2012) Bioremediation of polluted soil through the combined application of plants, earthworms and organic matter. J Environ Monit 14:2710–2717

78. Maiti RK, Pinero JLH, Oreja JAG, Santiago DL (2004) Plant based bioremediation and mechanism of heavy metal tolerance of plants: a review. Proc Indian Natn Sci Acad B70:1–12
79. Mao J, Luo Y, Teng Y, Li Z (2012) Bioremediation of polycyclic aromatic hydrocarbon-contaminated soil by a bacterial consortium and associated microbial community changes. Int Biodeterior Biodegrad 70:141–147
80. Malayeri BE, Chehregania A, Mohsenzadeha F, Kazemeinib F, Asgaric M (2013) Plants growing in a mining area: screening for metal accumulator plants possibly useful for bioremediation. Toxicol Environ Chem 95:434–444
81. Malekzadeh E, Alikhani HA, Sabaghebi-Firoozabadi GR, Zarei M (2012) Bioremediation of cadmium-contaminated soil through cultivation of maize inoculated with plant growth–promoting rhizobacteria. Bioremed J 16:204–211
82. Malkawi HI, Fatmi LM, Al-Deeb TM (2009) Mutational analysis of oil degrading genes in bacterial isolates from oil contaminated soil at the Jordan oil refinery. World Appl Sci J 6:208–220
83. Matin A, Little CD, Fraley CD, Keyhan M (1995) Use of starvation promoters to limit growth and select for trichloroethylene and phenol transformation activity in recombinant *Escherichia coli*. Appl Environ Microbiol 61:3323–3328
84. McGinn PJ, Dickinson KE, Park KC et al (2012) Assessment of the bioenergy and bioremediation potentials of the microalga *Scenedesmus* sp. AMDD cultivated in municipal wastewater effluent in batch and continuous mode. Algal Res 1:155–165
85. Mikeskova H, Novotny C, Svobodova (2012) Interspecific interactions in mixed microbial cultures in a biodegradation perspective. Appl Microbiol Biotechnol 95:861–870
86. Mishra VK, Upadhyay AR, Pathak V, Tripathi BD (2008) Phytoremediation of mercury and arsenic from tropical opencast coalmine effluent through naturally occurring aquatic macrophytes. Water Air Soil Pollut 192:303–314
87. Mohanty M, Pattnaik MM, Mishra Ak, Patra HK (2012) Bio-concentration of chromium—an *in situ* phytoremediation study at South Kaliapani chromite mining area of Orissa, India. Environ Monit Assess 184:1015–1024
88. Mohsenzadeh F, Nasseri S, Mesdaghinia A et al (2010) Phytoremediation of petroleum-pollutedsoils: application of *Polygonum aviculare* and its root-associated (penetrated) fungal strains for bioremediation of petroleum-polluted soils. Ecotoxicol Environ Safety 73:613–619
89. Moreira ITA, Oliveira OMC, Triguis JA et al (2011) Phytoremediation using Rizophora mangle L. in mangrove sediments contaminated by persistent total petroleum hydrocarbons (TPH's). Microchemi J 99:376–382
90. Moreira ITA, Oliveira OMC, Triguis JA et al (2013) Phytoremediation in mangrove sediments impacted by persistent total petroleum hydrocarbons (TPH's) using *Avicennia schaueriana*. Mar Pollut Bull 67:130–136
91. Morillo JA, Aguilera M, Antizar-Ladislo B, Fuentes S, Ramos-Cormenzana A, Russel NJ, Monteoliva-Sanchez M (2008) Molecular microbial and chemical investigation of the bioremediation of two-phase olive mill waste using laboratory-scale bioreactors. Appl Microbiol Biotechnol 79:309–317
92. Morris BEL, Henneberger R, Hubber H, Moissl-Eichinger C (2013) Microbial syntrophy: interaction for the common good. FEMS Microbiol Rev 37:384–406
93. Nam IH, Kim YM, Murugesan K, Jeon JR, Chang YY, Chang YS (2008) Bioremediation of PCDD/Fs-contaminated municipal solid waste incinerator fly ash by a potent microbial biocatalyst. J Hazard Mater 157:114–121
94. Nouri J, Khorasani N, Lorestani B, Karami M, Hassani AH, Yousefi N (2009) Accumulation of heavy metals in soil and uptake by plant species with phytoremediation potential. Environ Earth Sci 59:315–323
95. Orandi S, Lewis DM, Moheimani NR (2012) Biofilm establishment and heavy metal removal capacity of an indigenous mining algal-microbial consortium in a photo-rotating biological contactor. J Ind Microbiol Biotechnol 39:1321–1331

96. Paul D, Pandey G, Jain RK (2005) Suicidal genetically engineered microorganisms for bioremediation: need and perspectives. BioEssays 27:563–573
97. Peng S, Zhou Q, Cai Z, Zhang Z (2009) Phytoremediation of petroleum contaminated soils by *Mirabilis Jalapa* L. in a greenhouse plot experiment. J Hazard Mater 168:1490–1496
98. Perpetuo EA, Souza CB (2011) Engineering bacteria for bioremediation. In: Carpi A (ed) Progress in molecular and environmental bioengineering—from analysis and modeling to technology applications, 1st edn. In Tech, Rijeka, pp 605–632
99. Polti MA, Atjian MC, Amoroso MJ, Abate CM (2011) Soil chromium bioremediation: synergic activity of actinobacteria and plants. Int Biodeterior Biodegrad 65:1175–1181
100. Pous N, Puig S, Coma M, Balaguer MD, Colprim J (2013) Bioremediation of nitrate-polluted groundwater in a microbial fuel cell. J Chemi Technol Biotechnol 9:1690–1696
101. Prakken R (1959) Induced mutation. Euphytica 8:270–322
102. Rahaman MA, Hasegawa H (2011) Aquatic arsenic: phytoremediation using floating macrophytes. Chemosphere 83:633–646
103. Ratnaparkhe A, Tiwari A (2011) Bioprospecting approaches for biocatalysts from extremophiles. Int J Biomed Res 2:1–17
104. Robertson LA, Steer BA (2004) Recent progress in biocatalyst discovery and optimization. Curr Opin Chem Biol 8:141–149
105. Romeh AA (2010) Phytoremediation of water and soil contaminated with imidacloprid pesticide by Plantago major, l. Int J Phytoremed 12:188–199
106. Ronner-Holm SGE, Mennerich A, Holm NC (2006) Specific SBR population behaviour as revealed by comparative dynamic simulation analysis of three full scale municipal SBR wastewater treatment plants. Water Sci Technol 54:71–79
107. Ronner-Holm SGE, Holm NC (2009) Reduction of emissions using biomanipulations of activated sludge: selected full-scale examples. Environ Protect Eng 35:41–51
108. Ros M, Rodriguez I, Garcia C, Hernandez T (2010) Microbial communities involved in the bioremediation of an aged recalcitrant hydrocarbon polluted soil by using organic amendments. Bioresour Technol 101:6916–6923
109. Rosales E, Perez-Paz A, Vazquez S, Pazos A, Sanroman MA (2012) Isolation of novel benzo [a]anthracene-degrading microorganisms and continuous bioremediation in an expanded-bed bioreactor. Bioprocess Biosyst Eng 35:851–855
110. Rosenberg E, Bittan-Banin G, Sharon G, Shon A, Hershko G, Levy I, Ron EZ (2010) The phage-driven microbial loop in petroleum bioremediation. Microbial Biotechnol 3:467–472
111. Sakakibara M, Ohmori Y, Ha NTH, Sano S, Sera K (2011) Phytoremediation of heavy metal-contaminated water and sediment by *Eleocharis acicularis*. Clean Soil Air Water 39:735–741
112. Salinas-Martinez A, Santos-Cordova MDL, Soto-Cruz A et al (2008) Development of a bioremediation process by biostimulation of native microbial consortium through the heap leaching technique. J Environ Manage 88:115–119
113. Saslis-Lagoudakis CH, Savolinen V, Williamson EM et al (2012) Phylogenies reveal predictive power of traditional medicine in bioprospecting. Proc Nat Acad Sci USA 109:15835–15840
114. Selvin J, Ninawe AS, Kiran GS, Lipton AP (2010) Sponge-microbial interactions: ecological implications and bioprospecting avenues. Crit Rev Microbiol 36:82–90
115. Sharma S, Sundaram CS, Luthra PM, Singh Y et al (2006) Role of proteins in resistance mechanism of *Pseudomonas fluorescens* against heavy metal induced stress with proteomics approach. J Biotechnol 126:374–382
116. Shirdam R, Zand AD, Bidhendi GN, Mehrdadi N (2008) Phytoremediation of hydrocarbon-contaminated soils with emphasis on the effect of petroleum hydrocarbons on the growth of plant species. Phytoprotection 89:21–29
117. Sasmaz A, Sasmaz M (2009) The phytoremediation potential for strontium of indigenous plants growing in a mining area. Environ Exp Bot 67:139–144

118. Silva IS, Santos EDCD, Menezes CRD et al (2009) Bioremediation of a polyaromatic hydrocarbon contaminated soil by native soil microbiota and bioaugmentation with isolated microbial consortia. Bioresour Technol 100:4669–4675
119. Su-Qin C, Shan C, Liu DB, Xia HM (2006) An extracellular poly (3-hydroxybutyrate) depolymerase from *Penicillium* sp.DS9713a-01. World J Microbiol Biotechnol 22:729–735
120. Sizmur T, Palumbo-Roe B, Watts MJ, Hodson ME (2011) Impact of the earthworm *Lumbricus terrestris* (L.) on As, Cu, Pb and Zn mobility and speciation in contaminated soils. Environ Pollut 159:742–748
121. Smith VH, Graham DW, Cleland DD (1998) Application of resource ratio theory to hydrocarbon biodegradation. Environ Sci Technol 32:3386–3395
122. Smith WL, Wheeler WC (2006) Venom evolution widespread in fishes: a phylogenetic road map for the bioprospecting of piscine venoms. J Heredity 97:206–217
123. Solevic T, Novakovik M, Ilic M, Antic M, Vrvic MM, Jovancicevic B (2011) Investigation of the bioremediation potential of aerobic zymogenous microorganisms in soil for crude oil biodegradation. J Serb Chem Soc 76:425–438
124. Sprocati AR, Alisi C, Tasso F et al (2012) Effectiveness of a microbial formula, as a bioaugmentation agent, tailored for bioremediation of diesel oil and heavy metal co-contaminated soil. Process Biochem 47:1649–1655
125. Sriprapat W, Thiravetyan P (2013) Phytoremediation of BTEX from indoor air by *Zamioculcas zamiifolia*. Water Air Soil Pollut 224:1482
126. Stabili L, Schirosi R, Licciano M, Mola E, Giangrande A (2010) Bioremediation of bacteria in aquaculture waste using the polychaete *Sabella spallanzanii*. New Biotechnol 27:774–781
127. Stenuit B, Eyers L, Schular L, George I, Agathos SN (2009) Molecular tools for monitoring and validating bioremediation. In: Singh A et al (eds) Advances in applied bioremediation, soil biology, vol 17. Springer, Heidelberg, pp 339–353
128. Sun GD, Xu Y, Jin JH, Zhong PZ, Liu Y, Luo M, Liu ZP (2012) Pilot scale ex-situ bioremediation of heavily PAHs-contaminated soil by indigenous microorganisms and bioaugmentation by a PAHs-degrading and bioemulsifier-producing strain. J Hazard Mater 233–234:72–78
129. Sun H, Li J, Wang C, Wang L, Wang Y (2011) Enhanced microbial removal of pyrene in soils in the presence of earthworms. Soil Sediment Contamin 20:617–630
130. Sun H, Wang Z, Gao P, Liu P (2013) Selection of aquatic plants for phytoremediation of heavy metal in electroplate wastewater. Acta Physiol Plant 35:355–364
131. Sun Y, Zhou Q, Xu Y, Wang L, Liang X (2011) Phytoremediation for co-contaminated soils of benzo[a]pyrene (B[a]P) and heavy metals using ornamental plant *Tagetes patula*. J Hazard Mater 186:2075–2082
132. Suthar S (2008) Bioremediation of aerobically treated distillery sludge mixed with cow dung by using an epigeic earthworm *Eisenia fetida*. Environmentalist 28:76–84
133. Suthar S (2012) Vermistabilization of wastewater sludge from milk processing industry. Ecol Eng 47:115–119
134. Suthersan SS (ed) (1999) In situ bioremediation. Remediation engineering: design concepts. CRC Press LLC, Boca Raton
135. Thavamani P, Megharaj M, Naidu R (2012) Bioremediation of high molecular weight polyaromatic hydrocarbons co-contaminated with metals in liquid and soil slurries by metal tolerant PAHs degrading bacterial consortium. Biodegradation 23:823–835
136. Tang J, Feng T, Cui C, Feng Y (2013) Simultaneous biodegradation of phenanthrene and oxidation of arsenite by a dual-functional bacterial consortium. Int Biodeterior Biodegrad 82:173–179
137. Urgun-Demirtas M, Stark B, Pagilla K (2006) Use of genetically engineered microorganisms (GEMs) for the bioremediation of contaminants. Crit Rev Biotechnol 26:145–164
138. Vargas-Garcia MDC, Lopez MJ, Suarez-Estrella F, Moreno J (2012) Compost as a source of microbial isolates for the bioremediation of heavy metals: in vitro selection. Sci Total Environ 431:62–67

139. Virdis B, Robaey K, Yuan Z, Keller J (2008) Microbial fuel cells for simultaneous carbon and nitrogen removal. Water Res 42:3013–3024
140. Wang X, Wu N, Guo J, Chu X, Tian J, Yao B, Fan Y (2008) Phytodegradation of organophosphorus compounds by transgenic plants expressing a bacterial organophosphorus hydrolase. Biochemi Biophysi Res Comm 365:453–458
141. Wolicka D, Suszek A, Borkowski A, Bielecka A (2009) Application of aerobic microorganisms in bioremediation in situ of soil contaminated by petroleum products. Bioresour Technol 100:3221–3227
142. Wood TK (2008) Molecular approaches in bioremediation. Curr Opin Biotechnol 19:572–578
143. Xavier JB (2011) Social interaction in synthetic and natural microbial communities. Mol Syst Biol 7:483
144. Yamane K (1988) Introduction of experiments in genetic engineering. IX. Expression and regulation of cloned genes in microorganisms. Radioisotopes 37:423–433
145. Zamora-Castro J, Paniagua-Michel J, Lezama-Cervantes C (2008) A novel approach for bioremediation of a coastal marine wastewater effluent based on artificial microbial mats. Mar Biotechnol 10:181–189
146. Zand AD, Bidhendi GN, Mehrdadi N (2010) Phytoremediation of total petroleum hydrocarbons (TPHs) using plant species in Iran. Turk J Agric For 34:429–438
147. Zehnsdorf A, Seidel H, Hoffmann P, Schlenker U, Muller RA (2013) Conditioning of sediment polluted with heavy metals using plants as a preliminary stage of the bioremediation process: a large-scale study. J Soils Sediments 13:1106–1112
148. Zhang G, Zhang H, Zhang C, Zhnag G, Yang F, Yuan G, Gao F (2013) Simultaneous nitrogen and carbon removal in a single chamber microbial fuel cell with a rotating biocathode. Process Biochem 48:893–900
149. Zhang X, Zhao FJ, Huang Q, William PN, Sun GX, Zhu YG (2009) Arsenic uptake and speciation in the rootless duckweed *Wolffia globosa*. New Phytol 182:421–428
150. Zhao B, Poh CL (2008) Insights into environmental bioremediation by microorganisms through functional genomics and proteomics. Proteomics 8:874–881
151. Zhao B, Yeo CC, Poh CL (2005) Proteome investigation of the global regulatory role of sigma 54 in response to gentisate induction in *Pseudomonas alcaligenes* NCIMB 9867. Proteomics 5:1868–1876
152. Zhao B, Yeo CC, Tan CL, Poh CL (2007) Proteome analysis of heat shock protein expression in *Pseudomonas alcaligenes* NCIMB 9867 in response to gentisate exposure and elevated growth temperature. Biotechnol Bioeng 97:506–514
153. Zhou J, Qin F, Su J, Liao JW, Xu HL (2011) Purification of formaldehyde-polluted air by indoor plants of Araceae, Agavaceae and Liliaceae. J Food Agri Environ 9:1012–1018

Adv Biochem Eng Biotechnol (2015) 147: 185–224
DOI: 10.1007/10_2013_245

Published Online: 13 September 2013

Organisms for Biofuel Production: Natural Bioresources and Methodologies for Improving Their Biosynthetic Potentials

Guangrong Hu, Shiqi Ji, Yanchong Yu, Shi'an Wang, Gongke Zhou and Fuli Li

Abstract In order to relieve the pressure of energy supply and environment contamination that humans are facing, there are now intensive worldwide efforts to explore natural bioresources for production of energy storage compounds, such as lipids, alcohols, hydrocarbons, and polysaccharides. Around the world, many plants have been evaluated and developed as feedstock for bioenergy production, among which several crops have successfully achieved industrialization. Microalgae are another group of photosynthetic autotroph of interest due to their superior growth rates, relatively high photosynthetic conversion efficiencies, and vast metabolic capabilities. Heterotrophic microorganisms, such as yeast and bacteria, can utilize carbohydrates from lignocellulosic biomass directly or after pretreatment and enzymatic hydrolysis to produce liquid biofuels such as ethanol and butanol. Although finding a suitable organism for biofuel production is not easy, many naturally occurring organisms with good traits have recently been obtained. This review mainly focuses on the new organism resources discovered in the last 5 years for production of transport fuels (biodiesel, gasoline, jet fuel, and alkanes) and hydrogen, and available methods to improve natural organisms as platforms for the production of biofuels.

Keywords Biofuels · Biosynthetic potential · Methodologies · Natural bioresources

F. Li (✉)
Institute of Bioenergy and Bioprocess Technology, CAS, Songling Road No. 189, Qingdao 266101, China
e-mail: lifl@qibebt.ac.cn

G. Hu · S. Ji · Y. Yu · S. Wang · G. Zhou
Shandong Provincial Key Laboratory of Energy Genetics, Key Laboratory of Biofuels, Qingdao Institute of Bioenergy and Bioprocess Technology, Chinese Academy of Sciences, Qingdao 266101, China

Abbreviations

ACCase	acetyl-CoA carboxylase
ACP	acyl carrier
CoA	coenzyme A
DGAT	diacylglycerol acyltransferase
DHAP	dihydroxyacetone phosphate
ENR	enoyl-ACP reductase
FAT	fatty acyl-ACP thioesterase
G3PDH	gycerol-3-phosphate dehydrogenase
GPAT	glycerol-3-phosphate acyltransferase
HD	3-hydroxyacyl-ACP dehydratase
KAR	3-ketoacyl-ACP reductase
KAS	3-ketoacyl-ACP synthase
LPAAT	lyso-phosphatidic acid acyltransferase
LPAT	lyso-phosphatidylcholine acyltransferase
MAT	malonyl-CoA:ACP transacylase
PDH	pyruvate dehydrogenase complex

Contents

1 Introduction 187
2 Microalgae for Bioenergy 188
  2.1 Microalgae for Production of Transport Biofuels 188
  2.2 Microalgae for Production of Hydrogen 191
  2.3 Natural Strain Isolation, Screening, Selection, and Preservation 193
  2.4 Ways to Improve Microalgae Strains 195
3 Diversity of Plants Used for Energy Generation 199
  3.1 Definition and Classification of Energy Plants 199
  3.2 Exploitation and Utilization of Energy-Yielding Plants Around the World 200
  3.3 Looking for Novel Green Energy Plants 201
4 Yeasts for Production of Biofuels 202
  4.1 Ethanol Yeast 202
  4.2 Oleaginous Yeasts 203
  4.3 Yeast Biofuel Cells 204
5 Bacteria for Bioenergy Production 204
  5.1 Ethanol-Producing Bacteria 204
  5.2 Butanol and Longer Chain Alcohol-Producing Bacteria 208
  5.3 Biohydrogen-Producing Bacteria 211
6 Conclusion 212
References 213

1 Introduction

At present, primary fossil energy, such as coal, petroleum, and natural gas, accounts for more than 87 % of the energy supply [1]. According to the prediction of the IEA (International Energy Agency), global energy consumption will rise more than 50 % by 2035, from 5.33×10^{20} J in 2008 to 8.12×10^{20} J in 2035 [2]. Liquid fuels account for 34 % of total global energy consumption, and the need is increasing, especially due to demands from the emerging markets of developing countries, for example, China and India. Of the liquid fuels, 54 % are consumed by the transportation sector [2]. Most liquid fuels come from petroleum. In recent years, shale gas has been a star in fossil resources [3], and is expected to expand the global energy supply, reaching 46 % of the United States' natural gas supply. However, the prices of fossil fuels are expected to rise steadily in the long term because of decreasing supply and increasing demand. Pessimistic viewers predicted that crude oil production would reach a peak in the 2050s and decrease inevitably thereafter [4, 5]. Nevertheless, there is no doubt that fossil fuel resources are finite.

In addition, combustion of fossil fuels leads to the steady accumulation of greenhouse gases and global warming [2]. Global warming is considered to affect climate change, including increasing the average global temperature and sea level. Intense weather events around the world, such as El Niño-Southern Oscillation (ENSO), have been enhanced by as much as 60 % in the past half century [6, 7]. These changes not only have a negative impact on the Earth's ecosystem, but also threaten the sustainable development of human society. Therefore, it is imperative for humans to explore renewable energy as alternatives to fossil fuels.

At present, bioethanol and biodiesel account for almost 90 % of the biofuel market [8]. Bioethanol is traditionally produced through yeast or bacteria fermentation of sugars and starches from crops such as maize and sugar beet. In order to keep the global food supply secure, the next-generation bioethanol from non-food lignocellulosic biomass has been proposed. For many years photosynthetic autotrophic plants and microalgae have been considered as a possible biofuel feedstock, inasmuch as they can be harvested and use sunlight to convert CO_2 into a wide variety of metabolites [9].

Biomass is a biological material derived from living or recently living organisms. In this review, biomass refers specifically to plant- (algae-) based material. Plant and algae biomass are used as substrates for microbial fermentation to produce biofuels, including bioethanol [10], biodiesel [11], jet fuel [12], and hydrogen [13]. In the process, these organisms can be distinctly classified as energy-harvesting organisms (such as plants, algae, and cyanobacteria) and energy-converting organisms (such as yeast and bacteria). In the past decades, many researchers have devoted great effort to finding biological resources for bioenergy, including exploration of natural species and strain improvement through metabolic engineering and synthetic biology methods [14].

This review provides an overview of organism resources (microalgae, energy plant, yeast, and bacteria) discovered in the last 5 years for production of liquid biofuels (biodiesel, gasoline, jet fuel, and alkanes) and hydrogen, and summarizes efforts to improve organisms for bioenergy production.

2 Microalgae for Bioenergy

Eukaryotic microalgae and prokaryotic cyanobacteria are oxygenic photosynthetic microorganisms that live in a wide range of ecological habitats including fresh, brackish, and ocean water. Although over 40,000 species have been reported, there are more algae species yet to be identified [15]. The term "algae" commonly includes many organisms from different kingdoms of life, which can be single or multicellular and eukaryotic or prokaryotic. According to their morphology and biochemical characteristics, algae can be divided into three groups: (i) microalgae (unicellular eukaryotic organisms such as *Chlamydomonas reinhardtii* and *Chlorella vulgaris*); (ii) macroalage (seawood such as *Laminaria japonica* and *Porphyra dentate*); and (iii) cyanobacteria (*Arthrospira platensis* and *Aphanizomenon flos-aquae*).

Compared to ethanol from corn or biodiesel from soy, palm, and rape, biofuel production per acre from microalgae feedstock is higher [16] and does not compete with limited arable land and fresh water used for food production. In support of the Aquatic Species Program of the US Department of Energy in the 1970s, scientists analyzed about 3,000 different microalgae for their possibility of producing biofuels [17]. Although many species of microalgae have some traits that are ideal for biofuel production, most have serious disadvantages that blocked the emergence of a profitable microalgae-based biofuel industry. Several technical difficulties need to be resolved before microalgae can be used as an economical biofuel feedstock. These barriers include difficulties in obtaining high lipid productivity in a large-scale cultivation in outdoor conditions, harvesting microalgae cells in low-energy ways, and extracting biofuels from the microalgal feedstock by cost-effective methods. To speed up the utilization of microalgae in biofuel production, it is necessary to invent new methods to increase the productivity of microalgal culturing systems and push bioprospecting efforts to look for strains with as many ideal biofuel characteristics as possible. This section discusses the application of microalgae and cyanobacteria in biofuel production.

2.1 Microalgae for Production of Transport Biofuels

Under optimal conditions, the lipids are located in the membrane of algae cells, and constitute about 5–20 % of the dry cell weight (DCW). Among the membrane lipids, the glycosyl glycerides are enriched in the chloroplast, including

Table 1 Lipid contents in different classes of microalgae and cyanobacteria[1]

Microalgae	Average lipid contents (% dry cell weight)	
	Under normal conditions	Under stress conditions
Green microalgae	25.5	45.7
Diatoms	22.7	37.8
Cyanobacteria[2]	9.8	NA[4]
Other eukaryotic algal taxa[3]	27.1	44.6

[1] This table is edited according to the data of Hu et al. [15]
[2] To date large quantities of total lipids have not been found in cyanobacteria strains
[3] They include Chrysophytes, Haptophytes, Eustigmatophytes, Dinophytes, Xanthophytes, and Rhodophytes
[4] *NA* Not applicable

monogalactosyldiacylglycerol (MGDG), digalactosyldiacylglycerol (MGDG), and sulfoquinovosyldiacylglycerol (SQDG), whereas most phosphoglycerides are located in the plasma and endoplasmic membrane, mainly including phosphatidylcholine (PC), phosphatidylglycerol (PG), and phosphatidylethanolamine (PE) [18, 19]. Under stress conditions, many algae will synthesize and accumulate neutral lipids (20–55 % DCW), mainly in the form of triacylglycerols (TAGs). TAGs are the main storage molecules of carbon and energy, which are deposited in lipid bodies in the algal cell cytoplasm. Interestingly, the green alga, *ryococcus braunii* does not synthesize lipids but instead produces a great amount of hydrocarbons (C23–C40, up to 80 % DCW) under stress conditions [20, 21].

Over the past decades, hundreds of oleaginous microalgae species with high lipid content have been screened and characterized. Oleaginous algae are distributed in diverse taxonomic groups, and their lipid contents vary significantly (Table 1). From the prospective of statistic analysis, the intrinsic potential to produce lipid/oil is strain-specific, instead of genus-specific [15].

Microalgae synthesize fatty acids as precursors to be used for the assembly of lipids. The carbon chain length of fatty acids in the microalgal cells ranges from C14−C18 [22]. According to the number and position of double bonds on the carbon chain, the fatty acids can be divided into three classes: saturated, monounsaturated, and polyunsaturated. Comparatively speaking, the saturated and monounsaturated fatty acids are predominant in most algae species [23]. It is noted that the fatty acid compositions are also used to identify the algal taxa. Compared with higher plants, there are greater variations in the fatty acid profiles of algae. In some algal species, the predominant fatty acids are medium-chain length (i.e., C10–C14) [23], whereas others can synthesize very-long–chain fatty acids (>C20) [23–27]. The ability to produce large quantities of very-long–chain polyunsaturated fatty acids (PUFAs) is another good feature of some algal species [25, 28].

The microalgae cells synthesize the fatty acids in the chloroplast, whereas TAGs are mainly assembled at the ER (Fig. 1) [18, 22]. In the end, the de novo synthesis pathway of fatty acids produces a C16 and/or C18 fatty acid. Acetyl CoA carboxylase (ACCase) catalyzes the conversions of acetyl CoA to malonyl CoA, which is the first step in fatty acid biosynthesis. ACCase is an enzyme complex that

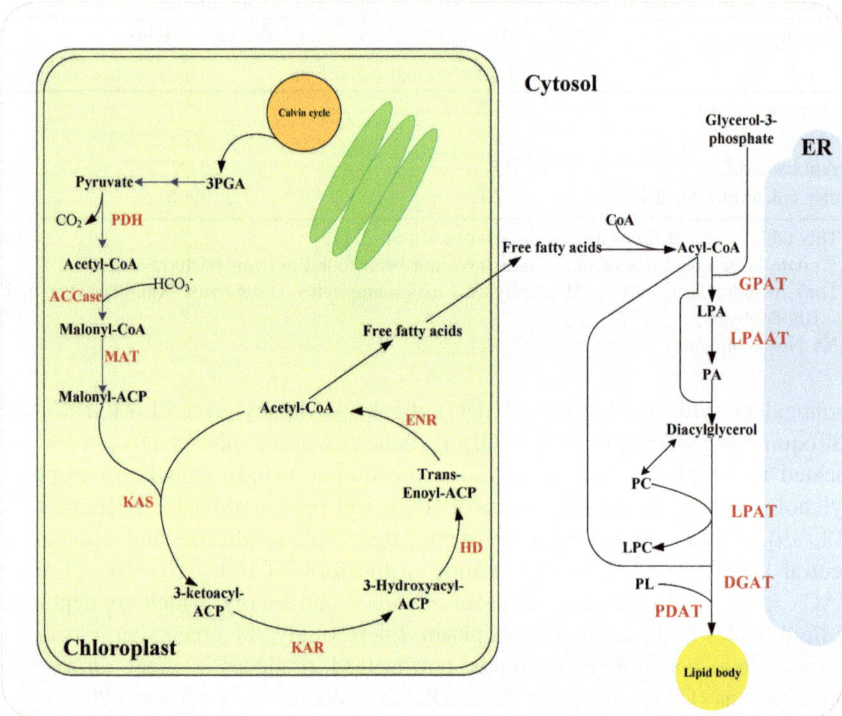

Fig. 1 Simplified overview showing the triacylglycerol biosynthesis pathway in a microalga cell. Fatty acids are produced in the chloroplast, whereas triacylglycerols are mostly synthesized at the ER and stored in the lipid body. *ACCase* acetyl-CoA carboxylase; *ACP* acylcarrier protein; *CoA* coenzyme A; *DGAT* diacylglycerolacyltransferase; *DHAP* dihydroxyacetone phosphate; *ENR* enoyl-ACP reductase; *FAT* fatty acyl-ACP thioesterase; *G3PDH* gycerol-3-phosphate dehydrogenase; *GPAT* glycerol-3-phosphate acyltransferase; *HD* 3-hydroxyacyl-ACP dehydratase; *KAR* 3-ketoacyl-ACP reductase; *KAS* 3-ketoacyl-ACP synthase; *LPAAT* lyso-phosphatidic acid acyltransferase; *LPAT* lyso-phosphatidylcholineacyltransferase; *MAT* malonyl-CoA:ACP-transacylase; *PDH* pyruvate dehydrogenase complex

exists in a eukaryotic and prokaryotic form [29]. The malonyl CoA is the carbon donor for fatty acid chain elongation through the 4 actions of condensation, reduction, dehydration, and reduction [22]. At the ER membrane, an enzyme named diacylglycerolacyltransferase (DGAT) catalyzes the final step of TAG assembly, in which a third fatty acid is transferred to the *sn-3* carbon of DAG [15, 22]. Another enzyme, phospholipid:diacylglycerolacyltransferase (PDAT), uses phospholipids as acyl donor and DAG as the acceptor to form the TAGs [30, 31].

The triglycerides in the algal lipids can be converted into different kinds of biofuels including biodiesel and jet fuel by chemical methods [32]. The fatty acid composition of triglycerides will affect the properties of biofuels. For example, the saturation and fatty acid chain length have an impact on the ignition quality, cold flow properties, and oxidative stability of biodiesel [33].

Oleaginous microalgae are a diverse group of microorganisms, many of which can produce novel feedstock for the production of renewable green biofuels. The algae strains used as candidates for biofuel production should have the following desirable traits: (i) high photosynthetic conversion efficiencies, (ii) rapid biomass productivity, (iii) substantial amounts of neutral lipids (>45 % DCW), (iv) the ability to thrive in different culture systems in the climatic zone, and (v) the ability to grow in a variety of wastewater in marginal lands.

2.2 Microalgae for Production of Hydrogen

Hydrogen gas is regarded as a promising energy because (i) it is renewable; (ii) it can produce a large amount of energy per unit molar and does not release CO_2 upon combustion; and (iii) it is simply converted to electricity in a power plant or by fuel cells. Although there are several ways to produce H_2 such as photovoltaics-electrolysis and gasification of biomass, biological H_2 production, especially using microalgae [13, 34, 35], has attracted considerable interest.

In the 1940s, Gaffron found the phenomenon of H_2 evolution in unicellular green microalgae [36]. After cultivation in dark and anaerobic conditions for a period of time, microalgae cells were induced to produce H_2 under light [37]. However, hydrogen evolution activity in microalgae cells was transient, and only lasted a few minutes at most. A hydrogenase in these green algae cells was discovered to be responsible for hydrogen evolution [38–40]. The gene was in the nucleus genome of microalgae cells, but its mature protein was located in the chloroplast stroma [40]. The hydrogenases of eukaryotic algae were monomeric proteins of about 45–50 kD, and belonged to a Fe hydrogenase family [37]. It was noted that the activity of Fe-hydrogenase was repressed by oxygen [41, 42].

Later, the expression of Fe-hydrogenase was found to be induced in the light under anaerobic conditions [42, 43]. In addition, sulfur deficiency can lead to reversible decrease of the oxygenic photosynthesis rate [44], but without changing the respiration [43]. Thus, a so-called "two-stage photosynthesis and H_2 production process" was proposed [43, 45]. First, green algae were cultivated in the light, then the algal cells were transferred into a sealed medium in which the sulfur supply was carefully limited. After the sulfur in the medium was consumed completely, the algal cells responded to S deficiency by changing photosynthesis and carbon metabolism to survive. Under S deficiency, the activity of photosystem II and oxygen evolution declined dramatically and the O_2 was depleted by cell respiration, which resulted in anaerobiosis in the culture. Subsequently, Fe-hydrogenase was induced and sustained H_2 produced in the light. In the period from 24 to 70 h, the rate of photosynthetic H_2 production was 2.0–2.5 $mL \cdot L^{-1} \cdot h^{-1}$[13]. Afterwards, the rate decreased slowly. At the same time, substantial amounts of cellular starch and protein were catabolized to support the hydrogen evolution [45].

The theoretical maximum yield of hydrogen by an algae cell should be 10 mol H_2 $m^{-2} \cdot d^{-1}$[13]. However, the real yields of H_2 production are only 10 %, due to some

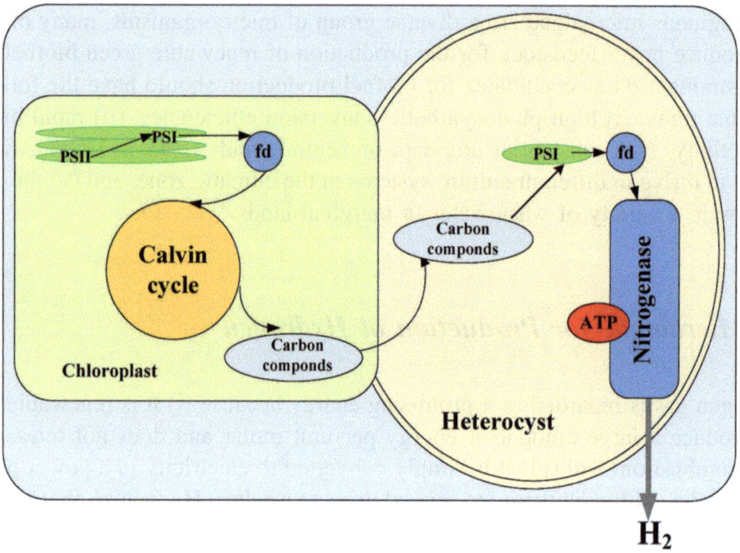

Fig. 2 Nitrogenase-mediated hydrogen production in the heterocyst of cyanobacteria

biological and engineering limitations, for example, solar energy conversion efficiency and the limited time of sustained hydrogen production in green algae [13]. Thus far, the microalgae species and mutants that proved to be able to generate hydrogen are: *C. reinhardtii* [43, 46, 47], *Scenedesmus obliquus* [48–50], *Chlorella fusca* [13], *Chlorella* sp. [51, 52], *Chlorella vulgaris* [53], *Chlorella protothecoides* [54], and *Platymonas subcordiformis* [55, 56].

Except for the eukaryotic microalgae, some prokaryotic nitrogen-fixing cyanobacteria are reported to be able to produce H_2 by nitrogenase. The filamentous cyanobacteria (e.g., *Plectonema boryanum* [57] and *Anabaena cylindria* [58]) can evolve H_2 using the endogenous carbohydrate as the electron donor under anaerobic conditions without N_2. Nitrogenase is responsible for nitrogen fixation in the presence of N_2. In the absence of N_2, nitrogenase catalyzes the reaction: [59]

$$2H^+ + 6\,e^- + 4ATP \rightarrow H_2 + 4ADP + 4Pi$$

The rate of hydrogen evolution catalyzed by nitrogenase is one-third to one-fourth that of nitrogen-fixation [35]. As is the hydrogenase, nitrogenase is also sensitive to oxygen. But the O_2 production by photosynthesis can be blocked by the N-depletion of the cultures [35, 57, 60].

Some cyanobacteria evolve a special kind of cell-heterocysts in their filamentous hyphae [61], which contain the nitrogenase and protect it from oxygen damage (Fig. 2). Photosynthesis and carbon fixation are conducted in vegetative cells of filamentous cyanobacteria; then the carbohydrates are transferred into heterocysts and decomposed to supply the nitrogenase with reducing power. Adenosine triphosphate (ATP) can be obtained from the PSI and anoxygenic

Table 2 Hydrogen production by some cyanobacteria species

Cyanobacteria species		References
Unicellular	*Aphanocapsa montana*	[64]
	Cyanothece sp.	[65, 66]
	Microcystis aeruginosa	[67]
	Gloeothece sp.	[68]
	Gloeobacter violaceus	[64]
	Gloeocapsa alpicola	[69]
	Synechocystis sp.	[68, 70]
	Synechococcus PCC602	[64]
Filamentous	*Leptolyngbya* sp.	[68]
	Lyngbya majuscula	[68]
	Anabaena cylindrical	[62]
	Anabaena variabilis	[68, 71]
	Anabaena sp.	[68, 72]
	Nostoc punctiforme	[68]
	Nostoc sp.	[68]
	Oscillatoria chalybea	[73]
	Plectonema boryanum	[74]
	Spirulina platensis	[75]

photosynthesis in heterocysts. In the presence of inert gas, for example, argon, such cyanobacteria can simultaneously produce H_2 and O_2 for a long time [62], even in outdoor conditions [63]. However, the efficiency of solar energy into H_2 by microalgae was <1–2 % in lab experiments at low light intensities, and declined dramatically to 0.3 % under outdoor conditions with sunlight illumination [35]. At present, hydrogen production is observed within at least 14 genera of cyanobacteria [59]. The most common cyanobacteria species that have nitrogenase and are able to generate hydrogen are listed in Table 2.

2.3 Natural Strain Isolation, Screening, Selection, and Preservation

The purposes of microalgae isolation and screening works are to find and maintain the promising microalgal strains for cultivation and improvement. Given the different culturing system and demands of biofuels and chemicals derived from microalgae, it is necessary to isolate new strains from a wide variety of habitats to provide the greatest metabolic biodiversity.

Generally, algae are isolated from different natural aqueous environments from fresh water to brackish water. Now, the algae that lived in extreme habitats, such as hypersaline, the polar environment, and thermal springs, are attracting more and more attention. In order to avoid duplication of efforts, large-scale sampling work should be organized in advance. All sampling locations should be planned carefully to include aquatic and terrestrial environments in different geographical zones to keep the gene pools at a maximum. Furthermore, except for the spatial distribution, temporal succession should be considered because the algae community in the habitats will vary with seasonal change.

Traditional methods to isolate new strains from natural environments have been elucidated in detail [76]. Generally, it will take a week or more to isolate some microalgae strains by traditional methods. Some automated isolation techniques such as flow cytometry have been used as high-throughput methods to sort fluorescence-activated cells of microalgae in large-scale isolation works [77–79]. After purification, strain identification is performed according to its morphological characteristics and molecular methods, for example, 18S rRNAgene sequence, ITS (internal transcribed spacer) sequence, and *rbcL* (ribulose-1,5-bisphosphate carboxylase/oxygenase large subunit) genes. Microalgae strains often are maintained in an agar medium in low temperature under low light although cryopreservation has been applied in microalgae maintenance [80].

Before screening oleaginous microalgae strains, it is necessary to give a clear definition of what a good oleaginous algae strain for biofuel production is. An ideal oleaginous microalgae strain would have advantages in three major areas: growth characteristics, lipid production, and strain robustness. The growth characteristics include a set of parameters such as growth rate, cell density, tolerance to environmental factors, and nutrient demands. Much work would be necessary to achieve these parameters by using traditional culturing techniques. Therefore developing an automated system to monitor all growth parameters of microalgae simultaneously would be significant. Screening for lipid production involves the determination of lipid contents and productivity, and fatty acid profiles of the lipid. For large-scale culture of a given oleaginous strain, its robustness must be considered carefully. The strain's robustness encompasses several parameters, including culture consistency, resilience, stability, and sensitivity to predators.

At present, the most difficult barrier in screening substantial oleaginous algal strains is the lack of high-throughput methods that can analyze many phenotypes at the same time (e.g., cell density, biomass concentration, lipid content, and composition) although many automated high-throughput analytical techniques have been applied in analysis of individual phenotypes. For example, the solvent extraction and weighing method is widely used to determine the lipid content in algae, but it requires a lot of biomass and time. Fluorescence semiquantification methods using dyes that need little biomass have been developed [81–83]. However, these methods are available in only a small range of algae strains. Lipidomics methodologies also accelerate the analysis of fatty acid profiles compared with the traditional GC–MS [84–87]. Recently, a new rapid screening procedure for oleaginous microalgae strains based on chlorophyll fluorometer and Nile-red dyes was developed [88].

Culture collections are important to preserve the diversity of strains, protect the genetic pool, and provide research materials. Up to now, the major algal collection agencies include UTEX (The Culture Collection of Algae at the University of Texas at Austin, Texas, about 3,000 strains, http://www.utex.org), CCMP (The Provasoli-Guillard National Center for Culture of Marine Phytoplankton at the

Bigelow Laboratory for Ocean Sciences in West Boothbay Harbor, Maine, more than 2,500 strains, http://ccmp.bigelow.org), CCAP (Culture Collection of Algae and Protozoa, about 2,000 strains, http://www.ccap.ac.uk), ANACC (The Australian National Algae Culture Collection, about 1,000 strains, http://www.csiro.au/places/Australian-National-Algae-Culture-Collection), and the FACHB collection (Freshwater Algae Culture Collection of the Institute of Hydrobiology, about 1,600 strains). Except for the maintenance of algal strains, these agencies can also perform research on identifying new algal species, phylogeny, determining physiological characteristics, and distributing information of algal strains.

2.4 Ways to Improve Microalgae Strains

Several microalgae species have some traits that are good for biofuel production [11, 17], but their lipid productivity and photosynthetic efficiency are much less than the theoretical value [15]. Even the modest improvements in photosynthetic efficiency and lipid productivity can cut down the cost and land for producing biofuels [11, 89]. By conventional breeding methods, microalgae strains with desired properties would be selected over time, as people did successfully with agricultural crops and microorganisms. Many mutants of *C. reinhardtii* were reported by conventional physical or chemical mutagenesis, such as X-rays, ultraviolet light, and EMS (ethylmethanesulphonate) and hybridization between mutants and/or wild-types were conducted through sexual reproduction [90]. Several starchless mutants of *C. reinhardtii* have been proved to be able to produce more lipids than their wild-types [91–93]. Since the 1970s, more and more genetic engineering technologies and mutageneses have been developed and applied successfully in the breeding of agriculture crops and industry microorganisms. Therefore, it is reasonable that using modern genetic engineering and mutagenesis would be feasible to obtain microalgae strains with desirable traits by bypassing the lengthy natural selection process.

Over the past 20 years, a more thorough understanding about microalgae physiology and central metabolism has been achieved with the significant advances in microalgal genome sequencing. Today, the huge size of information resources, such as the expressed sequences tag (EST), the mitochondrial, chloroplast, and nuclear genomes of many microalgae have been established and opened to all scientific and industrial communities via the Internet. The green alga *C. reinhardtii*is is a model organism and platform for phycological genetics study, and most methods for transgene expression and gene mutations were initially developed for this species. Subsequently, the tools are being applied in other algae. Since the genome of *C. reinhardtii* was published in 2007 [94], other algal nuclear genomes have been sequenced (Table 3). In addition, there are transcriptomes reported from some microalgae [95–99].

Gene transformation has been successfully applied in the nucleus and chloroplast of many microalgae strains belonging to *Chlorophyta, Rhodophyta*, and

Table 3 Sequenced genomes of algae[a]

Species	Status	References
C. reinhardtii	Published	[94]
Chlorella variabilis	Published	[100]
Cyanidioschyzon merolae	Published	[101]
Guillardia theta	Published	[102]
Micromonas pusilla	Published	[103]
Nannochloropsis gaditana	Published	[104]
Ostreococcus lucimarinus	Published	[105]
Ostreococcus tauri	Published	[106]
Phaeodactylum tricornutum	Published	[107]
Thalassiosira pseudonana	Published	[108]
Volvox carteri	Published	[109]
Botryococcus braunii	Incomplete	
Dunaliella salina	Incomplete	
Fragilariopsis cylindrus	Incomplete	
Galdieria sulphuraria	Incomplete	
Pseudo-nitzschia multiseries	Complete draft	
Porphyra purpurea	Incomplete	
Thalassiosira rotula	Incomplete	

[a] The genome information can be found on the websites of NCBI (http://www.ncbi.nlm.nih.gov/), JGI (http://genome.jgi.doe.gov/), and GOLD (http://www.genomesonline.org/cgi-bin/GOLD/index.cgi) [110]

Phaeophyta, diatoms, euglenids, and dinoflagellates, respectively [9]. According to statistical analysis, the efficiency of transformation is species-dependent. Thus, the transformation approaches must be carefully considered and optimized for each microalga. The ordinary transformation methods used to transfer exogenous genes into microalgal cells include glass beads, electroporation, microparticle bombardment, and *Agrobacterium tumefaciens*-mediated gene transfer [111]. In order to isolate the transformants efficiently, many selection markers/reporters are used, including antibiotic resistance and fluorescent/biochemical markers (Table 4). Because many microalgae are resistant to a wide range of antibiotics, the application of antibiotics in microalgal transformants is limited.

However, not all successful transformations can lead to stable expression of transgenes in either plastid or nucleus. The major problems related to foreign gene expression in microalgal cells include: no integration into the chromosome, recognition of the promoter sequence, biased codon usage, instability of mRNA, silencing by methylation, and epigenetic silencing mechanisms [9, 111]. Nuclear transformation of microalgae often results in random integrations of transgenes, whereas the chloroplast transformation can be achieved through homologous recombination [127]. RNA silencing has been used successfully to knock down gene expression in microalgae [128–130]. Compared to the RNA interference (RNAi), the more specific and stable artificial-micro–RNA (armiRNA) techniques for microalgae have been developed [131, 132].

Table 4 Marker/reporter genes expressed in microalgae transformants

Gene	Description	Marker/ reporter	Application	References
AadA	Adenylyl transferasespectinomycin resistance)	Marker	*Chlamydomonas* sp.	[112]
Als	Acetolactate synthase (resistance to sulphonylurea herbicides)	Marker	*Chlamydomonas* sp.	[113]
AphVIII	Aminoglycoside 3′phosphotransferase (resistance to paromomycin, kanamycin, and neomycin)	Marker	*Chlamydomonas* sp.	[112]
Ars	Arylsulphatase	Reporter	*Chlamydomonas* sp.	[114]
Ble	Bleomycin resistance protein (resistance to tallysomycin and related antibiotics)	Marker	*Chlamydomonas* sp.	[115]
Cat	Chloramphenicol acetyltransferase (chloramphenicol resistance)	Marker	*Chlamydomonas* sp.	[116]
Frustulin	Calcium-binding glycoprotein	Reporter	*Cylindrotheca fusiformis*	[117]
Cry1-1	Ribosomal protein S14	Marker	*Chlamydomonas* sp.	[118]
Gfp	Green fluorescent protein	Reporter	*P.tricornutum, Chlamydomonas* sp.	[119, 120]
Glut1	Glucose transporter	Marker or reporter	*P. tricornutum*	[119]
Gus	β-Glucuronidase	Reporter	*Amphidinium and Symbiodinium*	[121]
Hpt	Hygromycin B phosphotransferase	Marker	*Amphidinium and Symbiodinium*	[121]
Hup1	Hexose transporter	Reporter	*P. tricornutum, Cylindrotheca fusiformis*	[119, 117]
LacZ	β-Galactosidase	Reporter	*Haematococcus pluvialis*	[122]
Luc	Luciferase	Reporter	*P. tricornutum*	[123]
Nat	Nourseothricin resistance	Marker	*P. tricornutum*	[124]
NptII	Neomycin phosphotransferase II (resistance to G418)	Marker	*Chlamydomonas* sp.	[125]
Oee-1	Oxygen evolving enhancer protein	Marker	*Chlamydomonas* sp.	[126]

Inasmuch as lipids in the microalgae cells are the precursors for biofuel production, both improving the quantity and quality of lipids is the main target of microalgae breeding. To date, understanding of the lipid biosynthesis and catabolism, and the pathways that control carbon length and saturation of fatty acids comes from the investigation for terrestrial plants, for example, *Arabidopsis thaliana,* soy bean, and rapeseed. However, many homologs of genes that affected lipid metabolism in higher plants are found in the published microalgal genomes and EST database. It is postulated that the transgenic methodologies used to improve the lipid content in plants would be applied in microalgae.

The conversion of acetyl-CoA to malonyl-CoA is the first step in fatty acid biosynthesis, catalyzed by ACCase which is regarded as the limited step in organisms [22, 133]. Overexpression of ACCase in the seeds of *B. napus* only led to a minor rise in seed lipid content [134]. Similarly, 2–3-fold overexpression of ACCase activity in the diatom *C. cryptic* did not result in a significant increase of lipid content [135, 136]. However, overexpression of genes involved in TAGs formation can lead to increases in seed oil production, for example, glycerol-3-phosphate dehydrogenase (G3PDH) in the seeds of *B. napus* [137], glycerol-3-phosphate acyltransferase (GPAT) [138], lysophosphatidic acid acyltransferase [139], and diacylglycerolacyltransferase (DGAT) [140] in plants. In addition, the cellular lipid accumulations increased by cutting down the competitive metabolic pathways that synthesize storage compounds such as starch. Two starch-deficient mutants of *C. reinhardtii* can accumulate more TAGs than wild-type under high light and nitrogen deficiency [91, 92, 141].

It is important to improve the quality of the lipids: carbon chain length and degree of unsaturation of the fatty acids can affect the ignition quality (e.g., cetane number), cold-flow properties, and oxidative stability of biofuel. Generally, the chain length of fatty acids from microalgae is from 14 to 20, but ideal fatty acids for biodiesel feedstock should be 12:0 and 14:0. Fatty acids with a shorter chain length are good for the production of jet fuel. Acyl-ACP thioesterases are responsible for the releasing of the fatty acids chain from the fatty acid synthase and determining the chain lengths [133]. Expression of transgenic thioesterasecan can significantly change the fatty acid profile of lipids in plants [142, 143].

Microalgae are induced to produce and accumulate lipids in the cells when they face a stress condition, such as high light. An optimal intensity of light for most microalgal species growth is usually around 200–500 μmol · photons $m^{-2} \cdot s^{-1}$. Strong intensities of illumination will inhibit the microalgal growth rate, which is known as photoinhibition. Microalgae growth and proliferation are inhibited outside because the intensities of sunlight are often higher than 2,000 μmol · photons $m^{-2} \cdot s^{-1}$. Some researchers suggest a strategy to improve photosynthetic efficiency by reducing the number of chlorophyll antennae or light-harvesting protein complexes in the thylakoid of the chloroplast [13, 144, 145]. Some mutants with fewer chlorophyll and/or smaller light-harvesting complexes by random mutagenesis have been reported [88]. In addition, RNAi technologies are also used to knock down both LHCII and LHCI in *C. reinhardtii* [146].

As mentioned above, some genetic engineering approaches have been applied to several algal species aiming at improving lipid production, including *C. reinhardtii* [147, 148], *P. tricornutum* [124], *C. cryptic* [135, 136], and *Nannochloropsis* sp. [149]. However, only moderate success was achieved, mainly because the lipid metabolism, particularly the functions of key genes and enzymes involved in lipid synthesis and accumulation in these organisms, is not well understood. Furthermore, some transcription factors involved in the regulation of plant lipid accumulation were suggested as the second-generation targets to improve the lipid content [150–152].

Therefore, various physical mutagens, such as X-rays, β-rays, and heavy ion beam which have been successfully applied to crop and microorganism breeding, have been suggested to be used for microalgae trait improvement. Among them, heavy-ion beam induces a broad range of mutations, that is, base substitutions, small and large insertions/deletions, translocation, and inversions in the genomes of microalgae. Physical mutagens usually generate thousands of mutants, but it is a very laborious work to screen for desirable mutants. Establishment of a high-throughput screening method will be helpful for microalgal trait improvements. Based on the positive correlation between photosynthetic efficiency and lipid content, a new strategy has been developed to accelerate the screening of microalgae mutants with high yields of biomass and lipids [88].

3 Diversity of Plants Used for Energy Generation

3.1 Definition and Classification of Energy Plants

Energy plants refer to a variety of plants that can be effectively converted into energy, including plants rich in carbohydrates or oil. Except giving off heat when burning, energy plants can also be converted to solid, liquid, or gas fuels by physical and chemical methods [153].

Energy-yielding plants comprise a large number of species and are widely distributed, including trees, shrubs, herbs, and so on. To date, most energy plants in the world belong to Salicaceae, Euphorbiaceae, Leguminosae, Compositae, Myrtaceae, and Gramineae, including poplar, cassava, castor, *Hevea brasiliensis*, *Euphorbia tirucalli*, *Jatropha curcas*, *Euphorbia lathyris*, *Aleurites fordii*, soybean, *Cobaifera langadorffi*, *Sindora glabra*, sunflower, *Helianthus tuberosus*, *Eucalyptus* spp., rice, wheat, switchgrass, sugarcane, sorghum, maize, and *Miscanthus* among others. According to the chemistry component and application, energy plants can be categorized into several families:

(i) Saccharide-rich energy plants, including *Saccharum officinarum*, *Sorghum bicolor*, *Beta vulgaris*, and *H. tuberosus*, can produce fuel ethanol.
(ii) Starch-rich energy plants, including *Zea mays*, *Solanum tuberosum*, *Manihot esculenta*, *Lemna minor* Linn., and *Ipomoea batatas*, can also produce ethanol fuel.
(iii) Cellulose-rich energy plants, *Miscanthus*, *Populus* spp., *Panicum virgatum*, and *Eucalyptus* spp., can produce ethanol fuel and biogas (methane).
(iv) Ester-rich energy plants, including *Brassica napus*, *Elaeis guineensis*, *Helianthus annuus*, *Arachis hypogaea* L., and *Glycine max*, can produce biodiesel.
(v) Hydrocarbon-rich energy plants, including *E. lathyris*, *E. tirucalli*, *J. curcas*, and *Coriandrum sativum* L., can produce petroleum analogues.

3.2 Exploitation and Utilization of Energy-Yielding Plants Around the World

Since the 1970s, many countries, such as the United States, Brazil, Japan, and India, made plans to study and develop bioenergy [154–157]. After a large area of oil plants was successfully planted in California by Melvin Calvin [158], the studies on energy plants immediately stepped into a new era across the globe.

In America, researchers planted a large area of "oil plant," more than 1 million ha, and the yield of these energy plants was over 5×10^9 kg. In addition, America is the second largest ethanol producer by using maize as the raw material. In America, more than 90 % biodiesel is produced using soybeans. In recent years, American scientists made progress in cellulose ethanol production [159, 160]. In addition to the above progress, America also developed many other energy plants, such as switchgrass, *Pennisetum purpureum* Schumach, sorghum, poplar, *E. lathyris*, and *Ixeris chinensis*.

Brazil is a pioneer in ethanol fuel production. Making full use of their sugarcane resource, they have cultivated several high-yield sugarcane species and mastered skilled technologies in ethanol fuel production [161]. In addition to sugarcane, Brazil also developed sorghum to produce ethanol fuel. The GraalBio Company announced that they will build the first Brazilian cellulosic ethanol production factory at the end of 2013 [162]. To date, Brazil is the only country that does not supply pure petroleum for motor fuel. In addition to ethanol fuel, the Brazilian government also supports biodiesel production by using castor, sunflower, soybean, palm, and cotton. Brazil also owns plentiful wood resources, such as eucalyptus, *H. brasiliensis*, and *C. langadorffi*.

In Europe, France and Germany mainly use sugar beet and potato to produce ethanol fuel, respectively [163]. Germany and Austria constructed many biodiesel factories, and they mainly use rape as their raw material. Italy, Denmark, and Czech Republic also use rape to produce biodiesel [164]. Moreover, Europe also studied cellulosic ethanol plants, such as poplar, *Salix* spp., eucalyptus, *Liquidambar styraciflua*, *Miscanthus*, *P. virgantum*, *Phalaris arundinacea*, and *Arundo donax*.

Japan is the earliest country to develop biodiesel in Asia. Napier grass is widely cultivated in Japan, which is an ideal oil plant [165]. Japan also uses algae to produce biodiesel. In addition, Japan established a large farm, and planted 150,000 "oil plants" which can produce more than 100 barrels of oil. *Miscanthus giganteus*, originated in Japan, is now considered the energy plant with the most potential [166].

In Southeast Asia, many countries are developing energy plants. Malaysia uses palm to produce biodiesel [167]. The Philippines planted more than 10,000 ha of *Leucaena glauca* to produce biodiesel. Thailand uses cassava to produce ethanol and extracts petroleum analogues from *J. curcas*.

China is a country blessed with abundant natural resources. Although the study of energy plants started fairly late, great progress has been made. Chinese

Table 5 Primary noncrop energy plants

Name	Morphology	Location	Yield	Production
P. virgatum	Herb	America	5,000 $L\cdot ha^{-1}$	Ethanol
C. langadorffi	Tree	Brazil	~7,000 $kg\cdot ha^{-1}$	Diesel
J. curcas	Tree	China	$1.5-3 \times 10^3$ $kg\cdot ha^{-1}$	Diesel
E. lathyris	Tree	America and Europe	~7,000 $kg\cdot ha^{-1}$	Gasoline
P. purpureum	Herb	Japan	1.2×10^4 $kg\cdot ha^{-1}$	Petroleum
E. robusta	Tree	Australia	675 $g\cdot kg^{-1}$	Gasoline
H. tuberosus	Herb	North America	$2-4 \times 10^3$ $kg\cdot ha^{-1}$	Ethanol
I. chinensis	Herb	America	$1-6 \times 10^3$ $kg\cdot ha^{-1}$	Petroleum
Pittoaporumto bira	Shrub	Philippines	50 $g\cdot kg^{-1}$	Gasoline
Trachycarpus fortunei	Tree	Tropical rain forest	1×10^4 $kg\cdot ha^{-1}$	Fuel oil

scientists focus on a variety of energy plants, including maize, rice, soybean, rape, sugarcane, *Beta vulgaris*, sorghum, cassava, poplar, *J. curcas*, *Sindora glabra*, *Pistacia chinensis*, and *Cornus wilsoniana*.

3.3 Looking for Novel Green Energy Plants

Although great progress has been made in the exploitation and utilization of energy plants, some new problems have arisen. To date, ethanol fuel is mainly produced using sugarcane (Brazil), maize (America and China), sugar beet (France), and potato (Germany), which are main food and economic crops. The primary raw materials for producing biodiesel are soybean (America), rape (Austria), and palm (Malaysia), which are also important economic crops. To some extent, bioenergy converted from the above traditional crops eases the pressure of the energy demands. However, with economic development, the growing demand for energy is greatly increasing, which will lead to more and more large-scale use of crops and eventually result in the "competing for food with people, competing for arable land with food crops" problem.

The problem of crop yield and food security restricts the prospect of traditional energy plants. Therefore screening and developing novel noncrop energy plants is considered to be an effective approach to solve the energy crisis. The main noncrop plants for energy production are listed in Table 5. Recently, abundant and inexpensive lignocellulosic biomass energy plants are attracting more and more attention around the world, among which *Miscanthus* is one with the most potentiality.

Miscanthus is a C_4 perennial nonwood rhizomatous tall grass native to subtropical regions originating from eastern Asia and classified taxonomically with maize, sorghum, and sugarcane. The genus comprises around 17 species, such as *Miscanthus sinensis*, *M. sacchariflorus*, *M. floridulus*, and *Miscanthus giganteus*, and is widely adapted to a variety of environmental conditions, indicating its tolerant capability [168].

Since the 1980s, *Miscanthus* has been studied as a potential biofuel plant in Europe [169]. It is now planted widely in Europe and North America. As a candidate energy plant, *Miscanthus* has many advantages:

(i) High lignocellulosic biomass yield. *Miscanthus giganteus* yields have been reported to reach 27–44 $\times$ 10^4 kg $\cdot$ ha^{-1} in Europe.
(ii) Low production cost. Its low fertilizer and pesticide inputs reduce the cultivation cost.
(iii) Superior environment adaptability. It can grow on desolate, barren, marginal, or saline–alkali land, and survive low temperatures. This characteristic raises the utilization of wasteland.
(iv) Advantageous feedstock characteristics. Due to low ash content, *Miscanthus* has a high heating value ranging from 17 to 20 MJ $\cdot$ kg^{-1}. In addition, higher cellulose, semicellulose and lower lignin content make its biomass easily degraded.

To date, studies with respect to *Miscanthus* are primarily focused on genetic diversity analysis, identification of different quantitative trait loci (QTLs) and cultivation, but little is known about genome and genetic information of *Miscanthus* [170–172]. Recently, the genomic and small RNA of *M. giganteus* has been surveyed, providing the theoretical evidence for improvement of *Miscanthus* by biological engineering [173]. In summary, *Miscanthus* is a superior C_4 perennial grass with many advantageous characteristics which make it one of the most potentially novel energy plants.

To face and solve the energy crisis, a large number of energy plants have been screened and exploited to date, among which several crops have successfully achieved industrialization. With in-depth exploration, different countries developed their characteristic energy plants, for instance, the sugarcane of Brazil, soybean of America, sugar beet of France, potato of Germany, palm of Malaysia, and maize of China. In recent years, crop yield and food security have compelled researchers to develop noncrop energy plants, especially lignocellulosic biomass energy plants that are considered potential candidates as substitutes for fossil energy. Now energy plants are developing towards diversity.

4 Yeasts for Production of Biofuels

4.1 Ethanol Yeast

Ethanol as the first-generation biofuel is the most widely used transport fuel. The budding yeast *Saccharomyces cerevisiae* is the dominant microbial cell factory for ethanol production. However, native *S. cerevisiae* strains are not always adaptive to varied substrates and production environments. Other yeasts have been explored in bioethanol production. Thermotolerant yeasts are desirable in ethanol production for economic considerations. *Kluyveromyces marxianus* is well known for thermotolerance, and can even grow at 52 °C and ferment glucose to produce ethanol at 50 °C. *K. marxianus* can utilize various substrates, such as cellobiose, xylose, xylitol, arabinose, glycerol, lactose, and inulin. In ethanol fermentation from inulin, *K. marxianus* produces a significantly higher ethanol yield than native *S. cerevisiae* strains [174]. During ethanol fermentation from cassava starch hydrolysate by thermotolerant *Pichia kudriavzevii* strains at 40 °C, the highest ethanol concentration reached 7.86 % (w/v) after 24 h, with productivity of 3.28 $g \cdot L^{-1} \cdot h^{-1}$ and yield of 85.4 % of the theoretical yield [175].

Some nontraditional yeasts have been explored in bioethanol production from various biomasses. Seaweed biomass is a potential feedstock for biofuel production. Hydrolysates of seaweed contain a high salt concentration. Recently, the ability and efficiency of a marine yeast (*Candida* sp.) to grow and aerobically ferment seaweed polysaccharide-based hydrolysate to ethanol in the presence of 6.25–11.25 % salt concentration was validated [176]. Glycerol, the by-product of biodiesel production, is considered a waste by biodiesel producers. Ethanol fermentation from glycerol by the yeast *Pachysolen tannophilus* was tested in a recent study [177]. The highest ethanol production was 17.5 $g \cdot L^{-1}$ on 5 % (v/v) crude glycerol, corresponding to 56 % of the theoretical yield. A staged batch process achieved 28.1 $g \cdot L^{-1}$ ethanol, which is the maximum achieved so far for conversion of glycerol to ethanol in a microbial process.

4.2 Oleaginous Yeasts

The oily yeast genera include *Yarrowia*, *Candida*, *Rhodotorula*, *Rhodosporidium*, *Cryptococcus*, *Trichosporon*, and *Lipomyce* [178]. The oleaginous yeast *Y. lipolytica* is an attractive candidate for microbial oil production. It also has been found to be robust, able to grow on a variety of substrates, and has been used for lipid production on agro-industrial residues, industrial glycerol, and industrial fats. It has excellent lipid accumulation capacity, commonly accumulating up to 36 % of its dry cell weight in lipids [179]. With a fully sequenced genome and a growing body of tools, engineering of *Y. lipolytica* can be achieved with relative ease [180, 181]. Through engineering, lipid content of *Y. lipolytica* can reach up to 61.7 % [182].

Hydrolysates of lignocellulosic biomass contain glucose, xylose, and cellobiose among others. Effective utilization of these sugars remains challenging for microbial conversion, because most microorganisms consume such sugars sequentially with a strong preference for glucose. Efficient lipid production with simultaneous consumption of glucose and xylose was achieved in the yeast *Trichosporon cutaneum* [183]. The oleaginous yeast strain *T. cutaneum* AS 2.571 could assimilate glucose and xylose simultaneously, and accumulate intracellular lipid up to 59 % (DCW) with a lipid coefficient up to 0.17 $g \cdot g^{-1}$ sugar. The oleaginous yeast, *Lipomyces starkeyi*, was shown to consume cellobiose and xylose simultaneously and to produce intracellular lipids from cellobiose, xylose, and glucose [184]. Overall substrate consumption rates were close to 0.6 $g \cdot L^{-1} \cdot h^{-1}$, and lipid coefficients were 0.19 $g \cdot g^{-1}$ sugar, respectively. Novel oleaginous yeast species were identified, such as *T. cacaoliposimilis* sp. nov. and *T. oleaginosus* sp. nov. [185]. The draft genome of the red yeast *Rhodosporidium toruloides* MTCC 457 was also reported. The genome sequence will be valuable for molecular genetic analysis and manipulation of lipid accumulation in this yeast and for developing it as a potential host for biofuel production [186].

4.3 Yeast Biofuel Cells

Microbial fuel cells (MFCs) are devices that can use microbial metabolism to produce an electrical current from a wide range of organic substrates. The potential applications of MFC technology are involved in the production of electricity and degradation of wastes and toxic chemicals [187, 188]. Many bacteria possess the ability to transfer the electrons derived from the metabolism of organic matter to the anode. Marine sediment, soil, waste water, fresh water sediment, and activated sludge are all rich sources for these microorganisms [187]. Yeast microbial fuel cells have received little attention to date. Yeasts should be an ideal MFC catalyst because they are robust, easily handled, mostly nonpathogenic organisms with high catabolic rates and in some cases a broad substrate spectrum. The conventional yeast *S. cerevisiae* [189–192], *Hansenula anomala* [190, 193], *Candida melibiosica* [194], and *Arxulaa deninivorans* [195] have been evaluated as MFCs.

5 Bacteria for Bioenergy Production

Bacteria play an important role in the natural cycle of material and energy. The production of hydrogen, alcohols, and biogas is the main pathway for them to yield bioenergy. For bioenergy production, they often possess a broader substrate spectrum than yeast, and are capable of degrading pentoses, hexoses, disaccharides, and polysaccharides such as cellulose and alginate. Microbial bioconversion is a rapidly evolving subject. Novel bacteria with biofuel production ability are constantly being discovered in the environment, especially some bacteria from thermophilic conditions. This section introduces the latest advances in bacteria bioresource exploration and their application in alcohol and hydrogen production.

5.1 Ethanol-Producing Bacteria

The general ethanologenesis starts with the formation of pyruvate through several pathways (Fig. 3), namely the EMP (Embden–Meyerhof–Parnas), ED (Entner–Doudoroff), and PP (pentose phosphate) pathways [196]. The majority of microorganisms degrade hexoses through the EMP pathway or the ED pathway.

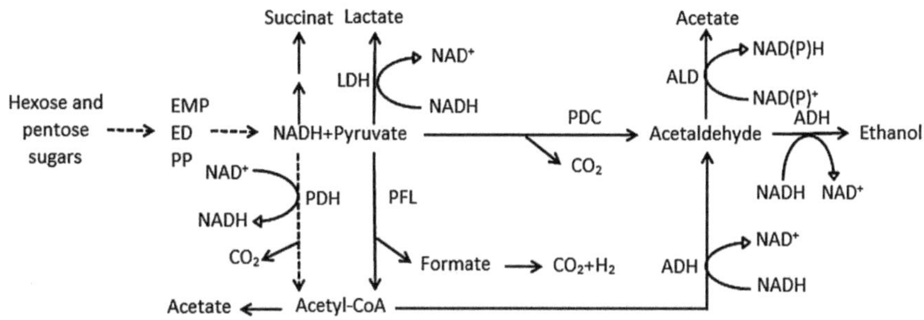

Fig. 3 Metabolic pathways for ethanol synthesis [196]. *EMP* Embden–Meyerhof–Parnas; *ED* Entner–Doudoroff; *PP* pentose phosphate; *ADH* alcohol dehydrogenase; *LDH* lactate dehydrogenase; *PDC* pyruvate decarboxlyase; *PDH* pyruvate dehydrogenase; *PFL* pyruvate-formate lyase; *ALD* acetaldehyde dehydrogenase

The catabolization of glucose with the EMP pathway generates 2 NADH and 2 pyruvates, together with the formation of 2 ATP [197]. The ED pathway has low distribution among anaerobic bacteria and is more restricted to Gram-negative bacteria and Archaea [197]. In some bacterial strains from *Clostridium* sp., *Caldicellulosiruptor* sp., *Thermoanaerobacter* sp., and *Caldanaerobacter* sp., the EMP pathway was utilized for conversion of glucose to phosphoenolpyruvate (PEP) [198]. PEP was then converted into pyruvate via an ATP-dependent pyruvate kinase (PPK) directly, or via an AMP-dependent pyruvate phosphate dikinase (PPDK) [198].

Pyruvate could be metabolized via several pathways to produce ethanol, lactate, succinate, formate, acetate, and so on (Fig. 3). One of the most important pathways to produce ethanol was the homoethanol fermentation pathway, which was catalyzed by PDC (EC 4.1.1.1) and ADH (EC 1.1.1.1). PDC catalyzes the decarboxylation of pyruvate to acetaldehyde, and acetaldehyde is then reduced to ethanol by ADH. PDC is common in yeast, but only few bacteria including *Zymomonas mobilis*, *Zymobacter palmae*, and *Acetobacter pasteurianus* possess the *pdc* gene [199–201].

Ethanol could be produced by a variety of bacteria. Some well-studied bacteria used for ethanol production include *Z. mobilis*, *Escherichia coli*, and *Klebsiella oxytoca* [202]. *Z. mobilis* is one of the well-studied bacteria due to its unique metabolism and ability to produce ethanol rapidly and efficiently as the main fermentation product from simple sugars. It shares a natively expressed homoethanol pathway similar to *S. cerevisiae*, which is controlled by the PET operon encoding PDC and ADH [203]. Unfortunately, *Z. mobilis* has a limited carbon substrate range, and lacks the ability to metabolize pentose which is abundant in hemicellulose. Therefore, *E. coli* and *K. oxytoca* are identified as choices for metabolic engineering towards homoethanol production either for the extensive understanding of its physiology and metabolism or for the wide range of substrates utilizing its ability [202].

Most of the well-studied ethanol-producing bacteria are mesophiles. In recent years, thermophilic bacteria have gained increased attention as candidates for ethanol production. Thermophilic bacteria are capable of producing various end-products including ethanol. There are several advantages in using thermophilic bacteria for ethanol production as concluded by Taylor et al. [204]. First, thermophiles are commonly able to ferment the pentose and hexose sugar fraction of the biomass, and some strains could degrade complex polycarbohydrates, such as lignocellulose. Second, they have the remarkable ability to tolerate fluctuations of environmental changes, such as pH and temperature. Third, ethanol could be removed or recovered more easily from the fermentation broth under high temperatures. Moreover, high temperature could reduce contamination risk, and eliminate low ethanol tolerance problems by growing bacteria at temperatures where "self-distillation" is possible.

These advantages are leading to a constant discovery of thermophilic bacteria with the capacity for ethanol production. Some bacterial strains from the genera of *Thermoanaerobacterium*, *Thermoanaerobacter*, and *Clostridium* have demonstrated good ethanol-producing capacities, such as *Thermoanerobacter ethanolicus* (65 °C), *Clostridium thermocellum* (60 °C), and *Thermoanaerobacterium* sp. AK17 (60 °C) [74, 78, 202, 205–207], *Geobacillus thermoglucosidasius* (70 °C), *Thermoanaerobacter saccharolyticum* (70 °C), and *Thermoanaerobacter mathranii* (70 °C) are also considered catabolically versatile hosts for ethanol production [204].

Bacteria could use a variety of substrates to produce ethanol. Industrial ethanol was mainly produced from starch derived from yellow corn and sucrose derived from sugarcane, which was referred to as "first-generation bioethanol" [204]. It was criticized for reducing food and land use for the diversion of food from the food chain into fuel production [208]. Thus, cellulosic materials were proposed to produce "second-generation bioethanol," and a series of novel ethanologenic bacteria with a broader substrate spectrum were discovered. However, reports of biofuel production from direct hydrolyzing lignocellulosic biomass by thermophilic bacteria are scarce. One example is *C. thermocellum*, which could rapidly decompose cellulosic materials and ferment the resulting sugars to ethanol, but its yield was low as a result of mixed acid fermentation [204]. In our recent work, an anaerobic and thermophilic bacterial community was enriched by coastal marine sediment. Most of the clones that accounted for 60 % of the clone library shared similarity with the type strain *C. thermocellum* ATCC 27405 [209]. Moreover, most of the clones represented by the clone library shared 16S rRNA similarities lower than 90 %, and all of them shared 16S rRNA similarities below 94 % (Table 6). They were demonstrated to be an untapped bacterial resource with both cellulose-degrading ability and ethanol-producing ability [209]. These results indicated that there were extensive unexplored bacterial resources existing in the ocean.

Except utilizing glucose, xylose, or other monosaccharide and biomass as substrates, there have been emerging studies on bioconversion into ethanol from substrates of carbon monoxide and glycerol in recent years [210, 211]. A few works utilizing such substrates are given in Table 7.

Table 6 Closest type strains of selected 16S rRNA clones from EzTaxon-e database

Clone no.	Length (bp)	Nearest type strain	Similarity (%)	Accession
2	1499	*Acetomicrobium faecale*	88.0	FR749980
4	1496	*Planifilum yunnanense*	88.0	DQ119659
6	1512	*Desulfotomaculum alkaliphilum*	91.2	AF097024
8	1498	*C. thermocellum*	89.4	CP000568
9	1500	*C. thermocellum*	87.6	CP000568
11	1505	*Clostridium straminisolvens*	88.8	AB125279
21	1624	*Caldicoprobacteroshimai*	89.8	AB450762
22	1505	*Desulfotomaculum halophilum*	90.6	U88891
23	1490	*D. halophilum*	86.6	U88891
27	1494	*Sporosalibacterium faouarense*	92.4	EU567322
39	1501	*Clostridium purinilyticum*	90.5	FR749894
41	1497	*S. faouarense*	86.7	EU567322
42	1492	*Planifilumfulgidum*	88.0	AB088362
46	1454	*Caloranaerobacter azorensis*	90.7	AJ272422
60	1487	*Bacillus thermolactis*	93.4	AY397764

Table 7 Some bacterial isolates with recently reported biofuel production ability

Isolates	Products	Growth temperature (°C)	Substracts	Reference
Geobacillus sp.XT15	Acetoin and 2,3-butanediol	55	Corn steep liquor powder	[212]
Kluyvera cryocrescens S26	Ethanol and 1,2-propanediol	30	Crude glycerol	[211]
Paenibacillus macerans N234A	Ethanol	37	Glycerol	[213]
Clostridium ljungdahlii	Ethanol	37	Sugars/syngas	[210]
Thermoanaerobacter J1	Ethanol	65	Glucose/xylose/	[214]
Thermoanaerobacter BG1L1	Ethanol	70	wheat straw/corn stover	[215, 216]
Thermoanaerobacterium AK17; *Thermoanaerobacter* Ak33; *Paenibacillus* AK25	Ethanol	60/70/50	Glucose/xylose/ cellulose/grass	[74]
Geobacillus sp. R7	Ethanol	70	Agricultural residues	[217]

The screening methods of ethanol-producing bacteria are mostly based on a culture-dependent method, which is achieved through the detection of ethanol production of the object strain. Several methods could be used for ethanol detection, including the hydrometer method, colorimetric method, gas chromatography, liquid chromatography, electrochemical and enzymatic assays, and so on [218]. Each method has advantages and disadvantages. Gas or liquid chromatographic analysis is the most commonly used method for quantifying ethanol yield, but the procedure is time consuming. Recently, researchers developed a method based on alcohol oxidase and peroxidase (AOP assay) for the high-throughput screening of thermophilic ethanol-producing bacteria [218]. They performed the assay in a 96-well microtiter plate and the method had high accuracy. Several isolates with high ethanol tolerance and ethanol yield were obtained by using this method [218].

Another approach for screening bacteria with potentials in ethanol production is to identify a few genetic biomarkers that play key roles in ethanol synthesis pathways by comparative meta-analysis of their genomes [198] and through analyzing these biomarkers, estimating the object strains' ethanol production abilities. This approach could also offer potential targets for metabolic engineering to increase ethanol yield.

5.2 Butanol and Longer Chain Alcohol-Producing Bacteria

Butanol could be served as a better biofuel with several advantages over ethanol. It has higher energy density, lower vapor pressure, and is less hygroscopic and therefore less corrosive [219]. It could be blended with conventional fuels at any ratio. Biological butanol is produced through the acetone–butanol–ethanol (ABE) fermentation process, and its production has a long history in the fermentation industry. In addition to acetone–butanol–ethanol, organic acids (acetic acid, lactic acid, and butyric acid) and gases (carbon dioxide and hydrogen) are also produced during the ABE fementation process (Fig. 4). Butanol could be produced by a variety of bacteria mainly from the genus of *Clostridia*. *C. acetobutylicum*, and *C. beijerinckii* are the most studied strains used in ABE fermentation. The metabolic pathway synthesizing the main products from glucose is shown in Fig. 4 [220].

In the past decades, butanol-producing bacteria including their mutants were constantly isolated and characterized. They were mostly from the genus of *Clostridium*, such as *Clostridium beijerinckii*, *C. saccaroperbutylacetonicum*, *C. saccharoacetobutylicum*, *C. aurantibutyricum*, *C. pasteurianum*, *C. sporogenes*, *C. cadaveris*, and *C. tetanomorphum* [221].

Although they had relatively high butanol yields, screening novel wild-type microbes with better butanol-producing ability is still of great importance for industrial strain breeding. There are several directions in the screening work, which include screening strains with higher butanol yields, broader substrate range, and higher butanol tolerance. Most of the screening work is based on testing physiological properties of the isolated strains.

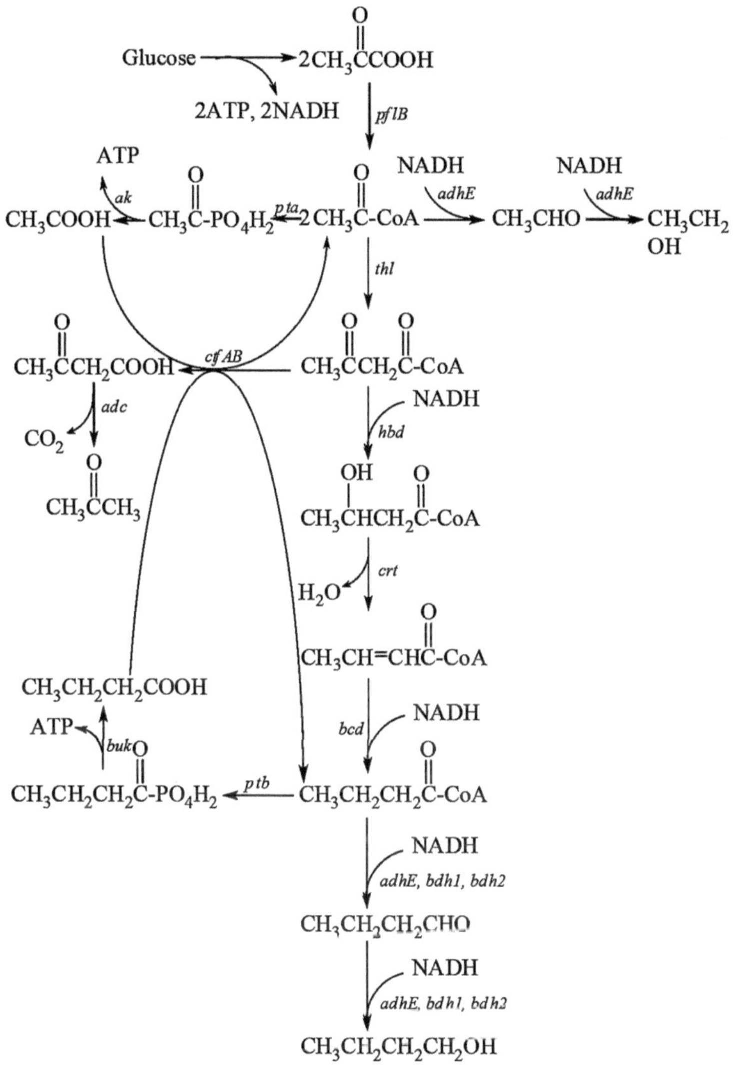

Fig. 4 Metabolic pathways of butanol production through the acetone–butanol–ethanol (ABE) fermentation process. Enzymes corresponding to each gene are: *pflB* pyruvate-ferrodoxin oxidoreductase gene, *pta* phosphate acetyltransferase gene, *ak* acetate kinase gene, *adhE* acetaldehyde/ethanol dehydrogenase gene, *thl* thilose gene, *ctfAB* acetoacetyl-CoA: acetate/butyrate CoA transferase gene, *adc* acetoacetate decarboxylase gene, *hbd* 3-hydroxybutyryl-CoA dehydrogenase gene, *crt* crotonase gene, *bcd* butyryl-CoA dehydrogenase gene, *ptb* phosphate butyltransferase gene, *buk* butyrate kinase gene, *bdh1*, *bdh2* butyraldehyde/n-butanol dehydrogenase gene

Recently, several interesting wild-type strains have been isolated from environmental samples. For example, *Clostridium* sp. G117 was isolated from soil samples from grassland in Singapore. This strain could produce dominant butanol that was 20 % higher than the butanol production by wild-type *C. acetobutylicum* ATCC 824 under similar conditions [222]. In addition, Strain G117 produced butanol and acetone as the main end-products, and only generated a negligible amount of ethanol through the AB (acetone–butanol) process [222].

Another example is a mesophilic *Clostridium* sp. strain BOH3. The strain has proved to be capable of utilizing cellulose and xylan to produce butanol and hydrogen. This feature distinguished BOH3 from most wild-type solventogenic strains [223]. In an earlier report of butanol-producing strain screening, a number of isolates classified as new strains of *C. acetobutylicum*, and *Clostridium* sp. NCP262 were reported to possess the ability to hydrolyze starch, carboxymethyl cellulose, xylan, inulin, chitosan, and so on. However, their performance on direct biofuel production from these polysaccharides is not substantiated in their study [224, 225]. Most of the studies converting polysaccharides to butanol were not direct processes, but utilized a prior hydrolysis pretreatment [226]. Screening strains with the ability of hydrolyzing polysaccharides would be a promising approach to produce biofuels on cheaper substrates, and would result in a simultaneous fermentation process [206]. Aiming at screening solvent-producing bacteria of the class Clostridia with cellulolytic activity, researchers developed a quick screening method [227]. First, bacteria isolates were screened based on their cellulolytic activity and butanol tolerance in selective media. Then these isolates were classified to the class Clostridia according to three selected criteria (endospore formation, sulfite-reducing ability, and metabolic products) . Last, the 16S rRNA gene of isolates was sequenced and the bacteria species were identified based on the phylogenetic tree based on the 16S rRNA gene.

A few researchers focused on searching for microorganisms with superior butanol tolerance. These studies would help to find suitable alternative hosts that could further conduct metabolic engineering for biofuel production [228]. Some salutary attempts have been made in this direction. For example, a bacterium identified as *Enterococcus faecium* capable of 2.5–3 % (w/v) butanol tolerance was isolated [229]. This isolate could produce butanol probably using different metabolic networks from the obligate anaerobe *C. acetobutylicum*. This isolate could also show tolerance to 10 % (w/v) ethanol and 3 % (w/v) isobutanol. With these distinct features, the isolate could be explored as a potential host for butanol production [229]. In order to isolate novel bacteria with the desired characteristics and with the potential for genetic engineering, more work should be done on strain screening.

With the full understanding of the metabolic pathways of butanol production, synthetic-biology approaches became popular to maximize fuel production [230]. To avoid some limitations of *Clostridium*, *E. coli*, *S. Cerevisiae*, *Pseudomonas putida*, *Bacillus subtilis*, *Lactobacillus brevis*, and *Synechococcus elongatus* were used to produce butanol after the introduction of a butanol pathway into these organisms [230]. In addition, isoprenoid-derived fuels, fatty-acid–derived fuels and polyketide-derived fuels are also becoming promising directions in biofuel production with the development of the synthetic-biology approach [230]. Apart from butanol, other C3–C5 normal and branched alcohols also possess similar chemical properties to butanol, including isobutanol, 2-methyl-1-butanol, 3-methyl-1-butanol, and isopentenol [231]. However, the production of these alcohols in native bacteria is not so efficient. To improve the titers and productivity, synthetic biology usually needs to be employed to do some synthetic pathway engineering [231]. For example, through modifying the amino acid biosynthetic pathways in *E.coli*, researchers have developed a metabolic engineering approach to produce higher alcohols including isobutanol, 1-butanol,2-methyl-1-butanol, 3-methyl-1-butanol, and 2-phenylethanol from glucose [232]. This nonfermentation strategy represents a major discovery for the production of C3–C5 alcohols [233]. In addition, C3 and C4 alcohols could also be produced from syngas. However, so far, the yield appears to be minor compared to ethanol produced by similar fermentations [210].

Screening of microorganisms capable of higher yield alcohols is underway. A new isolated strain XT15 from *Geobacillus* sp. was reported to be capable of producing 2,3-butanediol as one of the main products [234]. XT15 was found to be optimally growing between 45 and 55 °C, and capable of using glucose, galactose, mannitol, arabionose, and xylose as substrates. The yield of 2,3-butanediol could be up to 14.5 $g \cdot L^{-1}$, accompanied with 7.7 $g \cdot L^{-1}$ of acetoin. These results indicated its potential as a precious biological resource in a thermophilic fermentation process.

5.3 Biohydrogen-Producing Bacteria

Hydrogen is a clean and efficient fuel. It is known that photosynthetic organisms, for example, cyanobacteria and green algae can generate hydrogen gas [13, 43]. Hydrogen can also be generated by anaerobic fermentation of organic substrates, such as sugars, lingocellulosic biomass, and waste materials. Although H_2 production is common in bacteria in the process of anaerobic catabolism of organic compounds, it is only one of several electron sinks, inasmuch as other fermentation end-products are produced in addition to hydrogen [235]. A vast number of bacteria are known to produce hydrogen as an end-product, including strains from both mesophiles and thermophiles. Mesophilic bacteria such as *C. butyricum*, *Enterobacter aerogenes*, and *E. coli* have been extensively studied [236–238].

In recent years, H_2 production by thermophilic bacteria has began to attract increased interest. The most common thermophiles include *Pyrococcus furiosus*, *Thermococcus kodakaraensis*, and all *Thermotoga* and *Caldicellulosiruptor* species [239]. Most of these species do not encode adhE or aldH, and therefore produce negligible or no ethanol. The absence of ethanol-producing pathways makes more reducing equivalents that are disposed through H_2 production via hydrogenase [38]. Thermophilic bacteria have many advantages compared to mesophiles concerning H_2 production: higher H_2 yields and less variety of end-products [240].

Several thermophilic species have been reported as good producers with high H_2 yields in recent studies, including *Thermotoga neapolitana*, *Caldicellulosiruptor saccharolyticus*, and *C. owensensis* [241–245]. Some thermophilic bacteria from the genera *Clostridium* and *Thermoanaerobacterium* were also reported for hydrogen production, but the H_2 yield is lower than the species mentioned above [246, 247].

The fermentation for H_2 production using pure cultures is not feasible for large-scale prodution. In practice, mixed microbial cultures in sludge are frequently used in dark fermentation reactors to produce hydrogen [248, 249]. The dark fermentation can produce hydrogen from organic compounds constantly without the need for light [250]. Due to the adavatages of thermophilic bacteria in the process of fermentation, there are also increasing studies focusing on biohydrogen production by using thermophilic mixed cultures [250–253].

6 Conclusion

This review gives an outlook on organisms for biofuel production, including microalgae, energy plants, yeast, and bacteria. Plants and microalgae are primary producers that produce biomass via photosynthesis as biofuel feedstock utilized by yeast and bacteria. In the last several years, some species and strains in plants, microalgae, yeast, and bacteria have been reported to possess distinctive traits for biofuel production in the laboratory or on a pilot scale. Although great progress has been made, obstacles to industrial-scale production of biofuels remain. Isolation and characterization of organism resources from natural habitats should be a continuing effort around the world. The organisms used for biofuel production thus far are only a very small part of the identified species. Studies on additional species of plants, microalgae, yeast, and bacteria will help us to obtain novel insights into metabolism pathways in the organisms, which may provide more useful metabolites and more efficient biofuel production. With the assistance of omics technology, deep understanding of metabolic pathways and regulation networks will be possible. Benefiting from the fast-developing subjects of metabolic engineering, synthetic biology, and systems biology, many artificially synthesized "super-strains" with excellent characteristics for biofuel production may be created in the future.

Acknowledgments This work was supported by the National Basic Research Program of China (2011CB707404), the National Key Technology R&D Research Programs (2011BAD22B02-01, 2009AA10Z101), the Natural Science Foundation of China (U1232126), and the Key Research Program of Shandong Province (2012GGF01023).

References

1. Dudley B ed (2012) BP statistical review of world energy. June 2012, BP
2. Conti J (2011) International energy outlook 2011. IEA, Paris
3. Snow N (2013) BP: shale gas, tight oil to reshape global markets by 2030. Oil Gas J 111(2B):25–25
4. Shafiee S, Topal E (2009) When will fossil fuel reserves be diminished? Energy Policy 37(1):181–189
5. Sorrell S, Speirs J, Bentley R et al (2012) Shaping the global oil peak: a review of the evidence on field sizes, reserve growth, decline rates and depletion rates. Energy 37(1):709–724
6. Fedorov AV, Philander SG (2000) Is El Niño changing? Science 288(5473):1997–2002
7. Zhang Q, Guan Y, Yang H (2008) ENSO amplitude change in observation and coupled models. Adv Atmos Sci 25(3):361–366
8. Agarwal AK (2007) Biofuels (alcohols and biodiesel) applications as fuels for internal combustion engines. Prog Energy Combust Sci 33(3):233–271
9. Radakovits R, Jinkerson RE, Darzins A et al (2010) Genetic engineering of algae for enhanced biofuel production. Eukaryotic Cell 9(4):486–501
10. Balat M, Balat H, Oz C (2008) Progress in bioethanol processing. Prog Energy Combust Sci 34(5):551–573
11. Chisti Y (2007) Biodiesel from microalgae. Biotechnol Adv 25(3):294–306
12. Elmoraghy M, Farag IH (2012) Bio-jet fuel from microalgae: reducing water and energy requirements for algae growth. Res Inventy 1(2):22–30
13. Melis A, Happe T (2001) Hydrogen production. Green algae as a source of energy. Plant Physiol 127(3):740–748
14. Demirbas A (2007) Progress and recent trends in biofuels. Prog Energy Combust Sci 33(1):1–18
15. Hu Q, Sommerfeld M, Jarvis E et al (2008) Microalgal triacylglycerols as feedstocks for biofuel production: perspectives and advances. Plant J 54(4):621–639
16. Chisti Y (2008) Biodiesel from microalgae beats bioethanol. Trends Biotechnol 26(3):126–131
17. Sheehan J, Dunahay T, Benemann J et al (1998) US department of energy's office of fuels development, July 1998. A look back at the US department of energy's aquatic species program. Biodiesel from algae, close out report TP-580-24190. National Renewable Energy Laboratory, Golden
18. Riekhof WR, Benning C (2009) Glycerolipid biosynthesis in the Chlamydomonas sourecebook. In: Harris EH (ed). Elsevier Inc., Oxford, UK.41-61
19. Guckert JB, Cooksey KE (1990) Triglyceride accumulation and fatty-acid profile changes in chlorella (chlorophyta) during high ph-induced cell-cycle inhibition. J Phycol 26(1):72–79
20. Metzger P, Largeau C (2005) Botryococcus braunii: a rich source for hydrocarbons and related ether lipids. Appl Microbiol Biotechnol 66(5):486–496
21. Banerjee A, Sharma R, Chisti Y et al (2002) Botryococcus braunii: a renewable source of hydrocarbons and other chemicals. Crit Rev Biotechnol 22(3):245–279
22. Ohlrogge J, Browse J (1995) Lipid biosynthesis. Plant Cell 7(7):957–970
23. Cobelas MA, Lechado JZ (1989) Lipids in microalgae—a review.1. biochemistry. Grasas Y Aceites 40(2):118–145
24. Basova MM (2005) Fatty acid composition of lipids in microalgae. Int J Algae 7(1):33–57
25. Yuan C, Liu J, Fan Y et al (2011) Mychonastes afer HSO-3-1 as a potential new source of biodiesel. Biotechnol Biofuels 4:47
26. Bigogno C, Khozin-Goldberg I, Boussiba S et al (2002) Lipid and fatty acid composition of the green oleaginous alga Parietochloris incisa, the richest plant source of arachidonic acid. Phytochemistry 60(5):497–503

27. de Swaaf ME, de Rijk TC, Eggink G et al (1999) Optimisation of docosahexaenoic acid production in batch cultivations by Crypthecodinium cohnii. J Biotechnol 70(1–3):185–192
28. Cohen Z, Khozin-Goldberg I, Adlerstein D et al (2000) The role of triacylglycerol as a reservoir of polyunsaturated fatty acids for the rapid production of chloroplastic lipids in certain microalgae. Biochem Soc Trans 28:740–743
29. Sasaki Y, Konishi T, Nagano Y (1995) The compartmentation of acetyl-coenzyme a carboxylase in plants. Plant Physiol 108:445
30. Dahlqvist A, Stahl U, Lenman M et al (2000) Phospholipid: diacylglycerol acyltransferase: an enzyme that catalyzes the acyl-CoA-independent formation of triacylglycerol in yeast and plants. Proc Natl Acad Sci U S A 97(12):6487–6492
31. Yoon K, Han D, Li Y et al (2012) Phospholipid: diacylglycerol acyltransferase is a multifunctional enzyme involved in membrane lipid turnover and degradation while synthesizing triacylglycerol in the unicellular green microalga chlamydomonas reinhardtii. Plant Cell 24(9):3708–3724
32. Durrett TP, Benning C, Ohlrogge J (2008) Plant triacylglycerols as feedstocks for the production of biofuels. Plant J 54(4):593–607
33. Knothe G (2005) Dependence of biodiesel fuel properties on the structure of fatty acid alkyl esters. Fuel Process Technol 86(10):1059–1070
34. Benemann J (1996) Hydrogen biotechnology: progress and prospects. Nat Biotechnol 14(9):1101–1103
35. Benemann JR (2000) Hydrogen production by microalgae. J Appl Phycol 12(3):291–300
36. Gaffron H (1939) Reduction of carbon dioxide with molecular hydrogen in green algae. Nature 143:204–205
37. Happe T, Naber JD (1993) Isolation, characterization and N-terminal amino-acid-sequence of hydrogenase from the green-alga Chlamydomonas-reinhardtii. Eur J Biochem 214(2): 475–481
38. Voordouw G, Strang JD, Wilson FR (1989) Organization of the genes encoding fe hydrogenase in Desulfovibrio vulgaris subsp oxamicus monticello. J Bacteriol 171(7): 3881–3889
39. Adams MWW (1990) The structure and mechanism of iron-hydrogenases. Biochimica Et Biophysica Acta 1020(2):115–145
40. Happe T, Mosler B, Naber JD (1994) Induction, localization and metal content of hydrogenase in the green-alga Chlamydomonas-reinhardtii. Eur J Biochem 222(3):769–774
41. Ghirardi ML, Posewitz MC, Maness P-C et al (2007) Hydrogenases and hydrogen photoproduction in oxygenic photosynthetic organisms. Ann Rev Plant Biol 58(1):71–91
42. Ghirardi ML, Zhang JP, Lee JW et al (2000) Microalgae: a green source of renewable H-2. Trends Biotechnol 18(12):506–511
43. Melis A, Zhang LP, Forestier M et al (2000) Sustained photobiological hydrogen gas production upon reversible inactivation of oxygen evolution in the green alga Chlamydomonas reinhardtii. Plant Physiol 122(1):127–135
44. Wykoff DD, Davies JP, Melis A et al (1998) The regulation of photosynthetic electron transport during nutrient deprivation in Chlamydomonas reinhardtii. Plant Physiol 117(1): 129–139
45. Zhang L, Happe T, Melis A (2002) Biochemical and morphological characterization of sulfur-deprived and H2-producing Chlamydomonas reinhardtii (green alga). Planta 214(4): 552–561
46. Scoma A, Krawietz D, Faraloni C et al (2012) Sustained H-2 production in a Chlamydomonas reinhardtii D1 protein mutant. J Biotechnol 157(4):613–619
47. Show KY, Lee DJ, Tay JH et al (2012) Biohydrogen production: current perspectives and the way forward. Int J Hydrogen Energy 37(20):15616–15631
48. Florin L, Tsokoglou A, Happe T (2001) A novel type of iron hydrogenase in the green alga Scenedesmus obliquus is linked to the photosynthetic electron transport chain. J Biol Chem 276(9):6125–6132

49. Juantorena AU, Lastres O, Hernandez G et al (2012) Hydrogen production by microorganisms and its application in a PEMFC. Int J Energy Res 36(8):902–910
50. Papazi A, Andronis E, Ioannidis NE et al (2012) High yields of hydrogen production induced by meta-substituted dichlorophenols biodegradation from the green alga scenedesmus obliquus. Plos One 7(11):e49037
51. Song W, Rashid N, Choi W et al (2011) Biohydrogen production by immobilized Chlorella sp using cycles of oxygenic photosynthesis and anaerobiosis. Bioresour Technol 102(18):8676–8681
52. Chader S, Hacene H, Agathos SN (2009) Study of hydrogen production by three strains of Chlorella isolated from the soil in the Algerian Sahara. Int J Hydrogen Energy 34(11):4941–4946
53. Amutha KB, Murugesan AG (2011) Biological hydrogen production by the algal biomass Chlorella vulgaris MSU 01 strain isolated from pond sediment. Bioresour Technol 102(1):194–199
54. He M, Li L, Zhang L et al (2012) The enhancement of hydrogen photoproduction in Chlorella protothecoides exposed to nitrogen limitation and sulfur deprivation. Int J Hydrogen Energy 37(22):16903–16915
55. Guan YF, Deng MC, Yu XJ et al (2004) Two-stage photo-biological production of hydrogen by marine green alga Platymonas subcordiformis. Biochem Eng J 19(1):69–73
56. Ran CQ, Yu XJ, Jin MF et al (2006) Role of carbonyl cyanide m-chlorophenylhydrazone in enhancing photobiological hydrogen production by marine green alga Platymonas subcordiformis. Biotechnol Prog 22(2):438–443
57. Weare NM, Benemann JR (1974) Nitrogenase activity and photosynthesis in Plectonema boryanum. J Bacteriol 119(1):258–265
58. Weare NM, Benemann JR (1973) Nitrogen-fixation by Anabaena cylindrica.1. localization of nitrogen-fixation in heterocysts. Archiv Fur Mikrobiologie 90(4):323–332
59. Dutta D, De D, Chaudhuri S et al (2005) Hydrogen production by Cyanobacteria. Microbiol Cell Fact 4(1):36
60. Bergman B, Gallon J, Rai A et al (1997) N2 fixation by non-heterocystous cyanobacteria. FEMS Microbiol Rev 19:139–185
61. Fay P (1992) Oxygen relations of nitrogen fixation in cyanobacteria. Microbiol Rev 56:340–373
62. Weissman JC, Benemann JR (1977) Hydrogen production by nitrogen-starved cultures of Anabaena cylindrica. Appl Environ Microbiol 33(1):123–131
63. Hallenbeck P, Kochian L, Weissmann J et al (1978) Solar energy conversion with Hydrogen producing cultures of the blue green alga, Anabaena cylindrica. Biotechnol Bioeng Symp 8:283–297
64. Yu J, Takahashi P (2007) Biophotolysis-based hydrogen production by cyanobacteria and green microalgae. In: Méndez-Vilas A (ed) Communicating current research and educational topics and trends in applied microbiology. Formatex, Badajoz, Spain pp 79−89
65. Min H, Sherman LA (2010) Hydrogen production by the unicellular, diazotrophic cyanobacterium *Cyanothece* sp. Strain ATCC 51142 under conditions of continuous light. Appl Environ Microbiol 76(13):4293–4301
66. Bandyopadhyay A, Stöckel J, Min H et al (2010) High rates of photobiological H2 production by a cyanobacterium under aerobic conditions. Nat Commun 1(9):139
67. Rashid N, Song W, Park J et al (2009) Characteristics of hydrogen production by immobilized cyanobacterium Microcystis aeruginosa through cycles of photosynthesis and anaerobic incubation. J Ind Eng Chem 15(4):498–503
68. Schutz K, Happe T, Troshina O et al (2004) Cyanobacterial H-2 production—a comparative analysis. Planta 218(3):350–359
69. Antal TK, Lindblad P (2005) Production of H_2 by sulphur-deprived cells of the unicellular cyanobacteria Gloeocapsa alpicola and *Synechocystis* sp. PCC 6803 during dark incubation with methane or at various extracellular pH. J Appl Microbiol 98(1):114–120

70. Bernat G, Waschewski N, Roegner M (2009) Towards efficient hydrogen production: the impact of antenna size and external factors on electron transport dynamics in Synechocystis PCC 6803. Photosynthesis Res 99(3):205–216
71. Happe T, Schütz K, Böhme H (2000) Transcriptional and mutational analysis of the Uptake Hydrogenase of the Filamentous Cyanobacterium Anabaena variabilis ATCC 29413. J Bacteriol 182(6):1624–1631
72. Marques AE, Barbosa AT, Jotta J et al (2011) Biohydrogen production by Anabaena sp PCC 7120 wild-type and mutants under different conditions: light, nickel, propane, carbon dioxide and nitrogen. Biomass Bioenergy 35(10):4426–4434
73. Abdel-Basset R, Bader KP (2008) Hydrogen evolution in relation to PSI-reducible substrates in the cyanobacterium Oscillatoria chalybea assayed by means of mass spectrometry. Int J Hydrogen Energy 33(11):2653–2659
74. Huesemann MH, Hausmann TS, Carter BM et al (2010) Hydrogen generation through indirect biophotolysis in batch cultures of the nonheterocystous nitrogen-fixing cyanobacterium plectonema boryanum. Appl Biochem Biotechnol 162(1):208–220
75. Aoyama K, Uemura I, Miyake J et al (1997) Fermentative metabolism to produce hydrogen gas and organic compounds in a cyanobacterium, Spirulina platensis. J Fermentat Bioeng 83(1):17–20
76. Anderson RA Kawacbi M (2005) Traditional microalgae isolation techniques in Algal culturing techniques. In: Anderson RA (ed). Elsevier Academic Press, Amsterdam, pp 83–101
77. Crosbie ND, Pöckl M, Weisse T (2003) Rapid establishment of clonal isolates of freshwater autotrophic picoplankton by single-cell and single-colony sorting. J Microbiol Methods 55(2):361–370
78. Sinigalliano CD, Winshell J, Guerrero MA et al (2009) Viable cell sorting of dinoflagellates by multiparametric flow cytometry. Phycologia 48(4):249–257
79. Cellamare M, Rolland A, Jacquet S (2010) Flow cytometry sorting of freshwater phytoplankton. J Appl Phycol 22(1):87–100
80. Day JG, Brand JJ (2005) Cryopreservation methods for maintaining microalgal cultures. In: Anderson RA (ed) Algal culturing techniques. Elsevier Academic Press, Amsterdam, pp 165–188
81. Chen W, Zhang C, Song L et al (2009) A high throughput Nile red method for quantitative measurement of neutral lipids in microalgae. J Microbiol Methods 77(1):41–47
82. Chen W, Sommerfeld M, Hu Q (2011) Microwave-assisted Nile red method for in vivo quantification of neutral lipids in microalgae. Bioresour Technol 102(1):135–141
83. Govender T, Ramanna L, Rawat I et al (2012) BODIPY staining, an alternative to the Nile red fluorescence method for the evaluation of intracellular lipids in microalgae. Bioresour Technol 114:507–511
84. Dennis EA (2009) Lipidomics joins the omics evolution. Proc Natl Acad Sci U S A 106(7):2089–2090
85. Wenk MR (2010) Lipidomics: new tools and applications. Cell 143(6):888–895
86. Wu H, Volponi JV, Oliver AE et al (2011) In vivo lipidomics using single-cell Raman spectroscopy. Proc Natl Acad Sci U S A 108(9):3809–3814
87. Gross RW, Han X (2011) Lipidomics at the interface of structure and function in systems biology. Chem Biol 18(3):284–291
88. Hu G, Fan Y, Zhang L et al (2013) Enhanced lipid productivity and photosynthesis efficiency in a desmodesmus sp. mutant induced by heavy carbon ions. Plos One 8(4):e60700
89. Ghirardi ML, Dubini A, Yu J et al (2009) Photobiological hydrogen-producing systems. Chem Soc Rev 38(1):52–61
90. Harris EH (2009) The Chalmydomonas sourcebook. Academic Press, San Diego, CA
91. Li Y, Han D, Hu G et al (2010) Chlamydomonas starchless mutant defective in ADP-glucose pyrophosphorylase hyper-accumulates triacylglycerol. Metabolic Eng 12(4): 387–391

92. Work VH, Radakovits R, Jinkerson RE et al (2010) Increased lipid accumulation in the chlamydomonas reinhardtii sta7-10 starchless isoamylase mutant and increased carbohydrate synthesis in complemented strains. Eukaryotic Cell 9(8):1251–1261
93. Wang ZT, Ullrich N, Joo S et al (2009) Algal lipid bodies: stress induction, purification, and biochemical characterization in wild-type and starchless Chlamydomonas reinhardtii. Eukaryotic Cell 8(12):1856–1868
94. Merchant SS, Prochnik SE, Vallon O et al (2007) The Chlamydomonas genome reveals the evolution of key animal and plant functions. Science 318(5848):245–251
95. Nguyen AV, Thomas-Hall SR, Malnoe A et al (2008) Transcriptome for photobiological hydrogen production induced by sulfur deprivation in the green alga chlamydomonas reinhardtii. Eukaryotic Cell 7(11):1965–1979
96. Guarnieri MT, Nag A, Smolinski SL et al (2011) Examination of triacylglycerol biosynthetic pathways via de novo transcriptomic and proteomic analyses in an unsequenced microalga. Plos One 6(10):e25851
97. Rismani-Yazdi H, Haznedaroglu BZ, Bibby K et al (2011) Transcriptome sequencing and annotation of the microalgae Dunaliella tertiolecta: pathway description and gene discovery for production of next-generation biofuels. BMC Genomics 12:148
98. Rismani-Yazdi H, Haznedaroglu BZ, Hsin C et al (2012) Transcriptomic analysis of the oleaginous microalga Neochloris oleoabundans reveals metabolic insights into triacylglyceride accumulation. Biotechnol Biofuels 5:74
99. Wan L, Han J, Sang M et al (2012) De novo transcriptomic analysis of an oleaginous microalga: pathway description and gene discovery for production of next-generation biofuels. Plos One 7(4):e35142
100. Blanc G, Duncan G, Agarkova I et al (2010) The chlorella variabilis NC64A genome reveals adaptation to photosymbiosis, coevolution with viruses, and cryptic sex. Plant Cell 22(9):2943–2955
101. Matsuzaki M, Misumi O, Shin-I T et al (2004) Genome sequence of the ultrasmall unicellular red alga Cyanidioschyzon merolae 10D. Nature 428(6983):653–657
102. Curtis BA, Tanifuji G, Burki F et al (2012) Algal genomes reveal evolutionary mosaicism and the fate of nucleomorphs. Nature 492(7427):59–65
103. Worden AZ, Lee J-H, Mock T et al (2009) Green evolution and dynamic adaptations revealed by genomes of the marine picoeukaryotes micromonas. Science 324(5924):268–272
104. Radakovits R, Jinkerson RE, Fuerstenberg SI et al (2012) Draft genome sequence and genetic transformation of the oleaginous alga Nannochloropis gaditana. Nat Commun 3:686
105. Palenik B, Grimwood J, Aerts A et al (2007) The tiny eukaryote Ostreococcus provides genomic insights into the paradox of plankton speciation. Proc Natl Acad Sci U S A 104(18):7705–7710
106. Derelle E, Ferraz C, Rombauts S et al (2006) Genome analysis of the smallest free-living eukaryote Ostreococcus tauri unveils many unique features. Proc Natl Acad Sci U S A 103(31):11647–11652
107. Bowler C, Allen AE, Badger JH et al (2008) The Phaeodactylum genome reveals the evolutionary history of diatom genomes. Nature 456(7219):239–244
108. Armbrust EV, Berges JA, Bowler C et al (2004) The genome of the diatom Thalassiosira pseudonana: Ecology, evolution, and metabolism. Science 306(5693):79–86
109. Prochnik SE, Umen J, Nedelcu AM et al (2010) Genomic analysis of organismal complexity in the multicellular green alga volvox carteri. Science 329(5988):223–226
110. Liolios K, Mavromatis K, Tavernarakis N et al (2008) The genomes on line database (GOLD) in 2007: status of genomic and metagenomic projects and their associated metadata. Nucl Acids Res 36:D475–D479
111. León-Bañares R, González-Ballester D, Galván A et al (2004) Transgenic microalgae as green cell-factories. Trends Biotechnol 22(1):45–52

112. Sizova I, Fuhrmann M, Hegemann P (2001) A streptomyces rimosus aphVIII gene coding for a new type phosphotransferase provides stable antibiotic resistance to Chlamydomonas reinhardtii. Gene 277(1–2):221–229
113. Kovar JL, Zhang J, Funke RP et al (2002) Molecular analysis of the acetolactate synthase gene of Chlamydomonas reinhardtii and development of a genetically engineered gene as a dominant selectable marker for genetic transformation. Plant J 29(1):109–117
114. Ohresser M, Matagne RF, Loppes R (1997) Expression of the arylsulphatase reporter gene under the control of the nit1 promoter in Chlamydomonas reinhardtii. Curr Genet 31(3):264–271
115. Lumbreras V, Stevens DR, Purton S (1998) Efficient foreign gene expression in Chlamydomonas reinhardtii mediated by an endogenous intron. Plant J 14(4):441–447
116. Tang DKH, Qiao SY, Wu M (1995) Insertion mutagenesis of Chlamydomonas reinhardtii by electroporation and heterologous DNA. Biochem Mol Biol Int 36(5):1025–1035
117. Fischer H, Robl I, Sumper M et al (1999) Targeting and covalent modification of cell wall and membrane proteins heterologously expressed in the diatom Cylindrotheca fusiformis (Bacillariophyceae). J Phycol 35(1):113–120
118. Nelson JA, Lefebvre PA (1995) Targeted disruption of the NIT8 gene in Chlamydomonas reinhardtii. Mol Cell Biol 15(10):5762–5769
119. Zaslavskaia LA, Lippmeier JC, Shih C et al (2001) Trophic obligate conversion of an photoautotrophic organism through metabolic engineering. Science 292(5524):2073–2075
120. Fuhrmann M, Oertel W, Hegemann P (1999) A synthetic gene coding for the green fluorescent protein (GFP) is a versatile reporter in Chlamydomonas reinhardtii. Plant 19(3):353–361
121. Te, Lohuis MR, Miller DJ (1998) Genetic transformation of dinoflagellates (Amphidinium and Symbiodinium): expression of GUS in microalgae using heterologous promoter constructs. Plant J 13(3):427–435
122. Teng CY, Qin S, Liu JG et al (2002) Transient expression of lacZ in bombarded unicellular green alga Haematococcus pluvialis. J Appl Phycol 14(6):497–500
123. Falciatore A, Casotti R, Leblanc C et al (1999) Transformation of nonselectable reporter genes in marine diatoms. Marine Biotechnol 1(3):239–251
124. Zaslavskaia LA, Lippmeier JC, Kroth PG et al (2000) Transformation of the diatom Phaeodactylum tricornutum (Bacillariophyceae) with a variety of selectable marker and reporter genes. J Phycol 36(2):379–386
125. Hall LM, Taylor KB, Jones DD (1993) Expression of a foreign gene in Chlamydomonas reinhardtii. Gene 124(1):75–81
126. Chang M, Li F, Odom OW et al (2003) A cosmid vector containing a dominant selectable marker for cloning Chlamydomonas genes by complementation. Plasmid 49(1):75–78
127. Marin-Navarro J, Manuell AL, Wu J et al (2007) Chloroplast translation regulation. Photosynthesis Res 94(2–3):359–374
128. Casas-Mollano JA, Rohr J, Kim E-J et al (2008) Diversification of the core RNA interference machinery in Chlamydomonas reinhardtii and the role of DCL1 in transposon silencing. Genetics 179(1):69–81
129. Moellering ER, Benning C (2010) RNA interference silencing of a major lipid droplet protein affects lipid droplet size in Chlamydomonas reinhardtii. Eukaryotic Cell 9(1):97–106
130. Cerutti H, Ma X, Msanne J et al (2011) RNA-mediated silencing in algae: biological roles and tools for analysis of gene function. Eukaryotic Cell 10(9):1164–1172
131. Molnar A, Bassett A, Thuenemann E et al (2009) Highly specific gene silencing by artificial microRNAs in the unicellular alga Chlamydomonas reinhardtii. Plant J 58(1):165–174
132. Zhao T, Wang W, Bai X et al (2009) Gene silencing by artificial microRNAs in Chlamydomonas. Plant J 58(1):157–164
133. Ohlrogge JB, Jaworski JG (1997) Regulation of fatty acid synthesis. Ann Rev Plant Physiol Plant Mol Biol 48:109–136

134. Roesler K, Shintani D, Savage L et al (1997) Targeting of the Arabidopsis homomeric acetyl-coenzyme a carboxylase to plastids of rapeseeds. Plant Physiol 113(1):75–81
135. Dunahay TG, Jarvis EE, Roessler PG (1995) Genetic transformation of the diatoms Cyclotella cryptica and Navicula saprophila. J Phycol 31(6):1004–1012
136. Dunahay TG, Jarvis EE, Dais SS et al (1996) Manipulation of microalgal lipid production using genetic engineering. Biotechnol Appl Biochem 57–8:223–231
137. Vigeolas H, Waldeck P, Zank T et al (2007) Increasing seed oil content in oil-seed rape (*Brassica napus* L.) by over-expression of a yeast glycerol-3-phosphate dehydrogenase under the control of a seed-specific promoter. Plant Biotechnol J 5(3):431–441
138. Jain RK, Coffey M, Hi AKK et al (2000) Enhancement of seed oil content by expression of glycerol-3-phosphate acyl transferase genes. Biochem Soc Trans 28(6):959–960
139. Taylor DC, Katavic V, Zou JT et al (2002) Field testing of transgenic rapeseed cv. Hero transformed with a yeast sn-2 acyltransferase results in increased oil content, erucic acid content and seed yield. Mol Breed 8(4):317–322
140. Zheng P, Allen WB, Roesler K et al (2008) A phenylalanine in DGAT is a key determinant of oil content and composition in maize. Nat Genet 40(3):367–372
141. Wang ZT, Ullrich N, Joo S et al (2009) Algal lipid bodies: stress induction, purification, and biochemical characterization in wild-type and starchless chlamydomonas reinhardtii. Eukaryotic Cell 8(12):1856–1868
142. Voelker TA, Worrell AC, Anderson L et al (1992) Fatty-acid biosynthesis redirected to medium chains in transgenic oilseed plants. Science 257(5066):72–74
143. Yuan L, Voelker TA, Hawkins DJ (1995) Modification of the substrate-specificity of an acyl-acyl carrier protein thioesterase by protein engineering. Proc Natl Acad Sci U S A 92(23):10639–10643
144. Melis A (2009) Solar energy conversion efficiencies in photosynthesis: Minimizing the chlorophyll antennae to maximize efficiency. Plant Sci 177(4):272–280
145. Work VH, D'Adamo S, Radakovits R et al (2012) Improving photosynthesis and metabolic networks for the competitive production of phototroph-derived biofuels. Curr Opinion Biotechnol 23(3):290–297
146. Mussgnug JH, Thomas-Hall S, Rupprecht J et al (2007) Engineering photosynthetic light capture: impacts on improved solar energy to biomass conversion. Plant Biotechnol J 5(6):802–814
147. Li Y, Han D, Hu G et al (2010) Chlamydomonas starchless mutant defective in ADP-glucose pyrophosphorylase hyper-accumulates triacylglycerol. Metabolic Eng 12(4):387–391
148. Li Y, Zhang X, Hu Q et al (2011) A type-2 acyl-coa:diacylglycerol acyltransferase gene is essential for endoplasmic reticulum-based triacylglycerol synthesis in Chlamydomonas reinhardtii. J Phycol 47:S59–S59
149. Kilian O, Benemann CSE, Niyogi KK et al (2011) High efficiency homologous recombination in the oil-producing alga Nannochloropsis sp. Proc Natl Acad Sci U S A 108(52):21265–21269
150. Cernac A, Benning C (2004) WRINKLED1 encodes an AP2/EREB domain protein involved in the control of storage compound biosynthesis in Arabidopsis. Plant J 40(4):575–585
151. Ohto, M-a, Fischer RL, Goldberg RB et al (2005) Control of seed mass by APETALA2. Proc Natl Acad Sci U S A 102(8):3123–3128
152. Butelli E, Titta L, Giorgio M et al (2008) Enrichment of tomato fruit with health-promoting anthocyanins by expression of select transcription factors. Nat Biotech 26(11):1301–1308
153. Tuck G, Glendining MJ, Smith P et al (2006) The potential distribution of bioenergy crops in Europe under present and future climate. Biomass Bioenergy 30(3):183–197
154. Wilcox HA (1976) The Ocean Food and Energy Farm Project. J Marine Edu 16(19):76
155. Benemann J (1980) Proalcohol: the Brazilian alcohol program. Biosour Dig 2(3):156–179
156. Sangyōshō T (1979) Sunshine Project : new energy research and development in Japan. Tokyo: Ministry of International Trade and Industry, pp:1–54

157. Dasgupta B (1977) India's green revolution. Economic and political weekly, pp 241–260
158. Calvin M (1986) Renewable Fuels and Materials. Cell Biochem Biophys 9(1–2):189–210
159. Pimentel D, Patzek TW (2005) Ethanol production using corn, switchgrass, and wood; biodiesel production using soybean and sunflower. Nat Resour Res 14(1):65–76
160. Hertel TW, Golub AA, Jones AD et al (2010) Effects of US maize ethanol on global land use and greenhouse gas emissions: estimating market-mediated responses. BioScience 60(3):223–231
161. Goldemberg J (2007) Ethanol for a sustainable energy future. Science 315(5813):808–810
162. Sissell K (2013) Brazil's BNDES invests $293 million in cellulosic ethanol producer. Chemical Week
163. Panella L (2010) Sugar Beet as an Energy Crop. Sugar Tech 12(3):288–293
164. PRANKL H (2000) Standardisation of biodiesel. Final report of NTB-net Phase IV/Topic 1
165. Rahman M, Niimi M, Ishii Y et al (2006) Effects of season, variety and botanical fractions on oxalate content of napiergrass (Pennisetum purpureum Schumach). Grassland Sci 52(4):161–166
166. Stewart J, Toma Y, FERNÁNDEZ FG et al (2009) The ecology and agronomy of Miscanthus sinensis, a species important to bioenergy crop development, in its native range in Japan: a review. Global Change Biol Bioenergy 1(2):126–153
167. Sumathi S, Chai S, Mohamed A (2008) Utilization of oil palm as a source of renewable energy in Malaysia. Renew Sustain Energy Rev 12(9):2404–2421
168. Brosse N, Dufour A, Meng XZ et al (2012) Miscanthus: a fast-growing crop for biofuels and chemicals production. Biofuels Bioproducts Biorefining-Biofpr 6(5):580–598
169. Chung J-H, Kim D-S (2012) Miscanthus as a potential bioenergy crop in East Asia. J Crop Sci Biotechnol 15(2):65–77
170. Greef JM, Deuter M, Jung C et al (1997) Genetic diversity of European Miscanthus species revealed by AFLP fingerprinting. Genet Resour Crop Evol 44(2):185–195
171. Atienza SG, Satovic Z, Petersen KK et al (2003) Identification of QTLs influencing combustion quality in Miscanthus sinensis Anderss. II. Chlorine and potassium content. Theoret Appl Genet 107(5):857–863
172. Atienza SG, Satovic Z, Petersen KK et al (2003) Identification of QTLs influencing agronomic traits in Miscanthus sinensis Anderss. I. Total height, flag-leaf height and stem diameter. Theoret Appl Genet 107(1):123–129
173. Swaminathan K, Alabady MS, Varala K et al (2010) Genomic and small RNA sequencing of Miscanthus x giganteus shows the utility of sorghum as a reference genome sequence for Andropogoneae grasses. Genome Biol 11(2):R12
174. Hu N, Yuan B, Sun J et al (2012) Thermotolerant Kluyveromyces marxianus and Saccharomyces cerevisiae strains representing potentials for bioethanol production from Jerusalem artichoke by consolidated bioprocessing. Appl Microbiol Biotechnol 95(5):1359–1368
175. Yuangsaard N, Yongmanitchai W, Yamada M et al (2013) Selection and characterization of a newly isolated thermotolerant Pichia kudriavzevii strain for ethanol production at high temperature from cassava starch hydrolysate. Antonie Van Leeuwenhoek Int J General Mol Microbiol 103(3):577–588
176. Khambhaty Y, Upadhyay D, Kriplani Y et al (2013) Bioethanol from Macroalgal Biomass: Utilization of Marine Yeast for Production of the Same. Bioenergy Res 6(1):188–195
177. Liu XY, Jensen PR, Workman M (2012) Bioconversion of crude glycerol feedstocks into ethanol by Pachysolen tannophilus. Bioresour Technol 104:579–586
178. Ageitos JM, Vallejo JA, Veiga-Crespo P et al (2011) Oily yeasts as oleaginous cell factories. Appl Microbiol Biotechnol 90(4):1219–1227
179. Beopoulos A, Cescut J, Haddouche R et al (2009) Yarrowia lipolytica as a model for bio-oil production. Prog Lipid Res 48(6):375–387
180. Wang JJ, Zhang BR, Chen SL (2011) Oleaginous yeast Yarrowia lipolytica mutants with a disrupted fatty acyl-CoA synthetase gene accumulate saturated fatty acid. Process Biochem 46(7):1436–1441

181. Sabirova JS, Haddouche R, Van Bogaert IN et al (2011) The 'LipoYeasts' project: using the oleaginous yeast Yarrowia lipolytica in combination with specific bacterial genes for the bioconversion of lipids, fats and oils into high-value products. Microbial Biotechnol 4(1):47–54
182. Tai M, Stephanopoulos G (2013) Engineering the push and pull of lipid biosynthesis in oleaginous yeast Yarrowia lipolytica for biofuel production. Metabolic Eng 15:1–9
183. Hu C, Wu S, Wang Q et al (2011) Simultaneous utilization of glucose and xylose for lipid production by Trichosporon cutaneum. Biotechnol Biofuels 4(1):25
184. Gong Z, Wang Q, Shen H et al (2012) Co-fermentation of cellobiose and xylose by Lipomyces starkeyi for lipid production. Bioresour Technol 117:20–24
185. Gujjari P, Suh SO, Coumes K et al (2011) Characterization of oleaginous yeasts revealed two novel species: Trichosporon *cacaoliposimilis* sp nov and Trichosporon *oleaginosus* sp nov. Mycologia 103(5):1110–1118
186. Kumar S, Kushwaha H, Bachhawat AK et al (2012) Genome sequence of the oleaginous red yeast Rhodosporidium toruloides MTCC 457. Eukaryotic Cell 11(8):1083–1084
187. Du Z, Li H, Gu T (2007) A state of the art review on microbial fuel cells: a promising technology for wastewater treatment and bioenergy. Biotechnol Adv 25(5):464–482
188. Franks AE, Nevin KP (2010) Microbial Fuel Cells. Curr Rev Energies 3(5):899–919
189. Gunawardena A, Fernando S, To F (2008) Performance of a yeast-mediated biological fuel cell. Int J Mol Sci 9(10):1893–1907
190. Favre MF, Carrard D, Ducommun R et al (2009) Online monitoring of yeast cultivation using a fuel-cell-type activity sensor. J Ind Microbiol Biotechnol 36(10):1307–1314
191. Sayed ET, Tsujiguchi T, Nakagawa N (2012) Catalytic activity of baker's yeast in a mediatorless microbial fuel cell. Bioelectrochemistry 86:97–101
192. Hubenova YV, Rashkov RS, Buchvarov VD et al (2011) Improvement of yeast-biofuel cell output by electrode modifications. Ind Eng Chem Res 50(2):557–564
193. Prasad D, Arun S, Murugesan A et al (2007) Direct electron transfer with yeast cells and construction of a mediatorless microbial fuel cell. Biosensors Bioelectronics 22(11):2604–2610
194. Babanova S, Hubenova Y, Mitov M (2011) Influence of artificial mediators on yeast-based fuel cell performance. J Biosci Bioeng 112(4):379–387
195. Haslett ND, Rawson FJ, Barriere F et al (2011) Characterisation of yeast microbial fuel cell with the yeast Arxula adeninivorans as the biocatalyst. Biosensors Bioelectronics 26(9):3742–3747
196. Jarboe LR, Shanmugam KT, Ingram LO (2009) Ethanol. In: Schaechter M (ed) The desk encyclopedia of Microbiology, 2nd edn. Elsevier, Oxford and San Diego, pp 428–437
197. Sveinsdottir M, Sigurbjornsdottir MA, Orlygsson J (2011) Ethanol and hydrogen production with thermophilic bacteria from sugars and complex biomass. In: Shaukat S (ed) Progress in biomass and bioenergy production, In Tech, pp 359–394
198. Carere CR, Rydzak T, Verbeke TJ et al (2012) Linking genome content to biofuel production yields: a meta-analysis of major catabolic pathways among select H_2 and ethanol-producing bacteria. BMC Microbiol 12:295
199. Liu S, Dien BS, Cotta MA (2005) Functional expression of bacterial Zymobacter palmae pyruvate decarboxylase gene in Lactococcus lactis. Curr Microbiol 50(6):324–328
200. Kaczowka SJ, Reuter CJ, Talarico LA et al (2005) Recombinant production of Zymomonas mobilis pyruvate decarboxylase in the haloarchaeon Haloferax volcanii. Archaea 1(5):327–334
201. Chandra Raj K, Ingram LO, Maupin-Furlow JA (2001) Pyruvate decarboxylase: a key enzyme for the oxidative metabolism of lactic acid by Acetobacter pasteurianus. Arch Microbiol 176(6):443–451
202. Chen GQ, Jiang Y, Chen F (2008) Salt-induced alterations in lipid composition of diatom Nitzschia Laevis (*Bacillariophyceae*) under heterotrophic culture condition. J Phycol 44(5):1309–1314

203. Ohta K, Beall DS, Mejia JP et al (1991) Genetic improvement of *Escherichia coli* for ethanol production: chromosomal integration of Zymomonas mobilis genes encoding pyruvate decarboxylase and alcohol dehydrogenase II. Appl Environ Microbiol 57(4):893–900
204. Taylor MP, Eley KL, Martin S et al (2009) Thermophilic ethanologenesis: future prospects for second-generation bioethanol production. Trends Biotechnol 27(7):398–405
205. Balusu R, Paduru RR, Kuravi SK et al (2005) Optimization of critical medium components using response surface methodology for ethanol production from cellulosic biomass by Clostridium thermocellum SS19. Process Biochem 40(9):3025–3030
206. Dürre P (1998) New insights and novel developments in clostridial acetone/butanol/isopropanol fermentation. Appl Microbiol Biotechnol 49(6):639–648
207. Sveinsdottir M, Baldursson SRB, Orlygsson J (2009) Ethanol production from monosugars and lignocellulosic biomass by thermophilic bacteria isolated from Icelandic hot springs. Icelandic Agricul Sci 22:45–58
208. Tenenbaum DJ (2008) Food vs. fuel: diversion of crops could cause more hunger. Environ Health Perspect 116(6):A254–A257
209. Ji SQ, Wang SA, Tan Y et al (2012) An untapped bacterial cellulolytic community enriched from coastal marine sediment under anaerobic and thermophilic conditions. Fems Microbiol Lett 335(1):39–46
210. Kopke M, Mihalcea C, Bromley JC et al (2011) Fermentative production of ethanol from carbon monoxide. Curr Opinion Biotechnol 22(3):320–325
211. Choi WJ, Hartono MR, Chan WH et al (2011) Ethanol production from biodiesel-derived crude glycerol by newly isolated Kluyvera cryocrescens. Appl Microbiol Biotechnol 89(4):1255–1264
212. Xiao Z, Wang X, Huang Y et al (2012) Thermophilic fermentation of acetoin and 2,3-butanediol by a novel Geobacillus strain. Biotechnol Biofuels 5(1):88
213. Gupta A, Murarka A, Campbell P et al (2009) Anaerobic fermentation of glycerol in paenibacillus macerans: metabolic pathways and environmental determinants. Appl Environ Microbiol 75(18):5871–5883
214. Jessen JE, Orlygsson J (2012) Production of ethanol from sugars and lignocellulosic biomass by Thermoanaerobacter J1 isolated from a hot spring in Iceland. J Biomed Biotechnol 2012:186982
215. Georgieva TI, Mikkelsen MJ, Ahring BK (2008) Ethanol production from wet-exploded wheat straw hydrolysate by thermophilic anaerobic bacterium Thermoanaerobacter BG1L1 in a continuous immobilized reactor. Appl Biochem Biotechnol 145(1–3):99–110
216. Georgieva TI, Ahring BK (2007) Evaluation of continuous ethanol fermentation of dilute-acid corn stover hydrolysate using thermophilic anaerobic bacterium Thermoanaerobacter BG1L1. Appl Microbiol Biotechnol 77(1):61–68
217. Zambare VP, Bhalla A, Muthukumarappan K et al (2011) Bioprocessing of agricultural residues to ethanol utilizing a cellulolytic extremophile. Extremophiles 15(5):611–618
218. Qi X, Zhang Y, Tu R et al (2011) High-throughput screening and characterization of xylose-utilizing, ethanol-tolerant thermophilic bacteria for bioethanol production. J Appl Microbiol 110(6):1584–1591
219. Durre P (2007) Biobutanol: an attractive biofuel. Biotechnol J 2(12):1525–1534
220. Zheng YN, Li LZ, Xian M et al (2009) Problems with the microbial production of butanol. J Ind Microbiol Biotechnol 36(9):1127–1138
221. Huang H, Liu H, Gan YR (2010) Genetic modification of critical enzymes and involved genes in butanol biosynthesis from biomass. Biotechnol Adv 28(5):651–657
222. Chua TK, Liang DW, Qi C et al (2012) Characterization of a butanol-acetone-producing Clostridium strain and identification of its solventogenic genes. Bioresour Technol 135:372–378
223. Bramono SE, Lam YS, Ong SL et al (2011) A mesophilic Clostridium species that produces butanol from monosaccharides and hydrogen from polysaccharides. Bioresour Technol 102(20):9558–9563

224. Montoya D, Spitia S, Silva E et al (2000) Isolation of mesophilic solvent-producing clostridia from Colombian sources: physiological characterization, solvent production and polysaccharide hydrolysis. J Biotechnol 79(2):117–126
225. Montoya D, Arevalo C, Gonzales S et al (2001) New solvent-producing Clostridium sp strains, hydrolyzing a wide range of polysaccharides, are closely related to Clostridium butyricum. J Ind Microbiol Biotechnol 27(5):329–335
226. Ezeji TC, Qureshi N, Blaschek HP (2007) Bioproduction of butanol from biomass: from genes to bioreactors. Curr Opinion Biotechnol 18(3):220–227
227. Virunanon C, Chantaroopamai S, Denduangbaripant J et al (2008) Solventogenic-cellulolytic clostridia from 4-step-screening process in agricultural waste and cow intestinal tract. Anaerobe 14(2):109–117
228. Fischer CR, Klein-Marcuschamer D, Stephanopoulos G (2008) Selection and optimization of microbial hosts for biofuels production. Metabolic Eng 10(6):295–304
229. Ting CNW, Wu JC, Takahashi K et al (2012) Screened butanol-tolerant enterococcus faecium capable of butanol production. Appl Biochem Biotechnol 168(6):1672–1680
230. Peralta-Yahya PP, Zhang F, del Cardayre SB et al (2012) Microbial engineering for the production of advanced biofuels. Nature 488(7411):320–328
231. Lamsen EN, Atsumi S (2012) Recent progress in synthetic biology for microbial production of C3-C10 alcohols. Front Microbiol 3:196
232. Atsumi S, Hanai T, Liao JC (2008) Non-fermentative pathways for synthesis of branched-chain higher alcohols as biofuels. Nature 451(7174):86–89
233. Elshahed MS (2010) Microbiological aspects of biofuel production: current status and future directions. J Adv Res 1(2):103–111
234. Tabatabaei M, Tohidfar M, Jouzani GS et al (2011) Biodiesel production from genetically engineered microalgae: future of bioenergy in Iran. Renew Sustain Energy Rev 15(4): 1918–1927
235. Hallenbeck PC, Ghosh D (2009) Advances in fermentative biohydrogen production: the way forward? Trends Biotechnol 27(5):287–297
236. Kumar N, Das D (2000) Enhancement of hydrogen production by Enterobacter cloacae IIT-BT 08. Process Biochem 35(6):589–593
237. Chen WM, Tseng ZJ, Lee KS et al (2005) Fermentative hydrogen production with Clostridium butyricum CGS5 isolated from anaerobic sewage sludge. Int J Hydrogen Energy 30(10):1063–1070
238. Chittibabu G, Nath K, Das D (2006) Feasibility studies on the fermentative hydrogen production by recombinant *Escherichia coli* BL-21. Process Biochem 41(3):682–688
239. van Niel EWJ, Budde MAW, de Haas GG et al (2002) Distinctive properties of high hydrogen producing extreme thermophiles, Caldicellulosiruptor saccharolyticus and Thermotoga elfii. Int J Hydrogen Energy 27(11–12):1391–1398
240. van Niel EWJ, Claassen PAM, Stams AJM (2003) Substrate and product inhibition of hydrogen production by the extreme thermophile, Caldicellulosiruptor saccharolyticus. Biotechnol Bioeng 81(3):255–262
241. Munro SA, Zinder SH, Walker LP (2009) The Fermentation Stoichiometry of Thermotoga neapolitana and Influence of Temperature, Oxygen, and pH on Hydrogen Production. Biotechnol Prog 25(4):1035–1042
242. D'Ippolito G, Dipasquale L, Vella FM et al (2010) Hydrogen metabolism in the extreme thermophile Thermotoga neapolitana. Int J Hydrogen Energy 35(6):2290–2295
243. de Vrije T, Mars AE, Budde MAW et al (2007) Glycolytic pathway and hydrogen yield studies of the extreme thermophile Caldicellulosiruptor saccharolyticus. Appl Microbiol Biotechnol 74(6):1358–1367
244. de Vrije T, Bakker RR, Budde MAW et al (2009) Efficient hydrogen production from the lignocellulosic energy crop Miscanthus by the extreme thermophilic bacteria Caldicellulosiruptor saccharolyticus and Thermotoga neapolitana. Biotechnol Biofuels 2:12

245. Zeidan AA, van Niel EWJ (2010) A quantitative analysis of hydrogen production efficiency of the extreme thermophile Caldicellulosiruptor owensensis OLT. Int J Hydrogen Energy 35(3):1128–1137
246. Almarsdottir AR, Taraceviz A, Gunnarsson I et al (2010) Hydrogen production from sugars and complex biomass by Clostridium species, AK(14), isolated from Icelandic hot spring. Icelandic Agricul Sci 23:61–71
247. Cao GL, Ren NQ, Wang AJ et al (2010) Statistical optimization of culture condition for enhanced hydrogen production by Thermoanaerobacterium thermosaccharolyticum W16. Bioresour Technol 101(6):2053–2058
248. Ren NQ, Chua H, Chan SY et al (2007) Assessing optimal fermentation type for bio-hydrogen production in continuous-flow acidogenic reactors. Bioresour Technol 98(9):1774–1780
249. Cheong DY, Hansen CL, Stevens DK (2007) Production of bio-hydrogen by mesophilic anaerobic fermentation in an acid-phase sequencing batch reactor. Biotechnol Bioeng 96(3):421–432
250. Kongjan P, O-Thong S, Kotay M et al (2010) Biohydrogen production from wheat straw hydrolysate by dark fermentation using extreme thermophilic mixed culture. Biotechnol Bioeng 105(5):899–908
251. Koskinen PEP, Beck SR, Orlygsson J et al (2008) Ethanol and hydrogen production by two thermophilic, anaerobic bacteria isolated from icelandic geothermal areas. Biotechnol Bioeng 101(4):679–690
252. Koskinen PEP, Lay CH, Puhakka JA et al (2008) High-efficiency hydrogen production by an anaerobic, thermophilic enrichment culture from an icelandic hot spring. Biotechnol Bioeng 101(4):665–678
253. Kongjan P, Min B, Angelidaki I (2009) Biohydrogen production from xylose at extreme thermophilic temperatures (70 degrees C) by mixed culture fermentation. Water Res 43(5):1414–1424

Adv Biochem Eng Biotechnol (2015) 147: 225–229
DOI: 10.1007/978-3-662-45097-0

Index

A
ABE fermentation, 208
Acetyl CoA carboxylase (ACCase), 189
Actinobacteria, 41
Actinoflora, 111
Actinomycetes, 37, 40, 46, 111
Actinomycin D, 39
Acute inflammation, 63
African rice, 6
Agricultural biodiversity, 3
Agrobacterium, 24
Agrobiodiversity, 3
Air pollution, microbial bioremediation, 158
  phytoremediation, 169
Alcohols, 185
Alkaline protease inhibitor (API), 122
Alkaloids, 88
Allicin, 74
Amaranth, 17
Aminoglycoside, 42
Anthracenones, 39
Anthracophyllum discolor, 154
Anthracyclines, 39, 74
Anthraquinones, 85
Anti-inflammatory agents, 63
Antibiotics, 37, 111, 117
Anticancer activity, 59
Anticancer compounds, 37, 74
Antifungals, 117
Antimicrobial activity, 59, 79
Antineoplasic activity, 39
Antitumor activity, 37
Archazolides, 47
Aridicins, 46
Arsenic, 167
Arsenite, 154
Avermectins, 120
Avicennia schaueriana, 166

B
Bacteriome, 114
Bacterioplankton, 172
Barley, 4, 9
Benz[a]anthracene, 153
Benzene, toluene, ethylbenzene, and xylene (BTEX), 169
Berberine, 88, 90
Bioaugmentation, 139
Biodiesel, 185
  Japan, 200
Biodiscovery, 111
Biodiversity, 3, 37, 59, 137
Bioenergy, microalgae, 188
Biofuels, 185
Biological nitrification inhibition (BNI), 21
Bioprospecting, molecular, 142
  nonmolecular, 141
Bioremediation, 137, 147
Biosynthetic potential, 185
Bleomycin, 39
Brachialactone, 21
Bread wheat, 7
2,3-Butanediol, 211
Butanol, 185, 208

C
Cadmium, 154, 160–168
Caffeic acid, 85
Calcheamycin, 39
Camptotheca acuminata, 51

Camptothecin, 51
Cañahua (*Chenopodium pallidicaule*), 20
Carbohydrates, 185
Cardiac glycosides, 73
Cardiovascular diseases (CVDs), 59, 72
Cellulolytic fungi, 121
Central nervous systems (CNS), 62
Cereals, 4
  crop biodiversity, 1
Chaetomium globosum, 51
Chili peppers, 17
Chitinase, 122
Chlorella vulgaris, CO_2 absorption, 159
Chondrochloren, 48
Chondromyces crocatus, 48
Chondromyces pediculatus, 48
Cinnamaldehyde, 88
Climate change, 1
CNS, 59
Coptis chinensis, 90
Coptotermes lacteus, 127
Cornitermes cumulans, 123
Coumarins, 87
Crop genomics, 1
Crop improvement, 1
Cubitermes niokoloensis, 123
Cubitermes spp., 123
Curcumin, 74
Cymbidium goeringii, 73

D
Daptomycin, 52
Daunorubicin, 39
Dentigerumycin, 111
Diacylglycerolacyltransferase (DGAT), 190
Diallyl sulphide, 74
Dibenzofurans, 153
Digalactosyldiacylglycerol (MGDG), 189
Digitalis glycosides, 60
Digoxin, 73
Diterpenes, 119
Dopamine, 62
Doxorubicin, 39
Duckweed, 167
Dynemicin, 45

E
Earthworms, 170
Eco-taxonomy, 111
Eisenia foetida, 170
Eleutherine americana, 71
Emmer wheat, 7
Emodin, 72
Endosulfan, 155
Enediynes, 39
Energy plants, 199
Environmental tools, 137
Enzymes, oxidative, 121
Epirubicin, 39
Epothilones, 39
Essential oils, 87
Ethanol, 185, 187
  fuel production, 200
  synthesis, bacterial, 204
  yeast, 202
Etnangien, 49
Eugenol, 88
Euphorbia lathyris, 200
Evodia rutaecarpa, 70
Evodiamine, rutaecarpine, 70
Evolutionary clock, 142
Exploration, 137

F
Fatty acids, 189
Fatuoa pilosa, 87
Fe-hydrogenase, 191
Fidaxomicin, 52
Fish farm sediments, 171
Flavones, 85
Flavonoids, 85
Flavonols, 85
Fluoranthene, 170
Food security, 1
Functional genomics, 143
Fungal entomopathogens, 117
Fungi, 37
  ants, 117

G
Gamma-aminobutyric acid, 62
Gasoline, 185
Gastrointestinal (GI) ulceration, 89
Genetic engineering, 145
Genetic resources, 1
Genetic variation, crops, 5
Genistein, 74
Geosmin, 120, 128
Gifhornenolones, 120
Gigantol, 72, 73

β-1,3-Glucan recognition proteins (BGRPs), 122
Glucanases, 122
Glutamate, 62
Glycosyl glycerides, 188
Goca (*Oxalis tuberosa*), 17
Golden Rice, 25
Good agricultural and collection practices (GACP), 92
Goshuyuamide, 70
Greenhouse gases (GHG), 21, 169

H
Heavy metals, 155, 160
  phytoremediating plants, 145
Helicobacter pylori, 88
Hinkitiol, 72
Hodotermes mossambicus, 119
Human population pressure, 91
Hydrocarbons, 185
Hydrogen, bacteria, 211
  microalgae, 191
  production, nitrogenase-mediated, 192
Hydrogenases, 191
Hypericin, 51

I
Imidacloprid, 165
Inflammation, 63
Insect endosymbionts, 114
Insecticide, 165
Isobutanol, 211
Ixeris chinensis, 200

J
Jala, 9
Jasmonate, 62
Jet fuel, 185

L
Lantibiotics, 50
Lead, 164
Leiothrix spiralis (Eriocaulaceae), 86
Leucaena glauca, biodiesel, 200
Leymus racemosus (mammoth wild rye), 22
Lignin, degradation, 121
Lignocellulosic biomass, 185
Ligunstrazine, 72
Lindane, 155
Linezolid, 52
Lipids, 185
Lutein, 72
Luteolin, 70
Lycopene, 74

M
Macrotermes barneyi, 124
Maize, 4, 7, 10, 22, 24, 155, 164, 199
Mangifera indica, 86
Mentha suaveolens, 88
Menthol, 88
Mercury, 155, 159
Metagenomics, 143
Metal transporters, 145
Metarhizium anisopliae, 117
Metarhizium anisopliae, 121
Methodologies, 185
2-Methylisoborneol, 120, 128
Microalgae, bioenergy, 188
Microbial fuel cell (MFC), 158, 204
Millennium Ecosystem Assessment (MEA), 91
Millets, 4
Mining, 158, 160
Mithramycin, 39
Mitomycin C, 39
Miuraenamide, 48
Monogalactosyldiacylglycerol (MGDG), 189
Monoterpenes, 119
Municipal solid-waste incinerator fly ash, 153
Mycangimycin, 111
Myxobacteria, 37, 46
Myxochelins, 48

N
Nannochelins, 48
Naphterpins, 120
Nasutitermes corniger, 116, 124
Nasutitermes macrocephalus, 115
Nasutitermes triodiae, 120
Natural bioresources, 185
Natural resources, 137
New chemical entities (NCEs), 37
Niranthin, 71
Nitrate, 158
Nitrogen pollutants, Impact on biodiversity, 23
Nitrogen use efficiency (NUE), 24
Noradrenaline, 62
Norharmane, 118

O
Oat, 4

1-Octen-3-ol, 120
Odontotermes formosanus, 115, 118
Odontotermes mossambicus, 123
Oleaginous microalgae, 191
Oxaloterpins, 120

P
Paraliomyxa miuraensis, 48
Pathogen-associated molecular patterns (PAMPs), 122
Pattern recognition receptors (PRRs), 65
Pearl millet, 10, 22
Pedein, 48
Penicillin, 39, 50
Penicillium notatum, 50
Penicillium rubens, 39
Pennisetum purpureum, 200
Pennisetum spp., 16
Pentachlorophenol, degradation, 155
Pentostatin, 39
Pesticides, 154
Petroleum-contaminated soil, 165
Petroleum-degrading microorganisms, 148
Phalaris arundinacea, 165
Pharmaceutical bioprospecting, 90
Phenanthrene, 154
Phenolic acid polyphenols, 85
Phenols, 85
Pheromones, 118
Phialocephala fortinii, 51
Phosphatidylcholine (PC), 189
Phosphatidylethanolamine (PE), 189
Phosphatidylglycerol (PG), 189
Phosphoglycerides, 189
Phospholipid:diacylglycerolacyltransferase (PDAT), 190
Phyllanthin, 71
Physarum gyrosum, 52
Phytoremediation, 159, 165
Pinobatol, 71
Pirirubicin, 39
Plesiocystis/Enhygromyxa, 49
Podophyllotoxin, 74
Pollution, 137
  management, 137
Polychlorinated dibenzo-*p*-dioxins, 153
Polycyclic aromatic hydrocarbons (PAHs), 153
Polymorphisms, 6
Polyozellus multiplex, 70, 72
Polypeptides, 88
Polyphenols, 85
Polysaccharides, 185
Polysylvin, 70, 72
Polyzellin, 70, 72
Poplar, 200
Potato, 17
Proteins/peptides, antimicrobial, 89
Proteomics, 143
Pseudonocardia, 116
Psychoactive botanicals, 63
Puerarin, 72
PUFAs, 189
Pyridomycin, 46
Pyruvate, 204

Q
Quinoa, 1, 17
Quinones, 85

R
Rapamycin, 45
Resveratrol, 74
Retapamulin, 52
Reticulitermes chinensis, 118
Rhizophora mangle, 166
Rice, 4, 9

S
Saccharopolyspora spinosa, 119
Salimabromide, 49
Samanea saman, 166
Serotonin, 62
Slime molds, 51
Sludge, 156
Slurry, 156
Smalianthus sonchifolia, 70
Soils, bioremediation, 148, 160
  nitrification, 21
Solid waste treatment, 148
Sorangium cellulosum, 49
Sorghum, 4, 22, 200
Steroids, 48
Streptomyces avermitilis, 120
Streptomyces griseus, 39
Streptomycetes, 41
Streptomycin, 39
Streptozotocin, 39
Sulfoquinovosyldiacylglycerol (SQDG), 189
Symbiosis, 111, 114

T
Tagetes erecta, 71
Tagetes patula, 164
Transcription activator-like effector nucleases (TALENs), 24
Tannins, 86
Taxanes, 74
Taxol, 39
Taxomyces andreanae, 51
Teichoplanin, 46
Termicins, 117
Termites, 111
Terpenoids, 87, 120
Tetrachloroethylene, 156
Tetracyclines, 44
Thiolactone, 44
Transgenic crops, food security, 24
Transport biofuels, microalgae, 188
Transport fuels, 185
Treponema, 123
Triacylglycerols (TAGs), 189

V
Valrubicin, 39
Vancomycin, 46
Variecolortides, 51
Vermi-stabilization, 172
Verrucosispora gifhornensis, 120
VOCs, 120

W
Water hyacinth, 167
Water treatment, 156
Wheat, 4

Y
Yarrowia lipolytica, 203
Yeasts, biofuel, 202
  biofuel cells, 204
  oleaginous, 203

Z
Zinc, 160

MIX
Papier aus verantwortungsvollen Quellen
Paper from responsible sources
FSC® C105338

If you have any concerns about our products,
you can contact us on
ProductSafety@springernature.com

In case Publisher is established outside the EU,
the EU authorized representative is:
Springer Nature Customer Service Center GmbH
Europaplatz 3, 69115 Heidelberg, Germany

Printed by Libri Plureos GmbH
in Hamburg, Germany